CARBONATE PLATFORMS

CARBONATE PLATFORMS
FACIES, SEQUENCES AND
EVOLUTION

EDITED BY MAURICE E. TUCKER,
JAMES LEE WILSON, PAUL D. CREVELLO,
J. RICK SARG AND J. FRED READ

SPECIAL PUBLICATION NUMBER 9 OF THE
INTERNATIONAL ASSOCIATION OF SEDIMENTOLOGISTS
PUBLISHED BY BLACKWELL SCIENTIFIC PUBLICATIONS
OXFORD LONDON EDINBURGH
BOSTON MELBOURNE

© 1990 The International Association
of Sedimentologists
and published for them by
Blackwell Scientific Publications
Editorial offices:
Osney Mead, Oxford OX2 0EL
25 John Street, London WC1N 2BL
23 Ainslie Place, Edinburgh EH3 6AJ
3 Cambridge Center, Suite 208
    Cambridge, Massachusetts 02142, USA
107 Barry Street, Carlton
    Victoria 3053, Australia

First published 1990

Set by Setrite Typesetters, Hong Kong
Printed and bound in Great Britain
by William Clowes Ltd, Beccles and London

DISTRIBUTORS

Marston Book Services Ltd
PO Box 87
Oxford OX2 0DT
(*Orders*: Tel: 0865 791155
        Fax: 0865 791927
        Telex: 837515)

USA
Publishers' Business Services
PO Box 447
Brookline Village
Massachusetts 02147
(*Orders*: Tel: (617) 524−7678)

Canada
Oxford University Press
70 Wynford Drive
Don Mills
Ontario M3C 1J9
(*Orders*: Tel: (416) 441−2941)

Australia
Blackwell Scientific Publications
(Australia) Pty Ltd
107 Barry Street
Carlton, Victoria 3053
(*Orders*: Tel: (03) 347−0300)

British Library
Cataloguing in Publication Data

Carbonate platforms.
  1. Carbon strata
  I. Tucker, Maurice E.
  551.7'5

  ISBN 0−632−02758−4

Library of Congress
Cataloging in Publication Data

Carbonate platforms: facies, sequences, and
evolution/edited by
    Maurice E. Tucker . . . [*et al.*].
        p.   cm. − (Special publication
    number 9 of the International
    Association of Sedimentologists)
      ISBN 0−632−02758−4
      1. Rocks, Carbonate −
    Congresses.   2. Sedimentation and
    deposition − Congresses.   3. Tethys
    (Paleogeography) − Congresses.   I. Tucker,
    Maurice E.   II. Series: Special publication . . .
    of the International Association of
    Sedimentologists: no. 9. QE47.15.C3C35   1990
    552'.58 − dc20

Contents

Preface

This collection of papers on carbonate platforms is a companion to Special Publication 44, *Controls on Carbonate Platform and Basin Development*, published by the Society of Economic Paleontologists and Mineralogists in 1989. The latter was restricted in size for the sake of economy. Both volumes stem from the 1987 Los Angeles SEPM Research Symposium on carbonate platforms and basins, where many of the papers were presented orally or in poster sessions. This multi-national symposium derived from an international Penrose Conference on the subject held on the Isle of Capri in 1981 and organized by Professors Bruno D'Argenio and James L. Wilson.

Six of the papers in this book, those by Cocozza & Gandin, Elmi, Bice & Stewart, Ellis *et al.*, García-Mondéjar and Watts & Blome, were presented at the Los Angeles meeting and the others are recent additions.

The papers in this book are concerned with carbonate platforms and sequences from the Cambrian to Tertiary, and many derive from studies in the Tethyan region. Cocozza & Gandin compare Cambrian and Triassic carbonates from Sardinia and Tuscany in Italy, where carbonate platforms developed in rift-related settings. Gawthorpe & Gutteridge describe platform-margin bioclastic sand shoals from the Dinantian (Mississippian) of Derbyshire, central England, where facies, sequences and sand body lateral progradation are strongly controlled by sea-level changes. Based on studies of Carboniferous cycles, Walkden & Walkden present four computer models to simulate cycles in shallow-water carbonate, deep-water carbonate and mixed carbonate−clastic (deltaic environments. Middle Triassic (Muschelkalk) carbonate ramps in the Catalan Basin, Spain, are interpreted by Calvet, Tucker & Henton in terms of depositional systems tracts, but a strong tectonic control is argued for, rather than a simple eustatic forcing mechanism. Elmi describes the Ardèche carbonate platform (Triassic−Jurassic) from the western margin of the Subalpine Basin in France, bringing together many decades of field observations from a classic area of limestone scenery. From the Apennines of Italy, Bice & Stewart describe the formation of isolated platforms and their later drowning as a result of tectonism and ecological factors. A tectonic, as well as eustatic control on Jurassic carbonate buildups is documented from the Lusitanian Basin of Portugal by Ellis, Wilson & Leinfelder. Arnaud-Vanneau & Arnaud apply a sequence stratigraphy−depositional systems tracts approach to the lower Cretaceous (Urgonian) carbonate platform of the Jura and northern Subalpine chains and demonstrate the importance of major eustatic sea-level changes as well as tectonic influences of basement faults. The effects of basement structure on carbonate facies patterns are also documented by Wilson from the Mesozoic of northeastern Mexico. Finally, Watts & Blome describe the consequences of closure of an ocean basin on an adjacent carbonate platform margin from the Cretaceous of Oman.

The papers in this book clearly show the diversity of carbonate platforms and their facies in the Phanerozoic and many illustrate the large-scale roles of tectonics and eustatic sea-level changes (second and third order) in controlling broad facies patterns and platform evolution. One major problem of course is distinguishing between these two overriding controls. An uncritical application of the Vail-Haq *et al.* sea-level curve is unacceptable these days, particularly for deducing eustasy. Many of the carbonate platform sequences described in this book contain small-scale cycles, commonly shallowing-upward, and the topical issue here is whether they are the result of orbital forcing in the Milankovitch band producing sea-level changes (fourth and fifth order) or some other mechanism (tectonic or sedimentary). Critical field observations and imaginative computer modelling are likely to make important contributions to this problem.

MAURICE E. TUCKER, *Durham, UK*
JAMES L. WILSON, *New Braunfels, Texas, USA*
PAUL D. CREVELLO, *Littleton, Colorado, USA*
J. RICK SARG, *Midland, Texas, USA*
J. FRED READ, *Blacksburg, Virginia, USA*

Spec. Publs. int. Ass. Sediment. (1990) **9**, 9−37

Carbonate deposition during early rifting: the Cambrian of Sardinia and the Triassic−Jurassic of Tuscany, Italy

T. COCOZZA* *and* A. GANDIN

Dipartimento di Science della Terra, University of Siena, Via delle Cerchia 3, 53100 Siena, Italy

ABSTRACT

The platform carbonates of early Palaeozoic age in Sardinia and of the early Mesozoic age in Tuscany were deposited in analogous tectono-sedimentary situations, the former linked to the Iapetus, the latter to the Tethyan, early stages of continental rifting. Both units represent isolated pericratonic platforms initiated and developed in an intra-cratonic rift. Deposition began with dolomites and evaporites and ended with limestones. Persistence of shallow-water deposition documents a continuous balance between carbonate production, subsidence and/or sea-level fluctuations. The synsedimentary tectonic control during deposition resulted in high rates of sedimentation and differential subsidence. Three stages of platform collapse are recognized, which terminated shallow-water deposition. During the first stage the platforms were dissected into blocks and their differential movements produced contemporaneous tilting, uplift and drowning. During the second stage of collapse marginal plateaus were formed and neritic−pelagic sediments of various lithofacies onlapped sunken platform blocks. Nodular limestones were deposited on top of the more unstable blocks. The third collapse stage resulted in general deepening of the basins. The subsequent evolution of the basins reflects the different geodynamic development of the early Palaeozoic and early Mesozoic continental rifting. The former did not reach a spreading stage whereas the latter evolved into an open ocean. Climatic changes occurred during platform growth in both cases, and they provide further evidence of divergent movements connected with continental rifting.

INTRODUCTION

The Cambrian of southwestern Sardinia and the Triassic−Jurassic of Tuscany (Fig. 1) are characterized by sequences which show a similar pattern of depositional evolution. In both cases, a carbonate platform developed on siliciclastic deposits and after drowning was buried by red, nodular limestones and then other pelagic deposits (Fig. 2).

The sedimentary evolution of early Mesozoic sequences in Tuscany has been related to the early phases of continental rifting associated with the opening of Tethys (Bernoulli & Jenkins, 1974; Bernoulli *et al.*, 1979). The Lower Palaeozoic sequences of Sardinia are related to early stages of continental break-up connected with the opening of

the Iapetus Ocean. However, this extensional phase in Sardinia gave way to oblique transform rifting that resulted in pull-a-part basins and local gentle folding (Vai & Cocozza, 1986).

The aim of this paper is to compare the sedimentological characters of the two sequences in order to demonstrate that:

1 the Lower Palaeozoic and Lower Mesozoic carbonate platforms of Sardinia and Tuscany were tilted, isolated platforms, established, developed and drowned under a synsedimentary tectonic regime related to early stages of intra-cratonic rifting;

2 the climatic change from arid to humid conditions recorded within the platform stage of both sequences suggests the shifting of the continental plates from arid to humid tropical areas as a consequence of divergent movements during continental rifting.

* Deceased.

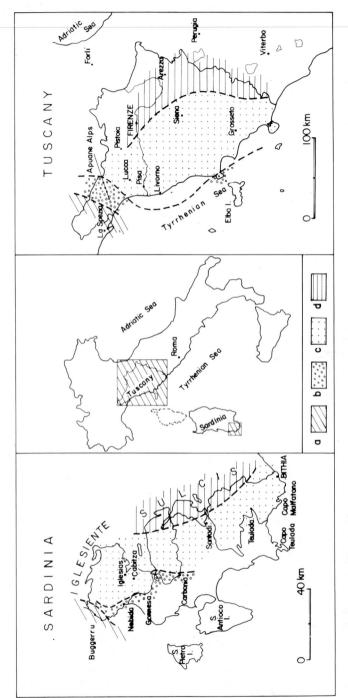

Fig. 1. Location maps and palinspastic reconstruction of Sardinian and Tuscan carbonate platforms. (a) basin; (b) western rimmed margin; (c) platform interior; (d) eastern margin.

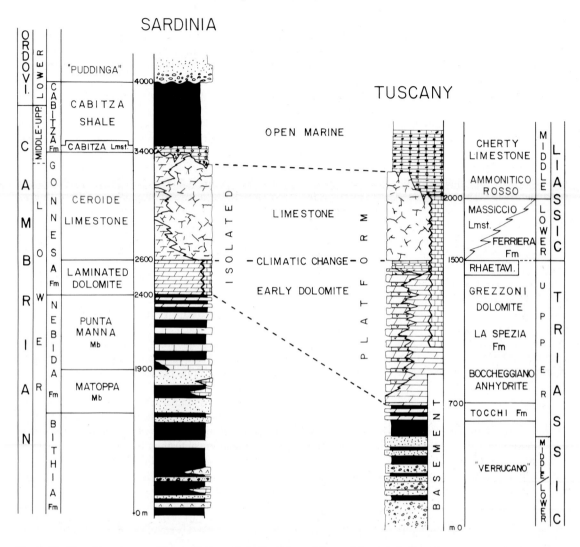

Fig. 2. Stratigraphic columns of Lower Palaeozoic in Sardinia and Lower Mesozoic in Tuscany.

GEOLOGICAL SETTING

Sardinia

Sardinia is the segment of the Meso-European Hercynian chain where the most complete Palaeozoic sequence in Italy is exposed. The Iglesiente–Sulcis region in southwestern Sardinia corresponds to the External Zone of the Hercynian chain, Central Sardinia to the Nappe Zone and northern Sardinia to the Axial Zone (Carmignani *et al.*, 1986b). In Iglesiente–Sulcis two sedimentary cycles are distinguished: Lower Cambrian to Lower Ordovician

and Upper Ordovician to Lower Carboniferous separated by the Sardic angular unconformity of Arenig age. In Central Sardinia, early Cambrian carbonates are missing, and the sequence consists of Middle Cambrian–Lower Ordovician basinal siliciclastic deposits (Carmignani *et al.*, 1986b).

In southwestern Sardinia, the Sardic phase, interpreted as a transpressional episode of a Caledonian oblique transform rifting (Vai & Cocozza, 1986), produced gentle E–W trending folds. During the Hercynian orogeny, three deformational phases gave rise to tighter N–S trending folds and local overthrusts, so that the present tectonic pattern of

Iglesiente−Sulcis is a dome and basin structure. To the north and southeast this structure is overlain by overthrust Cambrian (?) Lower Ordovician rocks of the Nappe Zone and surrounded by granite intrusions (Carmignani *et al.*, 1986a, b). The Caledonian and Hercynian structures were later re-activated during the Alpine orogeny by exclusively tensional movements resulting in the Campidano Graben and in the fracture along the Tyrrhenian coast.

The Lower Palaeozoic sequences of Sardinia were deposited on a proximal segment of the Gondwana continental shelf. The Cambrian−Lower Ordovician cycle starts with an Atdabanian (?) deltaic sequence (Bithia Formation), transgressive on a metamorphic, poorly-exposed basement (Gandin *et al.*, 1987). During Lower Cambrian times a shallow-water car-bonate platform developed and in the Middle Cambrian this collapsed, to be replaced by an intra-continental basin lasting until Tremadocian (Gandin, 1980; Carannante *et al.*, 1984; Courjault-Radè & Gandin, 1986). The deeper part of this basin is at present exposed in the Nappe Zone (eastern Sardinia) beyond the Campidano Graben (Carmignani *et al.*, 1986b; Gandin *et al.*, 1987).

Tuscany

Tuscany is part of the northern Apennine chain built during the Alpine orogeny. Triassic−Jurassic sediments occur in different tectonic units that have been detached from their basement and thrust over each other in an eastern direction. The more internal Ligurian Nappe comprises Upper Jurassic−Cretaceous oceanic deposits with ophiolites. The Tuscan Nappe that has been locally sheared off along Upper Triassic evaporites, consists of: (1) an external Tuscan Domain of Palaeozoic metamorphic basement and Mesozoic−Tertiary cover and (2) an internal Tuscan Domain (Apuan Alps Zone and Massa Zone) of a metamorphosed Mesozoic cover on Palaeozoic basement (Decandia *et al.*, 1980). At the end of the compressive orogenic phases (Lower Miocene) extensional tectonics produced NW−SE trending graben-and-horst structures, later dissected by a NE−SW trending transcurrent-fault system.

As a consequence of their tectonic history, the Lower Mesozoic deposits of the Tuscan Nappe are poorly exposed. However, they maintain their orig-inal palaeogeographic relationships without the need for major palinspastic corrections, except for the Livorno−Sillaro dextral transcurrent fault that puts

the marginal facies in contact with the internal facies of the Triassic−Jurassic platform (Fig. 1). Palinspastic reconstructions of the Triassic−Jurassic sequences place the Tuscan Units on the European continental margin of the Ligurian Ocean that was a N−S trending segment of Tethys (Bernoulli *et al.*, 1979).

The Alpine depositional cycle starts with the Middle−Upper Triassic Verrucano arid-climate red-beds sedimented on a continental basement already affected by extensional tectonics and Middle Triassic 'abortive' rifting (Gandin *et al.*, 1982). From Norian to Hettangian a carbonate platform existed, but at the beginning of the Sinemurian it collapsed and an epicontinental basin was established, which progress-ively deepened towards the Ligurian Ocean (Decandia *et al.*, 1980).

CAMBRIAN OF SARDINIA

The Cambrian sequence is up to 3500−4000 m thick, with lateral variations, especially in the carbonate unit. Its base (Bithia Formation, Lower Cambrian−?Atdabanian) is exposed only in the southernmost part of Sulcis region because of the structural setting of the area and the occurrence of Hercynian granite intrusions (Carmignani *et al.*, 1986a). It is overlain with angular unconformity by post-Arenig conglom-erates and megabreccias ('Puddinga').

The Lower Cambrian to Lower Ordovician se-quence (Figs 2 & 3) was deposited on an epi-continental shelf where deltaic terrigenous sediments (Bithia Formation) were gradually replaced by car-bonates (Nebida Formation). The exclusively car-bonate Gonnesa Formation consists of two informal members: Laminated Dolomite (Dolomia rigata) at the base, and Ceroide Limestone (Calcare ceroide) at the top (Cocozza, 1979). The carbonate platform underwent fragmentation, uplift and drowning and was finally buried by further shelf carbonates (Cabitza Limestone−Calcare nodulare) and then by basinal siliciclastics (Cabitza Shales−Scisti di Cabitza). The upper part of the sequence has prob-ably been removed during the emergence and depo-sition of the Puddinga (Cocozza, 1969; Carannante *et al.*, 1984; Gandin, 1990a).

Biostratigraphy in Sardinian Cambrian is mainly based on trilobites and archaeocyathids and most of these are endemic species and genera so that the biostratigraphic zonation cannot be precisely

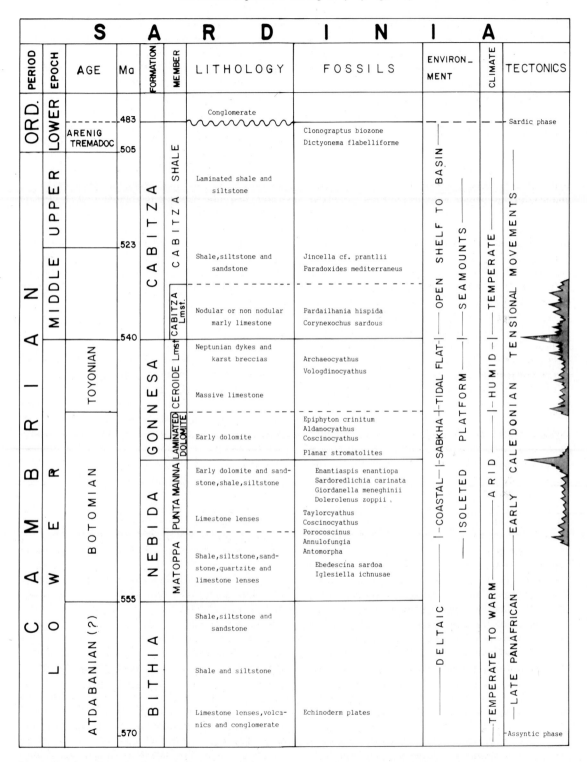

Fig. 3. Stratigraphy, depositional environments and tectonic events of the carbonate platform in Sardinia.

defined. Nevertheless Sardinian trilobite assemblages have been correlated with those of southern France (Montagne Noire), Spain and Morocco (Rasetti, 1972) and archaeocyathan assemblages with those of the Siberian platform and Altai−Sayan fold belt (Debrenne & Gandin, 1985).

Nebida Formation (Botomian)

This formation, up to 800 m thick, consists of a mixed carbonate siliciclastic sequence (Figs 2 & 3). In the lower part (Matoppa Member) green siltstones and sandstones include mounds built by *Epiphyton* algae and archaeocyathids, whereas the upper part (Punta Manna Member) has alternating sandstones, siltstones, shales and lenticular carbonate units of oolitic−oncolitic limestone with small archaeocyathid buildups. In the upper part of the sequence only early dolomite intercalations occur. In the eastern and southeastern areas, the limestone and the early-dolomite intercalations gradually decrease in thickness and number and in the easternmost sections they are absent. The intercalated siliciclastics are mostly fine-grained silts and shales with rare intercalations of quartz−sandstone.

The fauna is dominated by archaeocyathids in the limestones and trilobites in the siliciclastics. The archaeocyathids (*Taylorcyathus rectus, Coscinocyathus dianthus, Porocoscinus, Annulofungia* and *Anthomorpha*) are indicative of a general Botomian age (Rozanov & Debrenne, 1974). The trilobite fauna, mainly found in Punta Manna Member, includes *Hebediscina sardoa, Dolerolenus zoppii, Giordanella meneghinii* and *Enantiaspis enantiopa* and has been correlated by Rasetti (1972) to part of the Nevadella Zone of North America.

The formation is a shallowing-upward deltaic−marine mixed sequence evolving from prodelta (Matoppa Member) to delta plain−coast (lower Punta Manna Member) to sabkha (upper Punta Manna Member) (Gandin, 1990a). The facies distribution of lower Punta Manna Member documents an oolitic margin in the western edge of the presently exposed area (Bechstadt *et al.*, 1985). The interpretation of the depositional environment along the eastern edge is controversial. Bechstadt *et al.* (1985) interpreted it as a siliciclastic, backshoal-lagoon/tidal-flat complex flanking an eastern mainland. However, although the detailed analysis of this member is still in progress, some evidence suggests a deeper-water setting in the eastern and southern areas: (1) the siliciclastic fraction is dominated by

shales and silts, with rare fine sandstones; (2) channel sequences, ripple marks, flaser and cross-bedding are lacking in the east whereas they are common and well developed in the west; (3) palaeocurrents for the western area, indicate a provenance of the siliciclastic material from north−northwest. A similar source is suggested by the grain size of the sandstones which decreases from north to south and east (research in progress); and (4) the early-dolomitic sabkha facies at the top of the sequence in the western area is characterized in Iglesiente by intercalations of goethite-rich layers, interpreted as freshwater, bog-lake deposits (Carannante *et al.*, 1975). These intercalations are replaced towards the east and south by lenses of pseudomorphs of barite. Farther east both the tidal-flat early-dolomite and the barite lenses disappear. All these features suggest a deeper-water rather than a tidal-flat environment for the eastern outcrops of Punta Manna Member (Gandin, 1990a).

Laminated Dolomite (Botomian)

This unit consists of evenly bedded and laminated dolomite, 200−250 m thick (Figs 2 & 3). In northwestern Iglesiente (Buggerru area), archaeocyathids and *Epiphyton crinitum* are enclosed in oncoids and rare trilobite fragments and archaeocyathids in planar stromatolites. The poorly diversified archaeocyathid fauna is composed of typical Botomian genera such as *Aldanocyathus* and *Coscinocyathus* (Debrenne & Gandin, 1985).

In Iglesiente and northern Sulcis, shallowing-upward (intertidal to supratidal) cycles prevail; they are made up of: (1) banded dolomicrosparitic mudstone; (2) planar or wavy stromatolites (Fig. 6a); (3) stromatolites with fenestrae; and (4) vadose pisolite (*sensu* Esteban, 1976) (Fig. 4c) and/or desiccation breccias. Thin, discontinuous, locally silicified layers of oolitic grainstone occur as well as barite lenses and chert nodules which are pseudomorphic after anhydrite (Fig. 6b).

In northwestern Iglesiente (Buggerru area) the dolomite mudstone facies alternates with homogeneous black limestone (Bechstadt *et al.*, 1985); the planar stromatolites prevail and the supratidal facies are lacking. Oolitic facies contain archaeocyathids and skeletal algae (*Epiphyton*) and stromatolites entrapping archaeocyathids are intercalated in the upper part of the unit (Debrenne & Gandin, 1985). In eastern Sulcis this unit thins out; the early dolomite gives way to limestone and only the even-

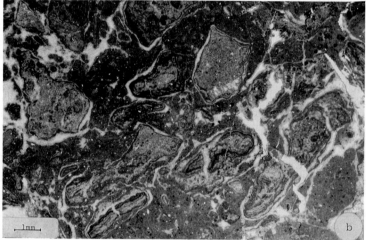

Fig. 4. Facies at the arid/humid climate transition: Laminated Dolomite/Ceroide Limestone boundary. (a) Vadose facies at the base of the platform consisting of limestone intraclasts and calcite cements. (b) Vadose facies at the top of the sabkha complex consisting of early dolomite intraclasts and dolomitic cements.

laminated stromatolitic facies persists (Gandin, 1990a).

The Laminated Dolomite represents a sabkha complex whose style of deposition has been related to the Trucial Coast model (Carmignani *et al.*, 1986a; Gandin, 1990a). However, in the Lower Cambrian sabkha of Sardinia, carbonate deposition prevailed over evaporitic sedimentation which is only recorded by small pseudomorphic chert nodules and barite lenses.

The areal distribution of the facies from west to east indicates a different morphology of the two ends of the platform (Fig. 12): (1) a western higher-energy margin, with patch-reefs built by archae-ocyathids and *Epiphyton*, whose remains are now found trapped in both the stromatolites and oncolites; and (2) an eastern ramp-like margin where, in a low energy shallow-subtidal environment, the stromatolites developed and were not dolomitized. This morphology as well as the complete absence of terrigenous material suggests that this sabkha developed on an isolated platform.

Ceroide Limestone (Upper Botomian to Lower − Middle Toyonian)

This unit is a massive body of pearl-grey limestone whose thickness ranges from 160 to 800 m (Figs 2 & 3). The transition from the Laminated Dolomite is gradational and marked by a continuous belt of black dolomitic limestone. By way of contrast, the upper boundary is an unconformity with pink, bedded marly limestone of Cabitza Formation. The Ceroide Limestone is locally affected by dolomitization (Grey Dolomite). In its upper part karstic features, calcite veins as well as breccia bodies are present.

Common fossils are echinoderm plates and trilobite fragments. Rare archaeocyathids and *Epiphyton* occur in the upper part of the limestone body, only in the western area. The archaeocyathids form assemblages with *Archaeocyathus altaicus*, *A.* cf. *grandis*, *A. kusmini* and *Vologdinocyathus tener* indicating a Lower–Middle Toyonian age (Debrenne & Gandin, 1985).

The Ceroide Limestone is noticeable for its great lateral variation in thickness in adjoining sections and complex facies mosaic (Fig. 5). The following facies have been recognized:

1 Tidal flat. Fenestral stromatolitic boundstone (Fig. 7f), and vadose pisolite (*sensu* Esteban, 1976) (Fig. 4a), with rare trilobite fragments.

2 Internal lagoon. Mudstone, peloidal wackestone and packstone.

3 Shoal. Three different facies associations have been found: type (1) consists of skeletal (*Epiphyton*) boundstone with local archaeocyathid mounds (Fig. 7b); type (2) is well-sorted ooid and oncoid grainstone (Fig. 7d). In both facies, trilobite and echinoderm fragments are common; and type (3) is represented by unsorted intraclast grainstone with lumps, peloids and rare trilobite and echinoderm fragments (Fig. 7h).

4 Open lagoon. Skeletal or poorly sorted granular wackestone, laminated mudstone and marly intercalations. The grains are oncoids, micritized grains and intraclasts, that are probably reworked calcrete clasts, and minor quartz silt and clay. Fossils are represented by trilobite and echinoderm fragments, commonly concentrated in small lenses.

These facies, indicative of a shallow-water platform, have a characteristic non-cyclic vertical distribution and are laterally discontinuous (Fig. 5). There is no suggestion of a regressive trend through this unit although there is clear evidence of exposure at the end of the Ceroide Limestone deposition. However, at the base of the sequence, everywhere in the central part of the platform, tidal-flat or lagoonal facies occur.

The areal distribution of the facies indicates an isolated platform bounded by epicontinental basins (Fig. 12). From west to east the following time equivalent facies can be recognized (Fig. 1):

1 Shelf. This occurs in a narrow belt along the coast of the Tyrrhenian sea (Buggerru area) where the well-stratified Ceroide Limestone has open-lagoon facies and in the upper part of a few sections in the platform interior where thin shale layers are intercalated with open-lagoon facies.

2 Platform rimmed margin. Along the western edge of the platform *Epiphyton* buildups with local archaeocyathid mounds interfinger with small oolitic–oncolitic shoals (type 1 and 2 shoal facies).

3 Platform interior. This consists of irregularly alternating restricted-lagoon and tidal-flat facies.

4 Platform ramp-like margin. In the eastern area, type 3 shoal facies interfingers with open-lagoon facies. Here the limestone decreases in thickness, is evenly bedded, and contains thin marly and shaly intercalations, slump structures (Cocozza, 1969) and channels.

Cabitza Limestone (Middle Cambrian)

This unit is made up of bedded, locally nodular, fossiliferous limestone alternating with red or green silty shales up to 80 m thick. It unconformably overlies the Ceroide Limestone with a sharp contact locally marked by a palaeokarstic surface and slope breccias. It contains rich assemblages of early Middle Cambrian age (Rasetti, 1972), including trilobites (*Corynexochus sardous*, *Pardailhania hispida*), echinoderms, *Chancelloria*, brachiopods, hyolithids and miscellaneous sponge spicules. Limestone and shale facies are highly variable in thickness and colour. The latter ranges from black to grey, green, pink, red or yellow. Three irregularly alternating different lithotypes can be found: (1) massive, evenly-bedded limestone with wavy shaly partitions; (2) nodular marly limestone (Fig. 8a); and (3) alternating calcareous and shaly planar laminae (Gandin, 1990a, b). The limestone facies are skeletal wackestones and packstones with a diverse open-marine fauna (Fig. 8b) and silt-sized quartz grains, pyrite and hematite. The latter gives the red colour to the intercalated red shales (Lecca *et al.*, 1983).

The sedimentary features and fossil content of this unit are on the whole indicative of a more or less restricted shallow-shelf environment, where pelagic siliciclastics and residual 'terra rossa' coming from the karstified platforms interfered with the *in situ* carbonate production. The genesis of the nodular structure of the limestone can be mainly related to compaction and dissolution of sediments with non-homogeneous density gradients, triggered by tectonic instability of the shelf. This facies was therefore

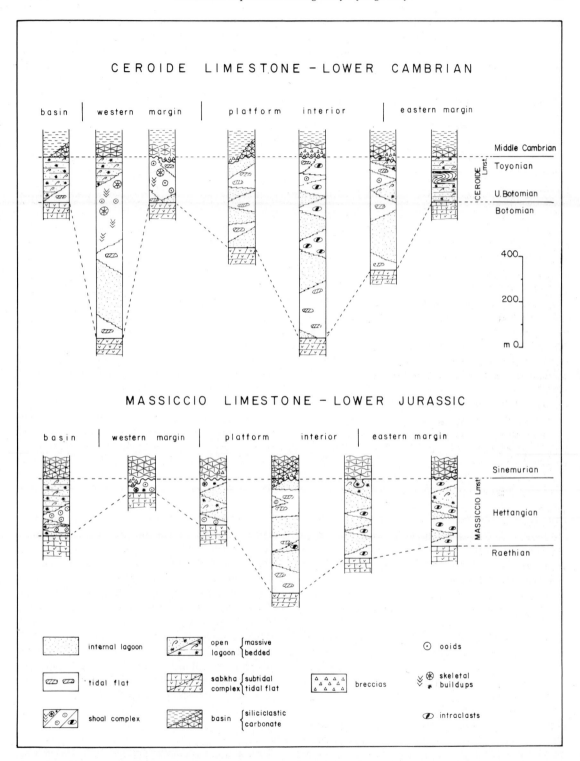

Fig. 5. Vertical and lateral facies distribution across Lower Cambrian Ceroide platform in Sardinia and Lower Jurassic Massiccio platform in Tuscany.

located on top of unstable highs whereas the massive and laminated facies correspond to more stable zones on the highs and in the troughs, respectively (Gandin, 1990b).

Cabitza Shales (Middle Cambrian to Arenig)

This unit consists of at least 600 m of siliciclastic deposits evolving from siltstones with fine-grained quartz−arenite intercalations to thinly laminated mudstones. Limestone nodules occur locally at the transition with the underlying Cabitza Limestone and disappear upwards. The laminated mudstone facies, characterized by rhythmical lamination, ripples, convolute lamination and slumps, documents deposition by turbidity currents, and indicates a gradual change towards a deeper-basin setting (Gandin, 1990a).

Fossils are very rare, commonly concentrated in small lenses. They are represented in the lower part by Middle Cambrian trilobite assemblages with *Paradoxides mediterranenus*, followed by *Jincella* cf. *prantli* assemblages (Rasetti, 1972) and in the upper part by *Dictyonema flabelliforme* and acritarchs (Tremadocian *Clonograptus* biozone; Barca et al., 1987).

Post-depositional processes

The Ceroide Limestone underwent extensive fresh water diagenesis as documented by blocky calcite cements. Marine cements are poorly developed and locally show traces of dissolution. In the upper part of the unit, palaeokarsts and neptunian dykes occur whereas at the boundary with the Cabitza Limestone and within it, breccias bodies are locally found (Fig. 2). Contorted beds, interpreted as cohesive slumps (Cocozza, 1969), occur along the eastern margin of the platform (Fig. 5).

Paleokarst

Karstic features consist of (1) cavities filled by breccias with hematitic matrix; (2) fractures filled by fibrous-radial and/or blocky calcite (Fig. 10b); and (3) cavities with smooth walls and pipes filled by shales of Cabitza lithologies or breccias and alternating carbonate laminae and black shales (Fig. 10a). Stable isotope analysis of the fibrous radial calcite of the karstic cavities as well as of the blocky cements of the Ceroide Limestone, support a meteoric origin for the diagenetic waters from which the calcite was

precipitated (δ O^{18} values = 12.57 PDB; Gandin & Turi, 1990). An extensive karstification of the platform, with the resulting formation of residual 'terra rossa' is also documented by the red hematitic material included in the breccia matrix and in the deposits overlying the massive bodies (Cabitza Limestone).

Neptunian dykes

The Ceroide platform was affected by extensive fracturing and most of the fractures were filled by fresh-water sparry calcite. However, there are small (up to 1 m wide) fractures filled by silicified marls with echinoderm and trilobite debris, equivalent to the homogeneous facies of the Cabitza Limestone.

Breccias

Two types of breccias occur at the boundary between platform and pelagic facies (Gandin, 1990b):
1 Karstic breccias of pebble-size, subrounded clasts of Ceroide Limestone that locally have pressure-dissolved contacts. The matrix is a red hematitic silt that also contains small calcite fragments and locally grades upward to pink marls of the Cabitza Limestone facies. The clasts are derived from lithotypes of the upper part of the platform, including archaeocyathid-rich facies that only occur along the western margin of the platform (Debrenne & Gandin, 1985; Gandin, 1990b). The breccias occur in palaeokarstic sinkholes within the upper part of the Ceroide Limestone or in pockets exactly at its boundary sealed by the Cabitza Limestone.
2 Slope breccias of blocks up to several meters in diameter; these are mainly derived from various facies of the Ceroide Limestone although some of them originated from the dolomitized equivalent of the Ceroide Limestone (Grey Dolomite). They occur as lenticular bodies up to 10 m thick and 100 m long, at the contact between Gonnesa and Cabitza Formations (Boni et al., 1981). Two different types of matrix occur: the former consists of marly facies of Cabitza Limestone, the latter of black shales and siltstones of Cabitza Shale. In the latter instance clasts of Cabitza Limestone can also be found among the blocks (Gandin, 1990b).

Slumping

Deformation structures and associated breccias are common along the eastern margin of the Ceroide

Limestone platform (Cocozza, 1969) and similar structures also occur along the western edge (Balassone *et al.*, 1985). Slump structures have also been reported from the Laminated Dolomite in the western and eastern margins of the platforms by Bechstadt *et al.* (1985). The deformation structures reported by them from the western stromatolitic — evaporitic facies are associated with that shown in Fig. 6a and similar to that of Fig. 6c and, therefore, are here interpreted as the result of plastic movements of now-vanished evaporites and connected with later tectonics.

The slump structures occurring in the eastern region and referred by Bechstadt *et al.* (1985) to the Laminated Dolomite, actually occur 20−50 m below the contact with the Cabitza Limestone and affect either the laminated or massive mudstone facies of the Ceroide Limestone in the ramp-like margin of the platform.

Ceroide platform evolution

Incipient platform

The first isolated platform stage is represented by the Laminated Dolomite deposited in a tidal-flat sabkha system flanked to the west and east by open-sea waters (Fig. 12 I). As a result of the present structural setting, the supposed basin sediments are not exposed. However, evidence of their existence is provided by: (1) lack of land-derived siliciclastics; (2) development of calcareous facies towards the western and eastern edges of the platform; and (3) the presence on the western edge of archaeocyathids and algal (*Epiphyton*) buildups. Moreover, the occurrence of normal marine organisms to the west is suggestive of a steeper margin there than in the east.

The depositional characters of the Laminated Dolomite and the rather uniform cyclic deposition and sediment thickness in the platform interior indicate regular, moderate subsidence of the platform and/or slow sea-level fall under an arid climate.

Ceroide Limestone platform

The vertical and lateral distribution of the facies and the lack of terrigenous material in the Ceroide Limestone are indicative of an isolated platform consisting of a wide, internal shallow-lagoon/tidal-flat system with asymmetric margins (Fig. 12 II). Throughout its growth the platform interior maintains a non-cyclic internal-lagoon/tidal-flat

deposition. This behaviour suggests that carbonate production kept up with irregular subsidence.

The western margin was built by organic reefs (algae and minor archaeocyathids) and oolitic shoals. Its growth appears to have been controlled by a rapid relative sea-level rise that favoured up-building and prevented lateral outgrowth. To the west the shoal was flanked by a deeper basin, whose existence can be inferred by the occurrence of a narrow belt of bedded, open-lagoon facies (Buggerru area).

The eastern margin, wider than the western one, had a ramp-like profile. It was mainly constructed by bedded marine lime mudstones containing a small amount of fine-grained siliciclastics. The limestone thickness gradually decreases toward the eastern Sardinia (Nappe Zone) siliciclastic basin. In this area too, although the transition to the basin is not exposed for structural reasons, the thickness reduction, the occurrence of siliciclastic material and the bedding of the Ceroide Limestone indicate a clear trend towards a deeper basin (Gandin *et al.*, 1987).

Within the platform the complex internal stratigraphy, the non-cyclic recurrence of the facies and the great thickness variability imply differential subsidence of the different areas. Rate of sedimentation can be estimated to be in the order of 30−160 mm 10^{-3} yr assuming that the time interval of deposition from Upper Botomian to Lower−Middle Toyonian was 5 Ma. (Toyonian is the upper stage of Lower Cambrian in the Russian subdivision of Cambrian (Rozanov & Sokolov, 1984). This stage was previously included in the Lenian whose duration is 15 Ma (Harland *et al.*, 1982). The Toyonian, which according to Rozanov & Sokolov (1984) is a short time interval is assumed here to have lasted 5 Ma). The high rates of sedimentation compare well with the values reported from the Tethyan Triassic platforms and the deposits of modern platforms (D'Argenio, 1970; Bernoulli, 1972; Wilson, 1975). The depositional behavior of the Ceroide platform deposits reflects an irregular and rapid subsidence balanced by carbonate production under a humid climate.

Platform exposure

The deposition of Ceroide Limestone ended when most of the platform emerged in late early Cambrian time (Fig. 12 III). Owing to the differential subsidence occurring during the previous depositional period, not all the platform sections were exposed

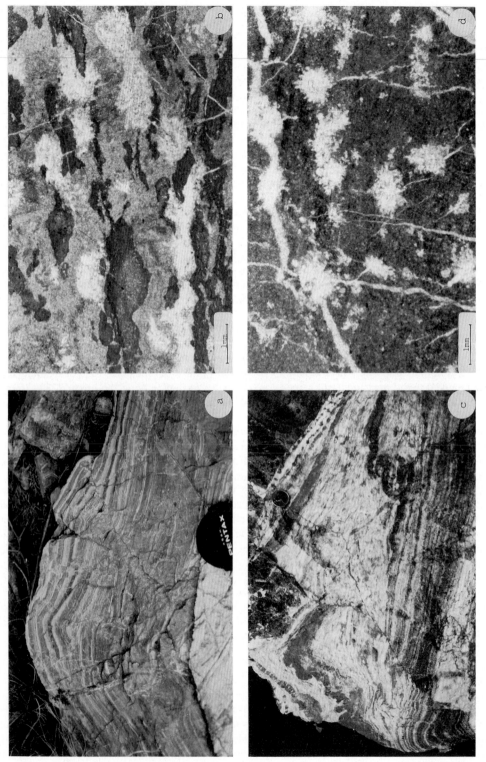

Fig. 6. Arid platform. Lower Cambrian Laminated Dolomite (above) and Upper Triassic Boccheggiano Anhydrite (below). (a) Stromatolitic laminites affected by tectonic deformation resulting from the former occurrence of sulphates. (b) Pseudomorphs of sulphate nodules in early-dolomitic mudstone. (c) Stromatolitic laminite and anhydrite layers tectonically deformed by the plastic behaviour of sulphates. (d) Anhydrite nodules and rosettes in early-dolomitic mudstone.

and the more distal marginal areas were never uplifted. Negative zones at the top of the Ceroide Limestone are characterized by restricted-lagoon, laminated facies including thin shaly layers in the marginal zones.

Positive zones were fractured and subjected to karstic dissolution producing cavities and clasts of the upper early-lithified sediments. The lack of well-developed regressive sequences attests to a rapid uplift of the blocks that was responsible for circulation of meteoric waters and consequent extensive fresh-water cementation of sediments not yet completely lithified in the marine environment, and local dolomitization.

Platform collapse

At the beginning of Middle Cambrian the previously uplifted platform began to collapse and drown (Fig. 12 IV). Biostratigraphic data and sedimentological evidence indicate that the drowning was differential and discontinuous. The drowning of the platform was achieved through two main stages leading to a marginal-plateau structure (Fig. 12).

A first drowning event in the early Middle Cambrian (*Pardailhania hispida* horizon) resulted in the deposition of Cabitza Limestone. Most of the platform was rather suddenly flooded to neritic depth. Exposure and shallow-water deposition persisted only on a number of blocks. The residual hematitic material, previously produced on the exposed platform, was in part swept out and mixed with pelagic clays and autochthonous carbonates. Fissures and karstic cavities, not yet filled by calcite or karstic breccias, were filled with marine unconsolidated sediment. Slope breccias accumulated at the fault scarps, mostly along the western and eastern margins of the platform (Boni *et al.*, 1981; Gandin, 1990b) and were covered by Cabitza lime muds. Most of the breccias occur at the Ceroide Limestone–Cabitza Limestone contact, but smaller breccia units occur also within the Cabitza Limestone. Their subrounded clasts of Ceroide Limestone lithofacies indicate local persistence of emergent Ceroide blocks. Deposition of Cabitza Limestone was characterized by carbonate production rhythmically alternating with siliciclastic material. During carbonate intervals a well-diversified benthic fauna developed. The block-and-basin topography of the seafloor resulted in two different facies of the Cabitza Limestone. The massive calcareous facies was formed on structural highs whereas the laminated marly facies was deposited in deeper-water settings. The nodular facies occurs only locally and at different stratigraphic levels of the sequence. Its genesis records instability of the blocks and their differential drowning.

A second drowning event took place later in the Middle Cambrian (*Paradoxides mediterraneus* to *Jincella* horizons). It led to the final foundering of the platform and to resumption of connections with the mainland. Quartz sands were at first transported in the Cabitza Shale basin in which only later exclusively siliciclastic distal turbidites were deposited. The few structural highs that were left exposed along the western margin during deposition of Cabitza Limestone also drowned and the karstic cavities were filled by black or grey Cabitza Shales. Megabreccias made up of Ceroide Limestone and locally Cabitza Limestone clasts accumulated at the fault scarps mostly along the western margin of the platform and were embedded in black muds and silt (Gandin, 1990b).

Climatic control

The Ceroide Limestone is a pure calcareous unit in which no traces of early dolomitization have been found in the widespread intertidal–supratidal facies. The *Epiphyton*-dominated buildups have been interpreted as indicative of normal marine water (Debrenne & Gandin, 1985). Moreover, the abundance of blocky calcite cements and the average limestone isotopic values ($\delta O^{18} = -8.9$ PDB) indicate a consistent supply of meteoric waters during the early diagenetic processes of the Ceroide Limestone (Gandin & Turi, 1990). These features imply humid tropical conditions and the same humid regime persisted when, with the uplift of the platform, the sediments underwent extensive fresh-water cementation, karstification and local dolomitization.

The lower boundary of the Ceroide Limestone platform is marked by a rather sharp transition from the underlying Laminated Dolomite. The latter was deposited in an arid, evaporitic tidal flat where the carbonates underwent early dolomitization.

A climatic change from arid to humid has been proposed to explain why Ceroide Limestone and Laminated Dolomite sediments went through such different early diagenetic processes (Fig. 4), since they were both laid down in the same tidal-flat conditions (Gandin, 1980).

MESOZOIC OF TUSCANY

The Upper Triassic—Middle Liassic carbonate sequence (≈ 2000 m thick) is underlain by continental deposits (Verrucano), which lie unconformably on the Palaeozoic basement or on Middle Triassic sediments. In the External Tuscan Domain this sequence records a major marine transgression (Figs 2 & 12) starting during the Carnian with lagoonal, mixed carbonate—siliciclastic—evaporitic deposits (Tocchi Formation), followed by siliciclastic-free evaporites and carbonates (Boccheggiano Anhydrite). The transition to the Lower Liassic Massiccio Limestone is marked by the deposition of carbonates with thin shale intercalations (*Rhaetavicula contorta* beds) in a wide metahaline lagoon.

In the Internal Tuscan Domain (Apuane Alps and Massa Zone), a carbonate platform developed during the Norian under an arid climate (Grezzoni Dolomite), locally directly on the basement. It interfingers to the west with a calcareous marly sequence laid down in a deepening-upward basin (La Spezia Formation) which persisted through Lower Liassic with the deposition of the Ferriera Formation. Elsewhere a carbonate platform developed under a humid climate (Massiccio Limestone) and this was later drowned and covered by open—marine neritic deposits (Ammonitico Rosso and Cherty Limestone).

Tocchi Formation (Carnian)

This formation consists of alternating sericitic and chloritic shales and thin dolomites (up to 40 m thick) passing upward to evaporite solution breccias (Figs 2 & 9). Rare and poorly preserved foraminifera (*Glomospira* cf. *kuthani*, *Diplotremina* sp. and *Glomospirella* spp.) suggest a Carnian age (Costantini *et al.*, 1980). They formed in restricted lagoons located along an arid shoreline (Costantini *et al.*, 1980).

Boccheggiano Anhydrite (Carnian to Norian)

A thick evaporitic sequence (up to 1000 m), known in Tuscany as Boccheggiano Anhydrite, Burano Formation or Cavernoso Limestone (Calcare cavernoso), overlies the Tocchi Formation and underlies the *Rhaetavicula contorta* beds (Figs 2 & 9). Boccheggiano Anhydrite (Anidriti di Boccheggiano) is an informal name commonly given to the Triassic anhydrite—carbonate unit that in the

subsurface of Boccheggiano (southern Tuscany) overlies the Palaeozoic Boccheggiano Formation (Costantini *et al.*, 1980). Burano Formation is an Umbrian unit. In Tuscany the authors commonly refer to the evaporitic sequence as the Burano Formation when found sheared off at the base or locally enclosed in the Ligurian nappes. Recent research on foraminiferal assemblages of both the Burano Formation and Boccheggiano Anhydrite, suggests different microfaunal provinces and a different age range (Norian to Rhaetian) for Burano Formation (Martini *et al.*, 1989). The Boccheggiano Anhydrite ranges from Carnian with *Triadodiscus eomesozoicus—Glomospirella capellinii* assemblages to Norian with *Glomospirella rosetta—Gandinella apenninica* assemblages (Zaninetti, pers. comm.). This unit consists of alternating layers of anhydrite and early dolomite. The Cavernoso Limestone is a dolomitic breccia with abundant calcite cements. It resulted from anhydrite—gypsum—anhydrite changes induced by Alpine tectonic stresses, and subsequent dissolution of the sulphates during exposure and weathering. It mostly represents the evaporitic sequence in outcrop whereas in the subsurface the sulphates and the stratigraphic order are better preserved (Costantini *et al.*, 1983; Martini *et al.*, 1989).

The carbonate facies are stromatolites and mudstone with anhydrite rosettes and nodules (Fig. 6c,d), wackestone with faecal pellets and skeletal grains, and fine-grained intraclast or oolitic grainstone. They indicate an arid low-energy intertidal—supratidal environment corresponding to a sabkha system (Martini *et al.*, 1989).

Grezzoni Dolomite (Carnian to Rhaetian)

Grezzoni Dolomite

This is a dolomite unit up to 300 m thick that overlies the Tocchi Formation (Montagnola Senese) or lies directly on the Palaeozoic basement (Apuane Alps). It grades upward to the Massiccio Limestone. Westward of the Apuane Alps, in the La Spezia region, Grezzoni Dolomite interfingers with La Spezia Formation (Ciarapica & Passeri, 1980). Early Jurassic Massiccio Limestone and Ferriera Formation are respectively superposed over Grezzoni Dolomite and La Spezia Formation (Figs 2 & 9).

La Spezia Formation consists of two members: the Norian Monte S. Croce Member and the

Rhaetian Portovenere Member. The facies are mostly mudstones associated with skeletal wackestone and packstone including low diversity bivalve assemblages and oolites in the lower part, and *Triasina hantkeni* and radiolarians in the upper part (Ciarapica & Zaninetti, 1984).

The Grezzoni Dolomite includes poorly diversified fossil associations of *Worthenia escheri*, megalodontids, dasycladacean algae (*Gyroporella*, *Physoporella*, *Diplopora*) and foraminifera. The last are found in two superposed assemblages, the first is Norian with *Agathammina australpina* and Glomospirellas, the second, Rhaetian with *Trasina hantkeni* and Involutinas (Ciarapica & Passeri, 1978).

The facies, consisting of oolitic and skeletal grainstone, wackestone with foraminifera, mudstone and stromatolites, document an arid setting in a shallow subtidal to supratidal depositional system (Ciarapica & Passeri, 1978). The Grezzoni Dolomite is bounded to the west by an oolitic shoal flanked by the deeper water of the La Spezia basin (Fig. 12). To the east the tidal flats graded to an hypersaline lagoon whose sediments probably passed into the Boccheggiano Anhydrite although evidence of this transition is missing because of the structural setting of the region (Ciarapica & Passeri, 1978; 1980).

Rhaetavicula contorta beds (Rhaetian)

This unit, 60–100 m thick, consists of black limestone, alternating with black shales which gradually disappear in the upper part (Figs 2 & 9). Gradational boundaries exist with the underlying evaporitic sequence and the overlying Massiccio Limestone. The black limestone facies consists of mudstone, commonly with anhydrite and celestite pseudomorphs, bioclastic packstone and grainstone, and stromatolites. Fossils, locally very abundant, form low-density assemblages dominated by bivalves (*Rhaetavicula contorta*) and foraminifera (*Triasina hantkeni*) diagnostic of Rhaetian.

In the Monte Cetona area of southeastern Tuscany the upper part of the *Rhaetavicula contorta* beds are characterized by skeletal banks with *Megalodon* and dasycladacean algae and mounds built by encrusting organisms and stromatolites (Ciarapica *et al.*, 1982). The depositional setting represents a confined, wide metahaline lagoon bounded to the east by low-relief shoals (Fig. 12).

The Upper Triassic sequences of Tuscany were deposited under an arid climate in a complex en-

vironmental setting. On the structural highs (Apuane Alps, Montagnola Senese) the Grezzoni Dolomite tidal-flat system developed; it was bounded to the west by an epicontinental basin and to the east it graded to a depressed, more subsident area, in which during Carnian and Norian, sulphate deposition prevailed over the carbonates (Boccheggiano Anhydrite), and during Rhaetian this was replaced by a wide metahaline lagoon (*Rhaetavicula contorta* beds). The areal distribution of the facies and the lack of terrigenous material in both the carbonate and evaporitic sequences document a wide platform detached from the emergent land, whose morphology was initially controlled by the basement structural pattern (Fig. 12).

Massiccio Limestone (Lower Liassic)

This unit consists of a massive body of pure white limestone whose thickness ranges from 50 to 1000 m. At the base it is bounded by the Grezzoni Dolomite or the *Rhaetavicula contorta* beds, and by Ammonitico Rosso at the top (Boccaletti *et al.*, 1975; Fazzuoli, 1974a, b) (Figs 2 & 9). While the lower boundary is gradational, the upper one is unconformable and is marked by an abrupt contact between the white massive limestone and the pink bedded marly limestone of Ammonitico Rosso.

The Massiccio Limestone is locally affected by recrystallization and minor dolomitization. In its upper part neptunian dykes, karstic features and calcite veins are common (Fazzuoli *et al.*, 1981). Common fossils are gastropods, bivalves, ostracods and arenaceous foraminifera. Solenoporacean and dasycladacean algae (*Palaeodasycladus mediterraneus* and *Thaumatoporella parvovesiculifera*), brachiopods, corals, echinoderms, sponge spicules and *Spirillina liassica* are locally common. They document normal-marine waters and a general Liassic age.

Westward of the northern Apuane Alps, the Massiccio Limestone interfingers with the Ferriera Formation that is an ammonite-rich calcareous unit containing *Schlotheimia angulata* whose age is Hettangian. Elsewhere in Tuscany, the age of the Massiccio Limestone is assumed to be Hettangian on the basis of the age of the underlying and overlying units (Giannini *et al.*, 1972). The Massiccio Limestone displays a greatly variable thickness in adjoining sections and a complex internal stratigraphy (Boccaletti *et al.*, 1975; Fazzuoli, 1974a, b) resulting from repeated vertical and lateral facies

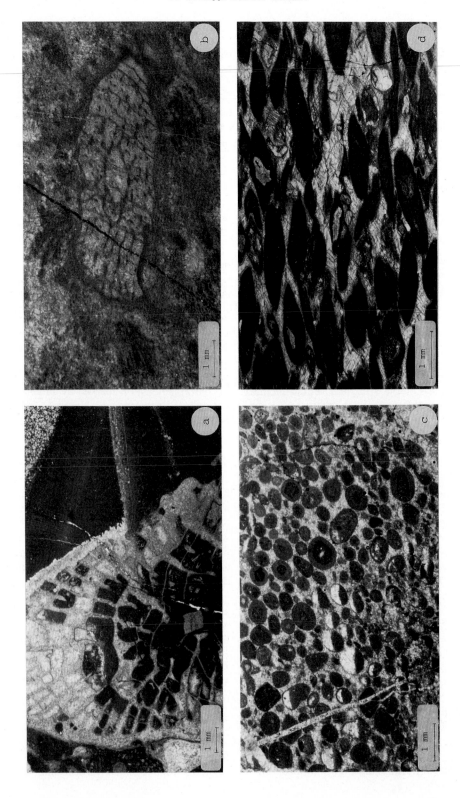

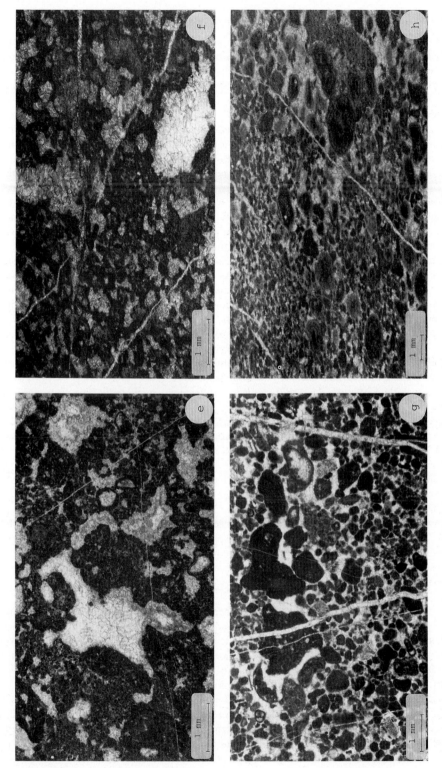

Fig. 7. Humid platform. Lower Cambrian Ceroide Limestone (right) and Lower Liassic Massiccio Limestone (left). (a) Mud mound with corals. (b) Buildup with algae (*Epiphyton*) and archaeocyathids. (c), (d) Oolitic grainstone; larger coated grains and skeletal remains occur in the Cambrian facies, whose grains are badly deformed. (e), (f) Fenestral fabric in stromatolitic boundstone, irregular fenestrae are filled by two generations of cement. (g), (h) Grainstone with calcrete grains and grapestone lumps; peloids and skeletal remains are associated with the Jurassic facies.

changes (Fig. 5). The following facies have been recognized:

1 Tidal flat. Fenestral stromatolitic boundstone (Fig. 7e), beach-rock and vadose pisolite (*sensu* Esteban, 1976). The stromatolitic boundstone is built by encrusting microbial organisms and algae (*Thaumatoporella*). Arenaceous foraminifera, gastropods and ostracods are rare.

2 Internal lagoon. Prevailing mudstone and peloidal—skeletal wackestone, with faecal pellets, ostracods and arenaceous foraminifera.

3 Shoal. Three different facies have been recognized: type (1) consists of well-sorted oolitic (Fig. 7c) or oolitic—oncolitic grainstone with lumps; type (2) corresponds to crinoid—coral mounds (Fig. 7a). Associated fossils in both facies are corals, solenoporacean and dasycladacean algae, molluscs, brachiopods and rare foraminifera. Type (3) is peloidal—intraclast or unsorted—intraclast grainstone (Fig. 7g). Molluscs, echinoderms and rare algae (*Paleodasycladus*, *Solenopora* and *Thaumatoporella*) are found in the intraclasts, as well as sponge spicules and open-marine foraminifera (*Spirillina liassica*).

4 Open lagoon. Skeletal wackestone with pelagic organisms such as ammonites, radiolarians and foraminifera, and locally reworked calcrete clasts and marly intercalations.

These facies have a non-cyclic vertical distribution and are laterally discontinuous (Fig. 5). A regressive trend at the top of the sequence occurs only in some sections of the platform interior. The areal distribution of the facies indicates an isolated platform bounded by epicontinental basins (Boccaletti *et al.*, 1975; Fazzuoli, 1980). From west to east the following facies equivalent zones can be recognized (Fig. 1).

1 Epicontinental basin. The Massiccio Limestone is here replaced by stratified, open-marine facies (Ferriera Formation). In northern Apuane Alps, where the Massiccio Limestone thins out, these strata interfinger in the slope zone off the Massiccio platform.

2 Platform-rimmed margin. This is built by oolitic shoals and locally by crinoidal—coral—mounds (type (1) and (2) shoal facies).

3 Platform interior. This consists of irregularly alternating, restricted-lagoon and tidal-flat facies.

4 Platform ramp-like margin. Type (3) shoal facies containing intraclasts with pelagic organisms interfingers with open-lagoon facies.

Ammonitico Rosso (Middle Lias)

The Ammonitico Rosso lies unconformably on the Massiccio Limestone. The contact is sharp and is locally marked by karstic breccias (Figs 2 & 9). This unit comprises three, irregularly alternating lithotypes: (1) pink, locally laminated limestone with local ammonite and/or crinoidal lumachelles (Calcare rosa a Crinoidi), commonly found at the base of the sequence; (2) red nodular marly limestone (Fig. 8c) with shaly partitions and local concentrations of ammonites (Calcare rosso ammonitico), recurring at different stratigraphic levels; and (3) grey cherty limestone (Cherty Limestone—Calcare selcifero) with thin argillaceous interbeds commonly prevailing in the upper part of the sequence (Fazzuoli & Pirini Radrizzani, 1981). Locally breccias and megabreccias of Calcare Massiccio blocks are found, enclosed in the Cherty Limestone (Fazzuoli, 1974b). The facies (Fig. 8d) are mudstone, skeletal wackestone, and packstone with crinoid plates, ammonites, radiolarians, sponge spicules and foraminifera which indicate an open-marine environment.

Ammonite assemblages document different stratigraphic levels: the older with *Arietites listeri* (*Arietites bucklandi* biozone) is Lower Sinemurian. Upper Sinemurian is documented by faunas with *Eoderoceras olenopthychum* whereas the younger assemblages with *Coeloceras psiloceroides*, *Amaltheus margaritatus* and *Protogrammoceras algovianum* record Pliensbachian (Federici, 1967). Higher in the sequence the Cherty Limestone grades into marls with pelagic bivalves (*Posidonomia alpina* Marls, Toarcian—Aalenian) that in turn pass to basinal, turbiditic siliceous shales and cherts (Diaspri, Dogger—Malm). These in the Ligurian Domain are associated with ophiolites.

Post-depositional processes

The Massiccio Limestone underwent extensive freshwater cementation. Cements mostly consist of blocky calcite whereas marine cements are poorly developed and locally show traces of dissolution. In the upper part of the unit neptunian dykes (6—8 m wide) and paleokarstic features are common. They are well developed along the western margin of the Massiccio platform and less common, but larger along the eastern margin. Breccia bodies are enclosed in karstic cavities or in the Cherty Limestone (Fazzuoli, 1980; Fazzuoli *et al.*, 1981).

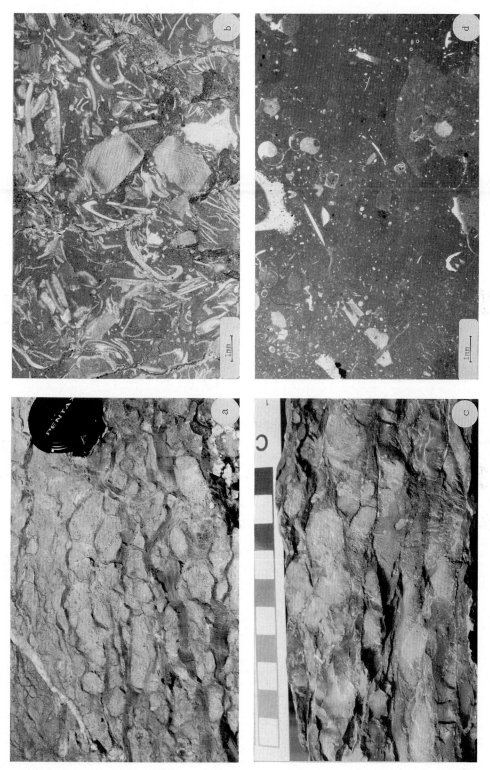

Fig. 8. Nodular facies. Middle Cambrian Cabitza Limestone and Middle Liassic Ammonitico Rosso (below). (a), (b) Limestone consisting of skeletal packstone with trilobite and echinoderm debris. (c), (d) Limestone consisting of wackestone with ammonites, echinoderms and sponge spicules.

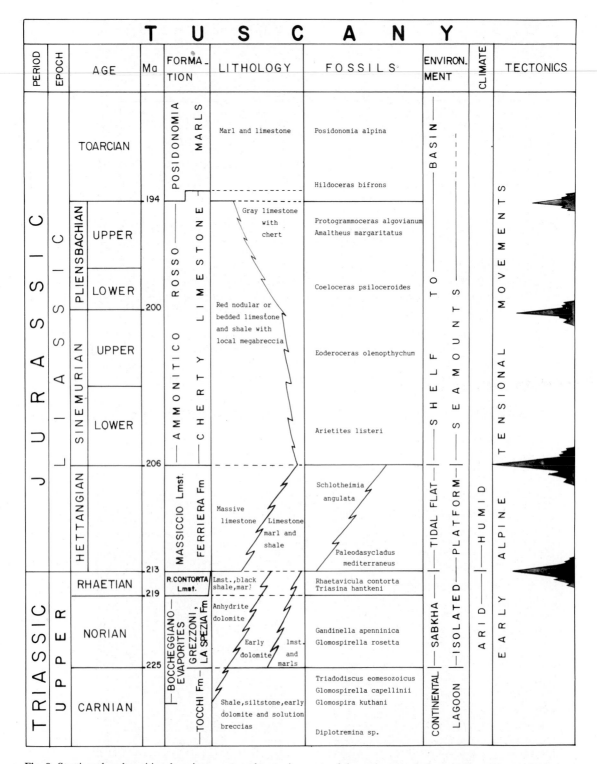

Fig. 9. Stratigraphy, depositional environments and tectonic events of the carbonate platform in Tuscany.

Fig. 10. Karstic cavities in the Ceroid Limestone. (a) Vadose, laminated internal sediment consisting of breccia (base outlined in ink), and carbonate and black shale laminae. (b) Fresh-water calcite filling of fractures.

Paleokarsts

These are cavities and sinkholes with smooth walls, cut in the Massiccio Limestone. The infills consist of (1) laminated vadose silt (Fig. 11a); (2) collapse breccias; and (3) pink, laminated or massive lithofacies of the Ammonitico Rosso. Smaller cavities and pipes are filled by radial fibrous and or blocky calcite (Fazzuoli *et al.*, 1981; Fig. 11b).

Neptunian dykes

These are filled by the pink lithofacies of Ammonitico Rosso that locally encloses angular clasts of Massiccio Limestone or ammonite lumachelles of Lower Sinemurian age.

Breccias

Two different types of breccia can be distinguished.

Type (1) occur within the Massiccio Limestone in sink-holes or pockets sealed by evenly bedded Ammonitico Rosso lithofacies. It includes polygenic breccias with angular to subangular clasts (up to 1 m in size) derived from various facies of the Massiccio Limestone, and from the lithology of the neptunian-dyke fills. The matrix consists of laminated calcareous silt. Type (2) is intercalated in the Cherty Limestone. It has been recognized rarely in Tuscany; however, blocks of the marginal facies (Crinoidal mounds) of the Massiccio Limestone are recorded along the western margin of the platform (Fazzuoli, 1974b).

Platform evolution

Incipient platform

During the Carnian a major transgression occurred on a structurally-controlled continental area (Fig. 12

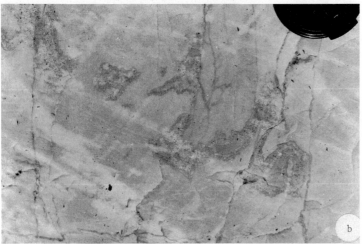

Fig. 11. Karstic cavities in the Massiccio Limestone. (a) Vadose, crudely-laminated internal sediment. (b) Fresh-water calcite filling of karstic pipes. Photographs courtesy of M. Fazzuoli.

I). Directly on the basement or on Middle Triassic red beds (Verrucano) an evaporitic platform developed. The first deposits, consisting of fine-grained siliciclastics alternated with early dolomite and sulphates and were laid down in coastal lagoons (Tocchi Formation). In Norian time on the structural highs, purely carbonate deposition commenced, building an oolitic sand shoal (Grezzoni Dolomite) which to the west graded into the La Spezia shallow-water confined basin (Monte S. Croce Member) and to the east to a rapidly-subsiding, wide tidal flat—sabkha system, where sulphates and early dolomite were produced (Boccheggiano Anhydrite).

During the Rhaetian the western La Spezia basin, deepened and radiolarian-rich limestones were de-posited (Portovenere Member). To the east of the persistent oolitic margin of the Grezzoni Dolomite the subsident sabkha system was replaced by a confined, metahaline lagoon (*Rhaetavicula contorta* beds), eastward (Monte Cetona) rimmed by a low-relief margin built by encrusting organisms. The occurrence of this margin suggests the individual development of an eastern, probably shallow-water negative zone during this time that at present is not exposed for structural reasons. At the end of Rhaetian the Upper Triassic Tuscan platform was to the west rimmed by an oolitic shoal facing a deeper basin and to the east the platform sloped down with a ramp-like configuration, to a shallow intrashelf basin.

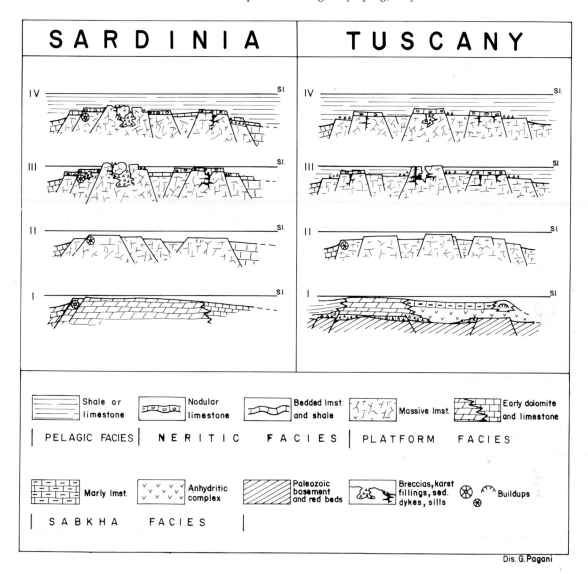

Fig. 12. Tectono-sedimentary evolution of the Cambrian and Upper Triassic—Liassic carbonate platforms of Sardinia and Tuscany: Stage I: Isolated platform established under arid climate on an inherited morphology (Botomian/Carnian—Rhaetian). Stage II: Isolated platform evolved under humid climate in an extensional tectonic regime which induced block faulting and differential subsidence (Upper Botomian—Toyonian/Hettangian). Stage III: Differential movements during drowning caused exposure of a number of blocks and resulted in karstic activity, slope breccias and neptunian dykes (Middle Cambrian/Lower Sinemurian). Stage IV: Drowning of the platform and onlap of pelagic sediments. The nodular facies formed on top of the more unstable blocks (Middle Cambrian—Lower Ordovician/Sinemurian—Pleinsbachian).

Massiccio Limestone platform

The vertical and lateral distribution of the facies and the lack of terrigenous material are indicative of an isolated platform consisting of a wide tidal-flat system with internal lagoons and asymmetric margins (Fig. 12 II), similar to the present situation in the Bahamas (Boccaletti *et al.*, 1975). The platform interior maintains throughout its growth, a non-cyclic, tidal-flat, internal-lagoon deposition. This behaviour suggests that carbonate production kept up with irregular subsidence.

The western margin faced a deeper basin, where pelagic lime-muds were deposited (Ferriera Formation) and this was rimmed by an oolitic shoal with fewer crinoid−coral mounds. The shoal growth appears to have been controlled by a rapid relative sea-level rise that prevented lateral outgrowth and favoured up-building. The eastern margin had a ramp-like profile. It was built by non-oolitic lime-sand shoals that have a seaward prograding trend towards the east, probably caused by a slower relative sea-level rise. The non-cyclic recurrence, the contrasting facies and thickness of the platform deposits imply differential subsidence in its different areas.

Rates of subsidence vary in the different sections of the Massiccio platform and are in the order of $70-140$ mm 10^{-3} yr. These values are higher than those reported for the post-Triassic Tethyan platforms and for the ancient deposits of recent platforms but compare rather well with the Triassic platforms of southern Apennines (D'Argenio, 1970; Bernoulli, 1972; Wilson, 1975). The depositional behaviour of the platform deposits documents an irregular and rapid subsidence balanced by carbonate production in an humid-climate regime.

Platform exposure

Deposition of the Massiccio Limestone was interrupted at the end of the Hettangian. Most of the platform interior and also its margins were cut by extensional fractures and the resulting blocks reacted in different ways to the foundering trend (Fig. 12 III). Some blocks tilted and became emergent with different rates so that varying depositional and diagenetic processes resulted. Local, more gradual emergence produced shallowing upward sequences with calcrete and beach-rock. Rapid uplift prevented deposition and started early cementation and karstification. Dissolution breccias and internal vadose sediments filled the karstic cavities. Circulation of meteoric waters was responsible for extensive freshwater cementation of the platform sediments and local dolomitization.

Platform collapse

Most of the shallow-water Massiccio Limestone platform was suddenly flooded at the beginning of Sinemurian by a neritic sea (Fig. 12 IV). This event is also recorded in the western basin where on the Hettangian (Schlotheimia angulata Zone) deeper-water sediments of Ferriera Formation, Ammonitico Rosso deposition began during early Sinemurian (Atractites bucklandi Zone). Differential and discontinuous drowning of the different parts of the platform is attested to by both biostratigraphic and sedimentologic evidence. The drowning of the platform, leading to a marginal plateau morphology (Bernoulli et al., 1979) was achieved through three main stages.

The first stage, dated as Lower Sinemurian (Atractites bucklandi Zone) followed the dissection of the platform and led to the differential collapse of the blocks. Along the western margin and in the platform interior, some of the blocks were uplifted and subaerial processes produced early lithification and karstification. On the sunken blocks either the grey cherty facies or the pink crinoidal facies of the Ammonitico Rosso were deposited. The latter filled also the open fissures of the neptunian dykes.

The second stage, ranging from Upper Sinemurian to Pliensbachian (Eoderoceras assemblages) resulted in the deposition of the red and grey cherty facies of the Ammonitico Rosso on the newly formed plateau. The diachronous occurrence of the nodular facies records the irregular morphology of the seafloor and the instability of the blocks during drowning. During this stage the neptunian dykes were sealed by Cherty Limestone and most of the blocks previously emergent along the former western margin of the platform subsided. This is attested to by the shallow-water facies of the breccia-blocks interbedded in the Cherty Limestone. Conversely some of the previously drowned ones were newly uplifted, as demonstrated by the karstic dissolution affecting the filling of the neptunian dykes.

The third stage, at the end of the Pliensbachian (Amaltheus margaritatus Zone), records the general deepening of the plateau. It is marked by the widespread deposition of the Cherty Limestone that locally includes turbidites, slumps and debris flow deposits with clasts of shallow-water carbonates derived from persistent uplifted blocks (Bernoulli et al., 1979). From the beginning of Toarcian, with the deposition of marls with pelagic bivalves (Posidonomia alpina beds) depth increased and resulted in the deposition of radiolarian cherts in a deep-water basin (Diaspri), during Malm−Neocomian.

Climate control

The Massiccio Limestone is a pure calcareous unit in

which intertidal–supratidal facies are abundantly represented and never affected by early dolomitization processes. Organisms were scarce but consist of forms indicating normal salinity. These points indicate humid-tropical conditions of the depositional setting. Persistent humid conditions during platform exposure are documented by extensive fresh-water cementation, karstification and local dolomitization.

The lower boundary of the Massiccio Limestone platform is marked by the gradual but rapid transition from the underlying Triassic deposits (Grezzoni Dolomite or *Rhaetavicula contorta* beds). The uppermost Triassic sediments and the lowermost Massiccio deposits were laid down in equivalent physiographic settings although the water salinity was different: metahaline-hypersaline in the Triassic, normal marine in the Jurassic. Such a depositional behaviour has been ascribed to a climatic change from arid to humid conditions (Ciarapica *et al.*, 1982).

TECTONO-SEDIMENTARY SETTING

The Mediterranean Triassic and early Jurassic carbonates have been interpreted to result from deposition along continental margins subjected to extensional tectonic regimes in connection with the opening of the Tethyan Ocean (Bernoulli, 1972; Bernoulli & Jenkins, 1974). Early extensional tectonics along embryonic passive continental margins, resulted in the formation of Bahamian-type platforms, some of which persisted throughout the Jurassic and Cretaceous, surrounded by deep-water basins (D'Argenio, 1970; Bernoulli, 1972). The Tuscan–Ligurian Mesozoic sequences record the evolution of an intracratonic rift in a N–S trending branch of the Tethys into the opening of the Ligurian Ocean in the Middle Jurassic (Bernoulli *et al.*, 1979).

The Sardinian Lower Palaeozoic sequences have been considered as the record of an oblique intracratonic rifting within the eastern side of the Gondwana craton, at some distance from the zone where the Iapetus Ocean would later open. The compressional Sardic phase which deformed and uplifted the Cambro-Ordovician basin was the response to transpressional movements that prevented further evolution towards an oceanic opening (Carannante *et al.*, 1984; Vai & Cocozza, 1986). The tensional movements controlling sedimentation during early Palaeozoic in Sardinia and early Mesozoic in Tuscany were a response to major plate tectonic movements.

The southward drifting of southern Europe during the early Palaeozoic is documented by palaeomagnetic data (Scotese *et al.*, 1979), and suggested by the distribution of trilobite faunas (Jell, 1973). The northern drifting of Europe during early Mesozoic is documented by palaeomagnetic data (Barron *et al.*, 1981) and suggested by the southerly migration of the evaporitic belt from Triassic to Jurassic (Frakes, 1979).

TECTONO-SEDIMENTARY EVOLUTION

The depositional evolution of the Cambrian and early Mesozoic platforms in Sardinia and Tuscany is not directly comparable to the models that have been established for carbonate systems without the complication of syndepositional tectonics (Read, 1985). The sedimentological analysis of the Cambrian–Lower Ordovician and Upper Triassic–Jurassic sequences documents a similar tectonic control in the segregation and depositional behaviour of the two platforms. Both units represent isolated pericratonic platforms fringed by epicontinental basins. Their growth started under an arid climate and ended in a humid regime. The features of the sediments produced during the humid stage in the two platforms reflect facies distribution and climate similar to the present situation in the Bahamas. However, during their growth they were surrounded by shallow-water basins and only in the late sinking stage did some of the uplifted blocks achieve Bahamian relief. Persistence of shallow-water deposition throughout their growth document a continuous balance between carbonate production, subsidence and/or relative sea-level fluctuations. The drowning trend of the platforms is reflected by the non-cyclic depositional pattern of the calcareous facies.

Synsedimentary extensional tectonics is also documented by high rates of sedimentation resulting in upbuilding and differential subsidence. One difference can be found in the pure siliciclastic deposition of the pelagic Middle Cambrian–Lower Ordovician compared with the prevailing carbonate character of the pelagic Middle Jurassic. Siliciclastic basinal deposits are reported to be a common feature in the Palaeozoic (Read, 1985). The lack of calcareous planckton in the seas and lack of vegetation on the lands, during early Palaeozoic, can explain this difference. Nonetheless a climatic variation should also be taken into account for the Sardinian sequences since the occurrence of possible tillites and the bryozoan-dominated littoral faunas in the overlying Caradocian deposits suggest a climatic trend towards cold regimes.

The fracturing and drowning of the Massiccio and Ceroide platforms are regarded by the authors as the first sign of rifting (Bernoulli *et al.*, 1979; Carannante *et al.*, 1984). However, the depositional behaviour of the Lower Cambrian and Upper Triassic–Jurassic sequences suggests that earlier extensional movements controlled both the detachment of the platforms and their internal stratigraphy during growth. Therefore they can be interpreted as indicative of the very beginning of continental rifting. Moreover, the sharp transition from early dolomite to pure limestone, occurring within the two platforms shortly before their collapse, could be interpreted as evidence of divergent movements that caused the drifting of the continental plates from tropical arid to humid zones.

Lower Cambrian platform

The platform developed on a Botomian, mixed siliciclastic carbonate delta-plain-coastal system fringed by an oolitic shoal to the west, and gently sloping to the east. This morphology was probably inherited and controlled by the underlying basement. With the cessation of the siliciclastic influx the platform was isolated from the mainland. During the first stage of growth, the isolated platform developed under an arid climate; in the platform interior and in the eastern ramp-like margin, a regular and moderate subsidence controlled the development of a sabkha system. Conversely, on the platform western edge a pulsing subsidence resulted in the development of a more subsident basin seaward of a rimmed margin.

The second stage of the platform (Upper Botomian) began with an abrupt climatic change towards humid conditions and mild extensional tectonic movements that resulted in the dissection of the platform into blocks. As a consequence, differential subsidence and high rates of sedimentation, caused the up-building of a complex facies mosaic in a tidal-flat/lagoonal system.

At the Lower Cambrian–Middle Cambrian boundary a stronger pulse of extensional tectonics led to the drowning of the platform. During the first stage of collapse, differential movements of the blocks, probably related to isostatic compensation, resulted in tilting, uplifting or sinking of the different blocks. Breccias accumulated at the fault scarps along the more active western and eastern edges of the platform and on the exposed blocks karstic processes produced caves, hematitic residual material and breccias.

During Middle Cambrian the second stage of collapse resulted in the drowning of most of the previously emergent blocks and the establishment of a marginal plateau structure. The onlapping neritic sediments underwent differential compaction so that on top of the more unstable blocks a typical nodular structure developed. The residual red hematitic material derived from the exposed blocks was mixed with the carbonate muds giving the characteristic red colour. The third and stronger tectonic pulse led to the final drowning of the platform and the installation of pelagic and turbiditic deposition that was interrupted during Lower Ordovician by the Sardic transpressional phase.

Upper Triassic–Lower Jurassic platform

The platform developed on the Palaeozoic basement, already fragmented into horsts and grabens, where during the Middle Triassic continental red beds were deposited. During the Carnian a marine transgression flooded the region and on the structural highs pure carbonate deposition developed. In the subsident zone evaporites and carbonates were laid down. Since Norian time, the platform was bordered on the west by a margin facing an epicontinental basin (La Spezia basin) and only in late Rhaetian on the eastern side, did a low-energy marginal shoal form. During this stage of the platform growth, arid climate and regular subsidence, more rapid in the eastern zone of the platform, controlled the deposition of a wide evaporitic sabkha system.

The beginning of the second stage of the platform growth is marked by an abrupt climatic change towards humid conditions. Extensional tectonic movements of low intensity caused the dissection of the platform in blocks and consequently high rates of sedimentation and differential subsidence. Since carbonate production balanced subsidence, the up-building of a complex facies mosaic in a tidal-flat/lagoonal system resulted.

A stronger pulse of tectonic subsidence resulted in a further dissection and in the drowning of the platform. During early Sinemurian a first submergence stage led to the onlap of neritic sediments on the shallow-water platform deposits and the filling of open fissures (neptunian dykes). Afterwards, as a consequence of isostatic compensation, a number of blocks tilted or uplifted and consequently emerged. On the more active, western margins breccias accumulated at the fault scarps and karstic processes on the exposed blocks produced caves, vadose fillings and breccias. Others blocks sunk and the onlapping pelagic sediments underwent differential compaction so that on top of the more unstable blocks the typical nodular structure of Ammonitico Rosso developed. From Sinemurian to Upper Pleinsbachian the blocks readjusted several times so that the nodular facies were produced at different stratigraphic levels. The final collapse of the platform leading to the beginning of pelagic turbiditic deposition coincided with the first opening of the Ligurian Ocean.

CONCLUSIONS

The events controlling the evolution of the Cambrian platform are related to intracratonic rifting in distal regions of the Gondwana landmass, concomitant with the early phase of opening of the Iapetus Ocean. The epicontinental basin established after the platform collapse never evolved into an oceanic basin owing to the transpressional movements of the Sardic phase. Conversely, the evolution of the early Mesozoic platforms is directly related to the Alpine events in Tethys. After the carbonate platform collapsed, a basin was formed and gradually evolved into the Ligurian branch of Tethys.

The general setting of both Lower Palaeozoic and Lower Mesozoic carbonate sequences represents an early intra-cratonic rifting phase. In this regime, extensional tectonics produced horst and graben structures that gradually evolved into platforms and epicontinental basins. The control of extensional tectonics on the depositional evolution of the two platforms is documented by the following features:
1 detachment and tilting of the carbonate platforms from the mainland in relation to extensional movements induced by thinning of the continental crust;
2 dissection of the isolated platforms and differential subsidence of the resulting blocks during their growth as a consequence of the reactivation of basement structures;
3 drowning of the platforms with associated tilting and uplift of a number of blocks as a response to

persistent, but intensified extensional movements coupled with isostatic compensation;
4 establishment of a marginal plateau and progressive onlapping of pelagic sediments onto the drowned platform during acceleration of the tectonic subsidence.

Further evidence of the tectonic control on sedimentation can be recognized in the climate change which occurred during the platform growth. It can be interpreted as the response to divergent movements of the continental plates during rifting.

The results of this study indicate that the sedimentary response to extensional tectonics during Lower Palaeozoic in Sardinia and Lower Jurassic in Tuscany was analogous despite the different age and the different further evolution of the intra-cratonic rifting that evolved in a passive continental margin in Mesozoic Tuscany whereas in the early Palaeozoic Sardinia did not reach a spreading stage.

ACKNOWLEDGEMENTS

The authors are grateful to Milvio Fazzuoli for helpful suggestions and documentation. Also to James Lee Wilson, Fred Read and Paul Crevello, and an anonymous reviewer for the constructive remarks following their critical reading of the manuscript and for the improvement of its clarity. The authors are indebted to Anna Bellini and Giancarlo Pagani for their assistance in drafting and typing. This study has been supported by the Italian Department of Education (MPI 40% Research Program).

Sadly, Tommaso Cocozza died suddenly on the 26th July 1989. He did not have the pleasure of seeing this, his last paper, published.

REFERENCES

BALASSONE, G., BONI, M., DI MAIO, G., SAVIANO, G. & SAVIANO, N. (1985) Alcune particolarita nei litotipi della Formazione di Gonnesa nell'Iglesiente nord orientale e nel Fluminese: ulteriori possibili evidenze di margine di piattaforma carbonatica. *Rend. Soc. Geol. Ital.* **8**, 87–90.

BARCA, S., COCOZZA, T., DEL RIO, M., PILLOLA, L. & PITTAU DEMELIA, P. (1987) Datation de l'Ordovician inférieur par Dictyonema flabelliforme et Acritarches dans la partie supérieure de la formation "Cambrienne" de Cabitza (SW de la Sardaigne, Italie): conséquences géodynamiques. *Compt. Rend. Acad. Sci. Paris* **305**, 1109–1113.

BARRON, E.J., HARRISON, C.G.A., SLOAN, II, J.L. & HAY,

W.N. (1981) Paleogeography, 180 million years ago to the present. *Eclogae Geol. Helvet.* **74**, 443–470.

BECHSTADT, T., BONI, M. & SELG, M. (1985) The Lower Cambrian of SW Sardinia: from a clastic tidal shelf to an isolated carbonate platform. *Facies* **12**, 113–140.

BERNOULLI, D. (1972) North Atlantic and Mediterranean Mesozoic facies: a comparison. *Init. Rep. DSDP* **11**, 801–871.

BERNOULLI, D. & JENKINS, H.C. (1974) Alpine, Mediterranean and Central Atlantic Mesozoic facies in relation to the early evolution of Tethys. In: *Modern and Ancient Geosynclinal Sedimentation* (Eds Dott, R.M. & Shaver, R.H.) Spec. Publ. Soc. Econ. Paleontol. Mineral. **19**, pp. 129–160.

BERNOULLI, D., KALIN, O. & PATACCA, E. (1979) A sunken continental margin of the Mesozoic Tethys: the Northern and Central Apennines. In: *La Sédimentation Jurassique Ouest-éuropéen* (Eds Beaudoin, B. & Purser, B.H.) Publ. Spec. Assoc. Sedimentol. Français 1, pp. 197–210.

BOCCALETTI, M., FAZZUOLI, M. & MANETTI, P. (1975) Caratteri sedimentologici del Calcare Massiccio a nord dell'Arno. *Boll. Soc. Geol. Ital.* **94**, 377–405.

BONI, M., COCOZZA, T., GANDIN, A. & PERNA, G. (1981) Tettonica, sedimentazione e mineralizzazione delle brecce al bordo sub-orientale della Piattaforma carbonatica cambrica (Sulcis, Sardegna). *Mem. Soc. Geol. Ital.* **22**, 111–122.

CARANNANTE, G., COCOZZA, T. & D'ARGENIO, B. (1984) Late Precambrian–Cambrian geodynamic setting and tectono-sedimentary evolution of Sardinia (Italy). *Boll. Soc. Geol. Ital.* **103**, 121–128.

CARANNANTE, G., COCOZZA, T., D'ARGENIO, B. & SALVADORI, I. (1975) Caratteri deposizionali e diagenetici della "Dolomia rigata" del Cambrico inferiore della Sardegna. *Rend. Soc. Ital. Mineral Petrol.* **30**, 1159–1173.

CARMIGNANI, L., COCOZZA, T., GANDIN, A. & PERTUSATI, P.C. (1986a) The geology of Iglesiente. In: *Guide-book to the Excursion on the Palaeozoic Basement of Sardinia*, International Geological Correlation Project 5, Newsletter, Special issue, pp. 31–49.

CARMIGNANI, L., COCOZZA, T., GHEZZO, C., PERTUSATI, P.C. & RICCI, C.A. (1986b) Outlines of the Hercynian Basement of Sardinia. In: *Guidebook to the Excursion on the Palaeozoic Basement of Sardinia*, International Geological Correlation Project 5, Newsletter, Special issue, pp. 11–21.

CIARAPICA, G. & PASSERI, L. (1978) I Grezzoni del nucleo apuano: nascita, sviluppo e morte di una piattaforma carbonatica iperalina. *Boll. Soc. Geol. Ital.* **97**, 527–564.

CIARAPICA, G. & PASSERI, L. (1980) Tentativo di ricostruzione paleogeografica a livello del Trias nella Toscana a Nord dell'Arno e sue implicazioni tettoniche. *Mem. Soc. Geol. Ital.* **21**, 41–49.

CIARAPICA, G. & ZANINETTI, L. (1984) Foraminifères et biostratigraphie dans le Trias Supérieur de la serie de La Spezia (Dolomies de Coregna et Formation de La Spezia, nouvelles formations), Apennin septentrional. *Rev. Paléobiol.* **3**, 117–134.

CIARAPICA, G., CIRILLI, S. & PASSERI, L. (1982) La serie triassica del M. Cetona (Toscana meridionale) e suo confronto con quella di La Spezia. *Mem. Soc. Geol.*

Ital. **24**, 155–167.

COCOZZA, T. (1969) Slumping e brecce intraformazionali nel Cambriano medio della Sardegna. *Boll. Soc. Geol. Ital.* **88**, 71–80.

COCOZZA, T. (1979) The Cambrian of Sardinia. *Mem. Soc. Geol. Ital.* **20**, 163–187.

COSTANTINI, A., GANDIN, A. & MARTINI, R. (1983) Prima segnalazione di foraminiferi del Trias nelle evaporiti di Boccheggiano. *Mem. Soc. Geol. Ital.* **25**, 159–164.

COSTANTINI, A., GANDIN, A., MATTIAS, P.P., SANDRELLI, F. & TURI, B. (1980) Un'ipotesi per l'interpretazione paleogeografica della Formazione di Tocchi. *Mem. Soc. Geol. Ital.* **21**, 203–216.

COURJAULT-RADÈ, P. & GANDIN, A. (1986) Comparative Early Palaeozoic geodynamic evolution in Montagne Noire (France) and Sardinia (Italy) (Abstracts). *12th Int. Sedimentol. Congress*, Canberra, Australia, pp. 69–70.

D'ARGENIO, B. (1970) Evoluzione geotettonica comparata tra alcune piattaforme carbonatiche dei Mediterranei Europeo ed Americano. *Atti Accad. Pontiniana Nuova Serie* **20**, 1–34.

DEBRENNE, F. & GANDIN, A. (1985) La Formation de Gonnesa (Cambrien, SW Sardaigne): biostratigraphie, paléogeographie, paléoecologie des Archéocyathes. *Bull. Soc. Géol. France* **1**(8), 531–540.

DECANDIA, F.A., GIANNINI, E. & LAZZAROTTO, A. (1980) Evoluzione paleogeografica del margine appenninico nella Toscana a Sud dell'Arno. *Mem. Soc. Geol. Ital.* **21**, 375–384.

ESTEBAN, M. (1976) Vadose pisolite and caliche. *Am. Assoc. Petrol. Geol. Bull.* **80**, 2048–2057.

FAZZUOLI, M. (1974a) Facies di "Laguna interna" nel Calcare Massiccio della Toscana sud-orientale. *Boll. Soc. Geol. Ital.* **93**, 369–396.

FAZZUOLI, M. (1974b) Caratteri sedimentologici del Calcare Massiccio nell'area della Pania di Corfino (Provincia di Lucca). *Boll. Soc. Geol. Ital.* **93**, 735–752.

FAZZUOLI, M. (1980) Frammentazione ed "annegamento" della piattaforma carbonatica del Calcare massiccio (Lias inferiore) nell'area Toscana. *Mem. Soc. Geol. Ital.* **21**, 181–191.

FAZZUOLI, M. & PIRINI RADRIZZANI, C. (1981) Lithofacies characteristics of the "Rosso Ammonitico" Limestone in the south-western Tuscany (Italy). In: *Rosso Ammonitico Symposium Proceedings, Roma 1980* (Eds Farinacci, A. & Elmi, S.) pp. 409–417.

FAZZUOLI, M., MARCUCCI PASSERI, M. & SGUAZZONI, G. (1981) Occurrence of "Rosso Ammonitico" and paleokarst sinkholes on the top of the "Marmi Fm" (Lower Liassic), Apuane Alps, Northern Apennines. In: *Rosso Ammonitico Symposium Proceedings, Roma 1980* (Eds Farinacci, A. & Elmi, S.) pp. 399–407.

FEDERICI, P.R. (1967) Prima segnalazione di Lias medio nel calcare nodulare rosso ammonitico dell'Appennino Ligure e considerazioni cronologiche sulla stessa formazione in Toscana. *Boll. Soc. Geol. Ital.* **86**, 269–286.

FRAKES, L.A. (1979) *Climates throughout Geological Time*, Amsterdam, Elsevier, p. 310.

GANDIN, A. (1980) Tectono-sedimentary evolution of an epicontinental shelf: Lower–Middle Cambrian of Sardinia (Italy). *11th Int. Congress on Sedimentology,*

Hamilton, Ontario (Abstracts) Hamilton 43, p. 43.

GANDIN, A. (1990a) Depositional and Paleogeographic evolution of the Cambrian in southwestern Sardinia. In: *International Geological Correlation* (Eds Sassi, F.P. & Bourrouilh, R.) Project 5, Newsletter 7, pp. 151–166.

GANDIN, A. (1990b) Anomalies at the transition between Gonnesa Fm and Cabitza Fm (Lower to Middle Cambrian; southwestern Sardinia). In: *International Geological Correlation* (Eds Sassi, F.P. & Bourrouilh, R.) Project 5, Newsletter 7, pp. 52–53.

GANDIN, A. & TURI, B. (1988) Sedimentologic and isotopic analysis of the Sardinian Cambrian carbonates In: *International Geological Correlation* (Eds Sassi, F.P. & Bourrouilh, R.) Project 5, Newsletter, 7, p. 56.

GANDIN, A., MINZONI, N. & COURJAULT-RADE, P. (1987) Shelf to basin transition in the Cambrian–Lower Ordovician of Sardinia (Italy). *Geol. Runds.* **76**, 827–836.

GANDIN, A., TONGIORGI, M., RAU, A. & VIRGILI, C. (1982) Some examples of the Middle-Triassic marine transgression in south-western Mediterranean Europe. *Geol. Runds.* **71**, 881–894.

GIANNINI, E., LAZZAROTTO, A. & SIGNORINI, R. (1972) Lineamenti di Geologia della Toscana meridionale. In: La Toscana meridionale. *Rend. Soc. Ital. Mineral. Petrol.* **27**, 33–168.

HARLAND, W.B., COX, A.V., LLEWELLYN, P.G., PICKTON, C.A.C., SMITH, A.G. & WALTERS, R. (1982) *A Geological Time-scale*, Cambridge Earth Science Series, pp. 1–131.

JELL, P.A. (1973) Faunal provinces and possible planetary reconstruction of the Middle Cambrian. *J. Geol.* **82**, 319–350.

LECCA, L., PALMERINI, V. & ZUDDAS, P. (1983) Le peliti dei calcari nodulari di Gutturu Pala e di altri affioramenti dell'Iglesiente (Sardegna sud-occiendentale). *Period. Mineral.* **52**, 97–116.

MARTINI, R., GANDIN, A. & ZANINETTI, L. (1989) Sedimentology, stratigraphy and micropalaeontology of the Triassic evaporitic sequence in the subsurface of Boccheggiano and in some outcrops of southern Tuscany (Italy). *Riv. Ital. Paleont. Strat.* **95**, 3–28.

RASETTI, F. (1972) Cambrian Trilobite Faunas of Sardinia. *Mem. Accad. Nazionale Lincei* **11**, 1–100.

READ, J.F. (1985) Carbonate platform facies models. *Am. Assoc. Petrol. Geol. Bull.* **69**, 1–21.

ROZANOV, A. YU. & DEBRENNE, F. (1974) Age of archaeocyathid assemblages. *Am. J. Sci.* **274**, 833–848.

ROZANOV, A. YU. & SOKOLOV, B.S. (1984) Lower Cambrian stage subdivision. Stratigraphy. *Izd. Nauk. Acad. Sci. USSR* 5–19 (In Russian).

SCOTESE, C.R., BAMBACH, R.K., BARTON, C., VAN DER VOO, R. & ZIEGLER, A.M. (1979) Paleozoic base maps. *J. Geol.* **87**, 217–277.

VAI, G.B. & COCOZZA, T. (1986) Tentative schematic zonation of the Hercynian chain in Italy. *Bull. Soc. Géol. France* **2**, 95–114.

WILSON, J.L. (1975) *Carbonate Facies in Geologic History*, Springer Verlag, New York, p. 740.

Spec. Publs int. Ass. Sediment. (1990) **9**, 39–54

Geometry and evolution of platform-margin bioclastic shoals, late Dinantian (Mississippian), Derbyshire, UK

R. L. GAWTHORPE* *and* P. GUTTERIDGE[†]

**Department of Geological Sciences, University of Durham, Durham DH1 3LE, UK; and
†School of Earth Sciences, Thames Polytechnic, Bigland Street, London E1 2NG, UK*

ABSTRACT

A belt of large-scale bioclastic sand shoals developed at the northern margin of the Derbyshire carbonate platform in northern England during the late Asbian–Brigantian (Dinantian). This belt was at least 2 km wide and extended at least 3 km parallel to the shelf margin. These sand shoals form a shoal complex at least 50 m in thickness composed mainly of bioclastic grainstone. The shoal complex has been divided into shoal sequences by boundaries which show toplap, downlap and onlap configurations. Some of these sequences contain clinoforms which are interpreted as large scale bedforms. Changes in internal geometry of these clinoforms indicate a passage from vertical accretion to basinward progradation during bedform development. These bedforms commonly overlie emergence surfaces and probably developed following a relative rise in sea-level of 20–25 m. Abandonment of these bedforms occurred as a result of subaerial exposure, high energy conditions causing erosional truncation, or the establishment of low energy conditions. Bedforms nucleated in a progressively basinward position throughout the deposition of the shoal complex resulting in basinward progradation of the complex as a whole.

INTRODUCTION

Carbonate sand shoals occur in a number of carbonate depositional systems such as ramps, rimmed shelves and epeiric platforms where they commonly form hydrocarbon reservoirs and host base metal deposits in association with shorelines and shelf breaks (e.g. Tucker, 1985). Carbonate sands have been extensively studied in modern environments where it has been shown that tidal regime and windward–leeward aspect of the shelf break exert a significant influence on the geometry and architecture of such sand bodies (e.g. Ball, 1967; Hine, 1977; Hine *et al.*, 1981). Other factors which are important in influencing the character of carbonate sand shoals include: antecedent topography, tectonic subsidence, eustacy and the type of carbonate sediment available. In the literature on both ancient and modern carbonate sands there is a bias towards those with a predominantly oolitic composition, whereas bioclastic sand bodies have received relatively little attention.

The objective of this paper is to describe a bioclastic–carbonate sand body complex of late Dinantian (Mississippian) age which formed in association with the northern margin of the Derbyshire carbonate platform in northern England. Particular attention will be paid to the internal geometry, facies variations and controls on the evolution of these carbonate sand bodies.

GEOLOGICAL SETTING

During late Dinantian times, northern England was situated immediately to the south of the palaeo-equator, and as a result, shallow siliciclastic-starved seas were dominated by carbonate deposition. The Derbyshire carbonate platform was one of several

* Present address: Department of Geology, The University, Manchester M13 9PL, UK.
[†] Present address: Cambridge Carbonates, 22 George Street, Cambridge, CB4 1AJ, UK.

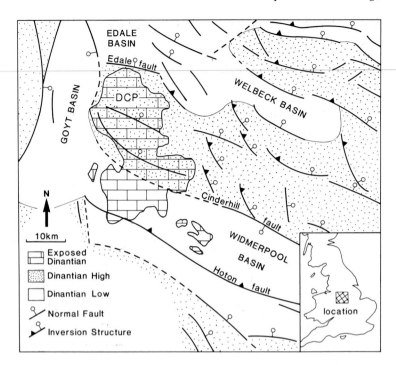

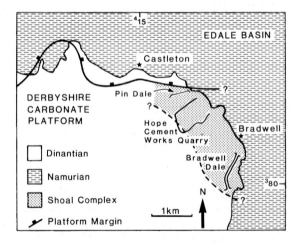

Fig. 1. Dinantian palaeogeography and structural setting of the Central Midlands showing the location of the exposed Derbyshire carbonate platform (DCP). The shelf-margin carbonate sand bodies discussed in this paper are located at the northern margin of the DCP, in the footwall to the Edale fault.

areas of shallow-water carbonate deposition which were located on a series of fault-controlled structural highs (Fig. 1). The Derbyshire carbonate platform formed in the footwall of the down-to-the-north Edale Fault which controlled the location and geometry of the Edale Basin in its hangingwall to the north. The southern platform margin is located on the hangingwall of the Hoton Fault which forms the major controlling fault of the Widmerpool Basin (Fraser *et al.*, 1990).

The bioclastic (mainly crinoidal) shoals described here formed during late Asbian to early Brigantian time along the northern margin of the Derbyshire carbonate platform, which is inferred to be a footwall carbonate margin (Lee, 1988; Gawthorpe *et al.*, 1989). These sand shoals are equivalent to the Monsal Dale and the lower part of the Eyam Limestones which were deposited in a variety of platform interior settings (Gutteridge, 1983, 1987, 1989). Scattered exposures of this bioclastic shoal complex occur between Castleton and Bradwell Dale (Fig. 2). Three large sections in Pin Dale, Hope Cement Works Quarry and Bradwell Dale provide excellent views of the internal geometry of this shoal complex.

These bioclastic shoals developed around the margin of the Derbyshire carbonate platform following a change in the nature of the platform margin

Fig. 2. Location map showing the main sections described in the text and the approximate limits of the shoal complex. Note·that the eastward extent of the shoal complex and the platform margin are masked by the younger Namurian cover.

from a high to a low angle rimmed margin during the late Asbian (Broadhurst & Simpson, 1973). The change coincided with a phase of active extension which is recognized over much of northern England (Gawthorpe *et al.*, 1989). This extensional episode caused reactivation of basement faults underlying the Derbyshire carbonate platform resulting in the initiation of an intrashelf basin (Gutteridge, 1987, 1989). Regional work on Dinantian basins in the subsurface and at outcrop shows that this late Asbian−early Brigantian episode of tectonism corresponds with the EC5 seismic sequence in the east Midlands (Fraser *et al.*, 1990).

The bioclastic shoal complexes have been referred to as 'flat-reefs' by previous workers (e.g. Shirley & Horsfield, 1940; Eden *et al.*, 1964; Stevenson & Gaunt, 1971). Such previous studies have concentrated on stratigraphical and palaeontological aspects of the bioclastic shoals whereas their sedimentology has not been discussed. The sequence boundaries identified in this study have been interpreted previously as major non-sequences and angular unconformities. The Asbian−Brigantian boundary has been correlated with one of the sequence boundaries described here from Pin Dale (e.g. Eden *et al.*, 1964; Stevenson & Gaunt, 1971). It follows from this study that many of these sequence boundaries have only local significance in terms of the evolution of the shoal complex and that the regional stratigraphical significance of the unconformities has probably been overstated.

INTERNAL GEOMETRY OF THE SHOAL COMPLEX

Application of a sequence stratigraphic approach in this study has enabled the recognition of a number of sequences within the shoal complex. The lateral and vertical relationships and the internal geometries of these sequences provide important information on the evolution of the shoal complex. Our approach is illustrated by Fig. 3.

The term *shoal complex* refers to the shelf margin bioclastic grainstone and associated facies as a whole. It is exposed for at least 3 km parallel to the shelf margin and forms a belt 1−2 m wide. The thickness of the complex is between 50 and 100 m. The shoal complex has been divided into a number of *shoal sequences* which are separated by *sequence boundaries* recognized by downlap, onlap, toplap and erosional truncation bedding configurations. The shoal sequences are several tens of metres thick and several hundred metres in lateral extent. Sub-sequences are recognized where there is evidence of a change of sedimentary style during deposition of a shoal sequence. This is usually seen as a change in clinoform geometry which implies a change in the style of accretion of the shoal sequence.

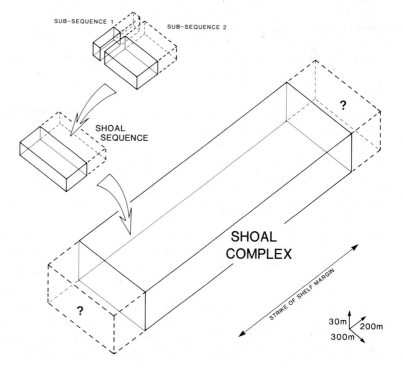

Fig. 3. Diagram illustrating the main elements of the shoal complex and their approximate dimensions. The principal building blocks of the shoal complex are shoal sequences which are bounded by unconformities, often paleokarst surfaces. Internally, shoal sequences can be subdivided into two main sub-sequences, often marked by a change in the clinoform geometry. Note that the dimensions of the shoal sequences and sub-sequences are relatively well constained, whereas the dimensions of the shoal complex, particularly parallel to the shelf margin, are poorly constrained.

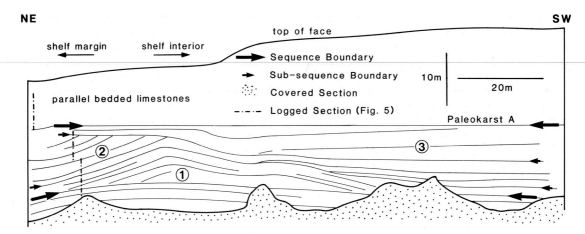

Fig. 4. Line drawing of part of the quarry at Pin Dale (SK159822) illustrating the main bedding plane configurations, sequence boundaries and sub-sequence boundaries. Note the mound-like form of sub-sequence 1 and downlap onto the lower sequence boundary. Sub-sequence 2 is characterized by well-developed clinoforms dipping towards the shelf margin and onlap onto the shelfward side of the lower strata. Sub-sequence 3 onlaps the stoss side of the bedform. The upper sequence boundary (paleokarst A) can be mapped into Hope Cement Works Quarry (see Fig. 6).

Pin Dale (Fig. 4)

Pin Dale is oriented approximately perpendicular to the shelf margin and shows the transition from the shelf break into the shelf-margin shoal complex over a lateral distance of ≈ 500 m. A shoal sequence some 10 m thick contains both shelfward and basinward dipping clinoforms. The base of the shoal sequence is marked by downlap and the upper boundary of the shoal sequence is marked by erosional truncation and onlap. The upper and lower sequence boundaries are associated with the development of multiple paleokarsts (Fig. 5).

The shoal sequence can be divided into two main sub-sequences. The lower sub-sequence has an asymmetric mounded form in cross-section with downlap at the base of the shoal sequence in both shelfward and basinward direction. Clinoforms dip (15–30°) in a basinward direction and (10–15°) in a shelfward direction. They are interpreted as the lee and stoss side of a bedform respectively. Internally, the vertical stacking of successive clinoforms indicates mainly vertical accretion of the shoal, with almost no progadation during the deposition of the sub-sequence.

The higher sub-sequence is characterized by well-developed clinoforms dipping basinwards at 25–30°. Onlap occurs on to the shelfward (stoss) side of the lower sub-sequence boundary. The basinward dipping clinoforms have a concave-up geometry and

pass basinward into bottomsets which extend over the shelf margin, a distance of ≈ 100 m to the northeast from the bedform crest. No downlap is observed. The foresets attain at least 10 m in height; however, the exact geometry of the top of the foresets is masked by the development of a paleokarstic surface at the top of the shoal. The development of this second sub-sequence is interpreted as a change from vertical accretion to basinward progradation of the bedform. A third minor sub-sequence displays onlap on to the stoss side of the bedform.

Hope Cement Works Quarry (Fig. 6)

This quarry provides a section ≈ 100 m thick through the shoal complex and ≈ 600 m in length perpendicular to the shelf margin. Three shoal sequences can be recognized within the exposed shoal complex here (Fig. 6), with several of the sequence boundaries marked by paleokarstic surfaces. The three shoal sequences are characterized by basinward-dipping clinoforms; shoal sequence 2 (Fig. 6) will be described in detail as this most clearly displays the evolution of the sand bodies.

As at Pin Dale, the shoal sequence is divisible into two main sub-sequences. The lower sub-sequence (Figs 6 & 7a) is characterized by sigmoidal, convex-up clinoforms. No bottomsets are developed and the foresets downlap at a high angle (up to 30°) on to the lower sub-sequence boundary. The beds underlying

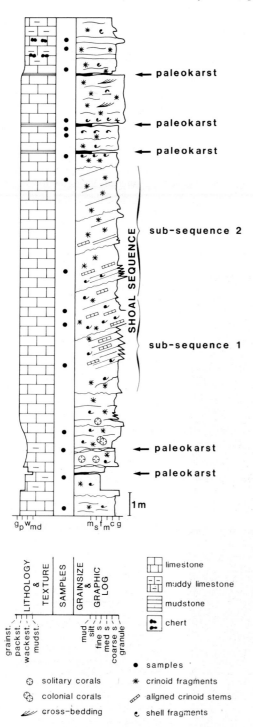

Fig. 5. Graphic log showing the sedimentological expression of the shoal sequence and sub-sequences in Pin Dale. For the location of the log see Fig. 4.

this sequence boundary contain rhizocretions and calcrete mottles indicating that the lower bounding surface was emergent prior to bedform development. The locus of the topset—foreset transition marked by successive clinoforms dips shelfward at ≈ 15° indicating both accretion and basinward progradation during deposition of this sub-sequence.

Clinoforms within the higher sub-sequence have a concave-up geometry and bottomsets are well-developed (Figs 6 & 7b). Several truncation surfaces are present within this sub-sequence. Small-scale cross-stratification superimposed on the clinoforms is locally developed with palaeocurrents indicating basinward transport of sediment down the clinoforms.

Bradwell Dale (Fig. 8)

Bradwell Dale provides a section through part of the shoal complex oriented perpendicular to the shelf margin. One shoal sequence is exposed, containing basinward- and shelfward-dipping clinoforms which represent a section perpendicular to a former bedform crest. Two sub-sequences are recognized within this shoal sequence on the basis of clinoform geometry.

The lower sub-sequence represents the early stages of bedform growth and is a symmetrical upbuilding of thinly-bedded (0.05–0.1 m) crinoidal bioclastic grainstone—packstone. The clinoforms dip 12–14° in a shelfward and basinward direction. The crest of the bedform trends 115–295°, which is sub-parallel to the shelf margin. Small-scale ripple cross-stratification is superimposed on these clinoforms indicating reworking of sediment by bimodal currents perpendicular to the bedform crest.

The higher sub-sequence is marked by a change from vertical aggradation to lateral accretion of the bedform with the development of tabular to convex-up shelfward and basinward dipping clinoforms up to 0.5 m in thickness. Basinward dipping clinoforms dip at 20–22° and contain superimposed tabular cross-stratification with cosets up to 0.2 m thick indicating basinward transport. Shelfward dipping clinoforms dip at 8–12° and contain superimposed tabular cross-stratification with cosets up to 0.2 m thick indicating alternate shelfward and basinward transport.

Form of the shoal sequences

The term 'form' is used to describe the shape of the

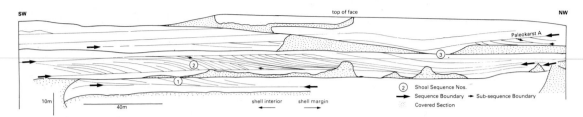

Fig. 6. Line drawing of the NW face of Hope Cement Works Quarry (SK159820) illustrating the main bedding plan configurations, sequence boundaries and sub-sequences. Note the position of paleokarst A which allows correlation with the Pin Dale section (Fig. 4). The section exposes three shoal sequences characterized by basinward dipping clinoforms. Note the two stage development of shoal sequence 2 with an initial phase of convex-up sigmoidal clinoforms and a later phase of oblique concave-up clinoforms with asymptotic bottomsets.

Fig. 7. Photographs illustrating the internal character of shoal sequence 2, Hope Cement Works Quarry (see Fig. 6 for location of shoal sequence 2). (a) Initial sub-sequence displaying convex-up clinoforms dipping to the right and downlapping onto the lower sequence boundary (arrowed). Person for scale (ringed). (b) Part of the second sub-sequence showing concave-up oblique clinoforms with well-developed bottomsets. Sequence boundaries arrowed, person (ringed) for scale.

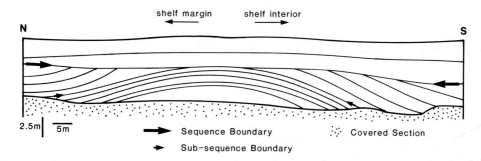

Fig. 8. Line drawing of the lower part of the quarry face at Bradwell Dale (SK173807), illustrating bedding plane configurations, sequence boundaries and sub-sequence boundaries. Note symmetrical upbuilding and mound-shaped form of lower sub-sequence and marked erosional truncation at the upper sequence boundary.

individual shoal sequences, the bedforms which produced them and the three-dimensional relationships between the shoal sequences which together constitute the architecture of the shoal complex. The main palaeokarstic surface developed at the top of the shoal sequence in Pin Dale and Hope Cement Works Quarry provides a means of correlation between the two exposures and allows constraints to be placed upon the three-dimensional form of the individual shoal sequences underlying this emergence surface. The absence of this shoal sequence in the southeastern face of Hope Cement Works Quarry places a minimum value for the along-strike dimension of the shoal sequence. These data, together with information from quarry faces, suggest that the shoal sequences have a sheet-like form ≈ 10 m thick, they are between 200 and 700 m wide measured perpendicular to the shelf margin and extend at least 300 m parallel to the shelf margin.

MICROFACIES

Shoal sequences at Pin Dale and Hope Cement Works Quarry (shoal sequence 3) were sampled for their microfacies. Four microfacies have been distinguished, their distribution is shown in Fig. 9.

Microfacies 1: Bioclast peloid intraclast grainstone (Fig. 10a)

Description

This microfacies is a grainstone with localized packstone texture. It contains a diverse allochem assemblage (Table 1) dominated by brachiopods, crinoids and peloids. Minor constituents include bryozoans, foraminifera, algae and oncolites. The following three intraclast types are also present: type A: sorted bioclast peloid grainstone (Fig. 10b); type B: calcretized bioclast grainstone−packstone (Fig. 10c); type C: bioclast wackestone (Fig. 10d).

Allochems in microfacies 1 are poorly to very well-sorted, grain-size varies from fine-sand to granule size. All bioclasts are disarticulated and fragmented and are moderately- to well-rounded. Grain roundness increases with increasing degree of micritization. Peloids, formed by micritization of bioclasts, are abundant and range in size from medium sand to granule. Oncolites (Fig. 11a) have a bioclast or peloid nucleus around which there is a complete, concentrically laminated coating of uniform thickness. Some specimens are normally graded and have preferential sub-horizontal alignment of shell fragments. There is no evidence of bioturbation.

An isopachous radial-fibrous calcite cement (Fig. 11b) is locally present as the initial phase of cementation. This precedes aragonite dissolution and is followed by the precipitation of a ferroan pore-filling sparry cement. This radial-fibrous calcite cement is also found within type A intraclasts.

Interpretation

This microfacies contains a diverse open marine bioclast assemblage dominated by brachiopods, crinoids, bryozoans and foraminifera. All the bioclasts show a high degree of fragmentation and rounding which, together with the predominant grainstone texture, suggest a high energy environment and frequent reworking. However, sedimen-

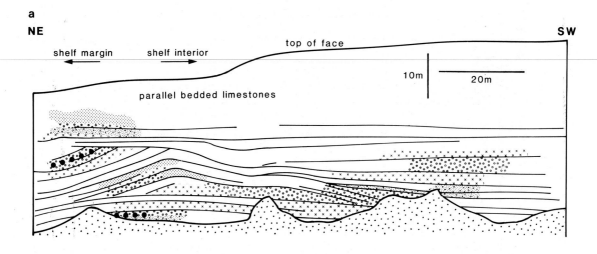

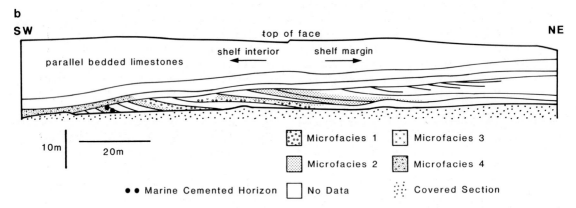

Fig. 9. Examples of microfacies distributions in shoal sequences. (a) Pin Dale (SK159822). (b) Sequence 3, Hope Cement Works Quarry (SK159820). Note the dominance of grainstones and packstones (microfacies 1 and 2) in the clinoforms and the occurrence of wackestones in parallel bedded carbonates above and below the shoal sequence boundaries and in onlapping sub-sequences (see Figs 4 and 6 for the location of sequence and sub-sequence boundaries).

tation rate was probably low as suggested by the high degree of micritization and abundance of peloids. Type A intraclasts were derived from a high energy, well-reworked fully marine environment. They were produced by erosion and reworking of cemented layers on the seafloor. Type B intraclasts are also interpreted as having been deposited originally in a high energy, fully marine environment. The micritic coating and pore bridging textures are interpreted as a pedogenic texture formed in a calcrete profile. This indicates that these intraclasts were reworked from parts of the shoal complex which were subaerially exposed or from shoal se-

quence boundaries. Type C intraclasts are similar to microfacies 4 which was deposited in sheltered areas within the shoal complex. The local preferential alignment of allochems is attributed to deposition by bedload processes or, in the case of layers of highly stacked shells, a high degree of winnowing. The radialfibrous cement is interpreted as a syndepositional marine phreatic cement because of its early diagenetic origin and its occurrence in the sorted grainstone intraclasts. The occurrence of cemented layers on the seafloor is attributed to periodic pauses in sedimentation.

Table 1. Qualitative analysis of the allochem assemblages, in the four main microfacies. The abbreviations for the intraclasts are; type A = sorted bioclast peloid grainstone, type B = calcretized bioclast grainstone–packstone, Type C = bioclast wackestone.

	Microfacies 1 Bioclast peloid intraclast gst.	**Microfacies 2** Bioclast intraclast oncolite gst.–pkst.	**Microfacies 3** Bioclast oncolite pkst.–wkst.	**Microfacies 4** Bioclast wkst.
Allochems in all samples				
Abundant	Brachiopod shell and spine fragments Crinoid ossicles Echinoderm plates Peloids	Crinoid ossicles and stems Brachiopod shell and spine fragments	Brachiopod shell and spine fragments	Molluscs (bivalves and gastropods) Calcitized spicules
Uncommon	Bryozoans (stick type and fenestrate) Foraminifera (endothyrids and archaediscids)	Bryozoans (stick type and fenestrate) Foraminifera (endothyrids and tetrataxids)	Bryozoans (stick type and fenestrate) Crinoid ossicles Echinoderm plates Foraminifera (tetrataxids)	Brachiopods Bryozoans (stick and fenestrate) Foraminifera (endothyrids)
Allochems in some samples				
Common	Foraminifera (tetrataxids, agglutinating types, earlandids) Algae (stachenids) Molluscs (bivalves and gastropods) Types A, B and C intraclasts	Algae (stachenids) Foraminifera (archaediscids) Oncolite Type C intraclast	Mollusc shell fragments Algae (kamaenids and ungdarellids) Foraminifera (earlandids, endothyrids and agglutinating types)	Ostracods Foraminifera (archaediscids) Algae (stacheoides)
Trace	Foraminifera (saccaminopsids) Oncolites Algae (kamaenids) Rugose and tabulate coral fragments Trilobite Calcitized monaxon spicules	Ostracod valves Algae (green, indet. and ungdarellid) Foraminifera (agglutinating types) Rugose coral Molluscs (bivalve and gastropod)	Algae (*Koninckopora* and stachendis) Foraminifera (saccaminopsids) Ostracods *Hexaphyllia* Peloids	Foraminifera (saccaminopsids) Green algae (indet.) Types A and C intraclasts

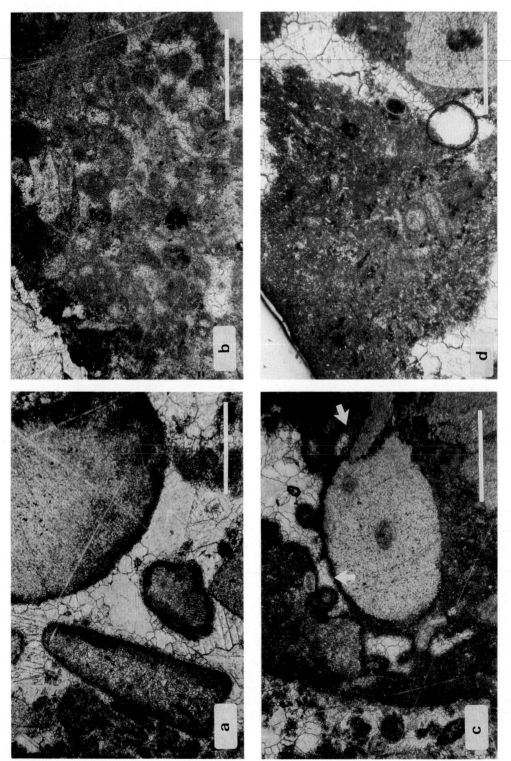

Fig. 10. Photomicrographs of microfacies 1 and intraclasts. (a) Bioclast peloid intraclast grainstone microfacies (microfacies 1). Note intensively micritized crinoid ossicles. Pin Dale, basinward dipping foresets sub-sequence 1. Photomicrograph of a peel. Scale bar = 250 μm. (b) Sorted bioclast grainstone intraclast. Pin Dale, basinward dipping foresets sub-sequence 1. Photomicrograph of a peel. Scale bar = 250 μm. (c) Calcretized bioclast packstone intraclast. Micrite lining intergranular porosity is arrowed. Pin Dale, basinward dipping foresets sub-sequence 1. Photomicrograph of a peel. Scale bar = 250 μm. (d) Bioclast wackestone intraclast. Pin Dale, basinward dipping foresets sub-sequence 1. Photomicrograph of a peel. Scale bar = 250 μm.

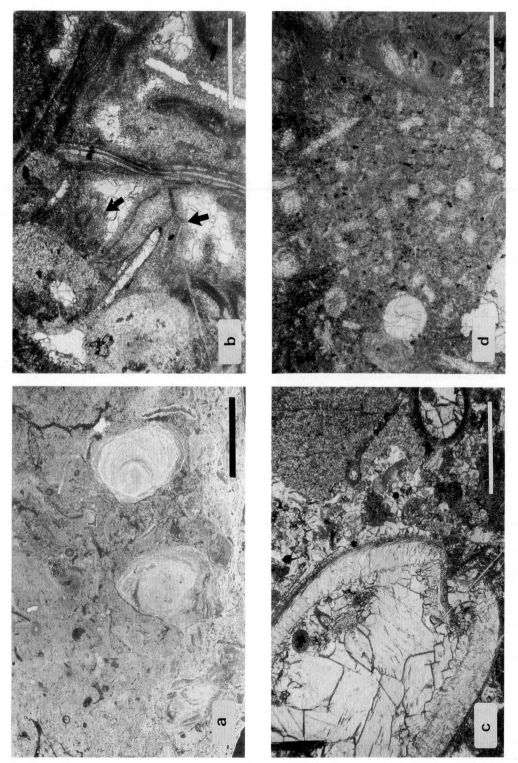

Fig. 11. Photomicrographs of microfacies 2, 3 and 4. (a) Bioclastic oncolite packstone–wackestone microfacies (microfacies 3). Oncolites show an initial complete concentric coating followed by an asymmetric coating. This implies an initial phase of continual agitation followed by a phase of decreasing agitation. Pin Dale immediately underlying lower shoal sequence boundary. Negative print of a peel. Scale bar = 1 cm. (b) Early marine isopachous radial fibrous-cement (arrowed). From upper part of foreset in shoal sequence 3 at Hope Cement Works Quarry. Photomicrograph of a peel. Scale bar = 250 μm. (c) Bioclast intraclast oncolite grainstone–packstone microfacies (microfacies 2). From shoal sequence 3 at Hope Cement Works Quarry. Photomicrograph of a peel. Scale bar = 250 μm. (d) Bioclast wackestone microfacies (microfacies 4). Bed immediately overlying shoal sequence 3 at Hope Cement Works Quarry. Photomicrograph of a peel. Scale bar = 250 μm.

Microfacies 2: Bioclast intraclast oncolite grainstone–packstone (Fig. 11c)

Description

This microfacies is a grainstone with development of local packstone texture. The allochem assemblage (Table 1) is dominated by brachiopods and crinoids. Minor constituents include bryozoans, foraminifera and algae. Peloids and micritized allochems are absent. Allochems are poorly to well-sorted. Bioclasts are disarticulated and fragmented although some whole brachiopod valves are present. Bioclasts show some evidence of sponge borings. Oncolites and bioclast wackestone intraclasts are also present. Oncolites have a bioclast nucleus with a complete coating which shows asymmetric thickening. There is local preferential alignment of shells and crinoid stems sub-parallel to bedding and normal size grading is present. An isopachous radial-fibrous calcite cement similar to that in microfacies 1 is locally present.

Interpretation

This microfacies contains a diverse allochem assemblage representing fully marine conditions. The grain-supporting texture, high degree of disarticulation, fragmentation and rounding of the allochems, together with the presence of oncolites indicate generally high energy conditions. The complete coating of the oncolite nuclei suggests a high frequency of rolling. However, the asymmetric thickening of the coating indicates alternating low energy conditions when the oncolites were stationary. The occurrence of packstone is attributed to infiltration carbonate mud through the sediment following a change to lower energy conditions. The bioclastic wackestone intraclasts indicates reworking of low energy sediment from sheltered areas adjacent to the sand body complex. A significant contrast with microfacies 1 is that the allochems within this microfacies have not been micritized. Some, however, have been bored by sponges which suggests a significant pause in sedimentation. This difference is interpreted in terms of differing sedimentation rates. In contrast to microfacies 1, microfacies 2 was subject to a higher sedimentation rate which did not allow sufficient time for micritization to take place. Bedload transport and winnowing are suggested by the orientation of bioclasts and stacking of shells. As in microfacies 1 the radial-fibrous calcite cement is interpreted as a marine phreatic cement.

Microfacies 3: Bioclast oncolite packstone–wackestone (Fig. 11a)

Description

This microfacies is transitional in texture between packstone and wackestone. The allochem assemblage (Table 1) is dominated by brachiopods with less abundant crinoids, bryozoans, foraminifera, algae, bioclast wackestone intraclasts, oncolites and local peloids. Allochems have generally not been micritized, although some have been partially micritized and some display sponge borings. Bioclasts are disarticulated and fragmented with a variable degree of rounding. Oncolites have a bioclast nucleus and the coatings show an early stage of complete accretion of uniform thickness followed by an asymmetric thickening of the coating. A further stage of incomplete partial accretion is shown in some oncolites. The coating of some oncolites contains several horizons which have been penetrated by sponge borings. In some specimens a mottled texture, picked out by a patchy distribution of carbonate mud and a preferential arcuate alignment of bioclasts, is present.

Interpretation

This microfacies contains a similar allochem assemblage to microfacies 1 and 2 and represents fully marine conditions. The disarticulation and fragmentation of bioclasts suggests a high-energy environment. However, the variable degree of rounding implies that allochems with differing histories of reworking are present and the packstone–wackestone texture suggests final deposition in a moderate- to low-energy environment. Bioturbation is indicated by mottling of the sediment and the occurrence of an arcuate alignment of bioclasts. The paucity of peloids and micritized and bored grains indicates that sedimentation rates were generally high, although episodic, as indicated by bored layers within oncolitic coatings. The change in the nature of the oncolitic coatings from symmetrical to asymmetrical can be explained either as a decrease in the energy of the environment, or to increasing size of the oncolite making it more difficult to be moved.

Microfacies 4: Bioclast wackestone (Fig. 11d)

Description

This wackestone has an allochem assemblage (Table

1) which differs from that of other microfacies in that it is dominated by molluscs and calcified spicules with brachiopods and crinoids as relatively minor components. Sorted bioclast peloid grainstone and bioclast wackestone intraclasts are also present. Bioclasts show a range of preservation states; the majority are disarticulated and fragmented with a high degree of rounding. However, whole unabraded bioclasts and fragmentary bioclasts with angular fracture surfaces are also present. This microfacies commonly displays a mottled texture which is picked out by a patchy distribution of carbonate mud and the development of an arcuate preferential alignment of bioclasts.

Interpretation

The dominance of molluscs and calcitized spicules suggest unusual salinity or a low oxygen content of the environment. The wackestone texture and the occurrence of whole bioclasts with minimal abrasion suggest deposition in a generally low energy environment. However, the presence of highly disarticulated, fragmented and rounded bioclasts suggests that these may have been reworked elsewhere and transported into the lower energy environment. The occurrence of intraclasts also indicates input of reworked sediment from surrounding areas. The common occurrence of sediment mottling and the arcuate preferential alignment of bioclasts is attributed to bioturbation.

EVOLUTION OF THE SHOAL COMPLEX

Bedform dynamics

Figure 9 shows that the clinoforms within shoal sequence 3 at Hope Cement Works Quarry and Pin Dale comprise alternations of microfacies 1 and 2 with the local occurrence of microfacies 3 in the Pin Dale bedform. This alternation is interpreted in terms of episodic bedform migration. During pauses in bedform migration, rates of sediment burial were low allowing micritization of grains and cementation of the bedform surface to occur. During bedform migration, rates of sediment burial were much higher and micritization was retarded allowing the deposition of 'fresh' bioclastic sediment. The occurrence of microfacies 3, particularly in bottomset beds, is interpreted as the establishment of low energy conditions allowing deposition and infiltration of micrite into the sediment of the bedform. Microfacies 4 is

generally restricted to areas adjacent to the main shoal sand bodies, and to the sequences overlying the shoal complex.

At all localities there is evidence of a two-stage development of the bedforms. This is shown schematically in Fig. 12, which is based on shoal sequence 2 in Hope Cement Works Quarry. This two-stage development is inferred to represent the initial upbuilding of the bedform during a phase when sedimentation rate was 'catching-up' with relative sea-level rise. The onset of basinward progradation reflects a limitation of vertical growth, possibly by sea-level stillstand, or that sedimentation had caught up with relative sea-level rise. Following this the bedform prograded basinwards with minimal vertical accretion. At the Hope Cement Works Quarry the bedform represented by shoal-sequence 2 prograded by 500−700 m basinwards. At Pin Dale progradation resulted in spill over the shelf break, and supply of sediment to the basinal area to the north.

Within the clinoforms, low angle erosion surfaces which truncate underlying clinoforms are interpreted as reactivation surfaces formed by high energy events during which the topography of the bedform was partly eroded. Re-establishment of the bedform after the development of these truncation surfaces is either by the conformable deposition of clinoforms, or downlap of clinoforms onto the reactivation surface (e.g. shoal sequence 3 at Hope Cement Works Quarry, Fig. 9b).

At Bradwell Dale, shelfward and basinward dipping clinoforms and bimodal smaller-scale cross-stratification indicates both shelfward and basinward sediment transport. This, together with the large scale of the bedform may suggest a tidal influence on the deposition of the shoal complex. However, the marked asymmetry of the shoal complex suggests a dominant ebb tidal current. The predominant basinward transport of sediment across the shelf margin along strike to the west suggests that prevailing offshelf wind or storm driven currents may also have been important along the shelf margin. Thus the shoal complex probably developed under a mixed storm−tide influence.

The extent of progradation may have been limited by sediment production or by extrinsic factors such as emergence or flooding due to relative sea-level change. The mechanism of nucleation, development and abandonment of these bedforms, as illustrated by the shoal sequences, has important consequences for the development and architecture of the shoal complex as a whole which is discussed in the next section.

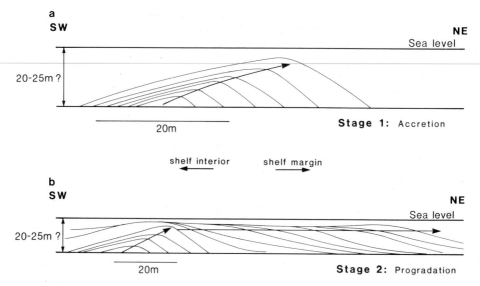

Fig. 12. Shoal sequence evolution based on examples from Pin Dale, Hope Cement Works Quarry and Bradwell Dale. (a) Stage 1 predominantly vertical accretion of bedform associated with relative rise of sea-level. Clinoforms are sigmoidal and convex-up, with downlap onto lower sequence boundary which is often a paleokarst. (b) Stage 2 stillstand of sea-level limits vertical accretion of the bedform with the result that it progrades basinward, under the influence of storm and tidal currents. Clinoforms have an oblique, concave-up geometry with asymptotic bottomsets.

Controls on shoal complex evolution

Where the lower sequence boundary of the shoal sequences is exposed, it is marked by evidence of emergence such as a paleokarst or calcrete. This implies that the bedforms developed during a relative sea-level rise which flooded an emergent platform. The bedform dynamics discussed above suggest an initially rapid relative transgression followed by still-stand. The dimensions of the bedforms suggest that episodic relative sea-level rises of at least 20–25 m took place during deposition of the shoal complex.

Features associated with the upper bounding surfaces of the shoal sequences suggest that abandonment of these bedforms occurred as a result of three processes: (1) erosional truncation, where the topography of the bedform was planed-off by a high energy (storm?) event; (2) development of low energy conditions over the bedform followed by onlap and burial of the bedform. This is seen at Pin Dale (Figs 4 & 9a) where the main bedform is onlapped by microfacies 4 which was deposited in a low energy, (deeper water?) setting; and (3) emergence, due to shoal up-building or relative fall in sea-level.

Hope Cement Works Quarry provides evidence

of the architecture of the shoal complex. The three shoal sequences (Fig. 6) form a total thickness of ≈ 50 m. Younger shoal sequences are progressively offset in a basinward direction relative to the immediately underlying sequence. The younger shoal sequences nucleate close to the crest of the bedform in the underlying sequence. The architecture of the shoal complex is thus one of progressive basinward-stacking of shoal sequences. The facies relationships parallel to depositional strike are unclear owing to poor exposure. These shoal sequences coalesce to produce a shoal complex which is wedge shaped, thinning away from the shelf margin towards the shelf interior. They form a belt varying from several hundred metres to several kilometres in width and several kilometres in length along depositional strike.

The predominant basinward sediment transport and bedform progradation is seen in modern carbonate platform margins of the windward-protected and leeward types (Hine *et al.*, 1981). There is no evidence of a persistent barrier parallel to the shelf margin in the form of a 'reef' or emergent area which could have afforded protection of the shelf margin. A leeward setting of the northern margin of the Derbyshire carbonate platform would be in agreement with the E−W or SE−NW prevailing Dinantian storm tracks deduced by Schofield (1982).

CONCLUSIONS

A complex of bioclastic−carbonate sand shoals some 2 km wide and extending for several kilometres along strike developed at the northern margin of the Derbyshire carbonate platform during the late Asbian and early Brigantian. This complex is divided into several shoal sequences separated by sequence boundaries recognized by toplap, onlap and downlap bedding configurations. These shoal sequences contain clinoforms which represent large-scale bedforms. The internal geometry of these shoal sequences indicates an initial stage of development dominated by vertical accretion which is followed by a phase of basinward progradation. The bedforms prograde by a maximum of 500−700 m and are seen to spill over the shelf margin. These bedforms developed on emergent surfaces, probably in response to a relative rise in sea-level of 20−25 m. Tidal currents and storm activity were both important in controlling the internal geometry of the shoals. The bedforms are made up mainly of bioclast grainstone with local packstone and wackestone, the latter reflecting lower energy conditions. The grainstone is a mixture of unmicritized 'fresh' sediment deposited during episodes of bedform migration and highly micritized peloidal sediment reflecting pauses in bedform migration.

Abandonment of the bedforms took place by high energy events planing-off the bedform topography, establishment of low energy conditions, or sub-aerial exposure. Early cementation of the bedforms occurred during pauses in migration and during meteoric diagenesis associated with subaerial exposure. The bedforms nucleated in a progressively basinward position throughout the deposition of the shoal complex, resulting in basinward progradation of the facies complex as a whole.

ACKNOWLEDGEMENTS

The authors would like to thank Mr Peter Dumenil, quarry manager of the Blue Circle Hope Cement Works Quarry for allowing us to work in the quarry and Al Fraser of BP and Maurice Tucker of Durham University for commenting on the contents of the paper. RLG acknowledges support from the Durham University Research Foundation and the Society of Fellows.

REFERENCES

BALL, M.M. (1967) Carbonate sand bodies of Florida and the Bahamas. *J. Sed. Petrol.* **37**, 556−591.

BROADHURST, F.M. & SIMPSON, I.M. (1973) Bathymetry on a Carboniferous reef. *Lethaia* **6**, 367−381.

EDEN, R.A., ORME, G.R., MITCHELL, M. & SHIRLEY, J. (1964) A study of part of the margin of the Carboniferous Limestone 'Massif' in the Pin Dale area of Derbyshire. *Bull. Geol. Surv. G.B.* **21**, 73−118.

FRASER, A.J., NASH, D.F., STEELE, R.P. & EBDON, C.C. (1990) A regional assessment of the Intra-Carboniferous play of northern England. In: *Classic Petroleum Provinces* (Ed. Brooks, J.) Spec. Publ. Geol. Soc. Lond. (in press).

GAWTHORPE, R.L., GUTTERIDGE, P. & LEEDER, M.R. (1989) Late Devonian and Dinantian basin evolution in northern England and North Wales. In: *The Role of Tectonics in Devonian and Carboniferous Sedimentation in the British Isles* (Eds Arthurton, R.S., Gutteridge, P. & Nolan, S.C.) Occ. Publ. Yorks. Geol. Soc. 6, pp. 1−23.

GUTTERIDGE, P. (1983) *Sedimentological study of the Eyam Limestone Formation in the east central part of the Derbyshire Dome.* Unpublished PhD thesis. University of Manchester.

GUTTERIDGE, P. (1987) Dinantian sedimentation and the basement structure of the Derbyshire Dome. *Geol. J.* **22**, 25−41.

GUTTERIDGE, P. (1989) Controls on carbonate sedimentation within a Brigantian intrashelf basin, Derbyshire. In: *The Role of Tectonics in Devonian and Carboniferous Sedimentation in the British Isles* (Eds Arthurton, R.S. Gutteridge, P. & Nolan, S.C.) Occ. Publ. Yorks. Geol. Soc. 6, pp. 171−187.

HINE, A.C. (1977) Lily Bank, Bahamas; history of an active oolite sand shoal. *J. Sed. Petrol.* **47**, 1554−1582.

HINE, A.C., WILBER, R.J. & NEUMANN, A.C. (1981) Carbonate sand bodies along contrasting shallow bank margins facing open seaways in Northern Bahamas. *Bull. Am. Assoc. Petrol. Geol.* **65**, 261−290.

LEE, A.G. (1988) Carboniferous basin configuration of central and northern England modelled using gravity data. In: *Sedimentation in a Synorogenic Basin Complex: The Upper Carboniferous of Northwest Europe* (Eds Besly, B.M. & Kelling, G.) pp. 69−84. Blackie, Glasgow.

SCHOFIELD, K. (1982) *Sedimentology of the Woo Dale Limestone Formation, Derbyshire.* Unpublished PhD thesis, University of Manchester.

SHIRLEY, J. & HORSFIELD, E.L. (1940) The Carboniferous Limestone of the Castleton − Bradwell area, North Derbyshire. *Q. J. Geol. Soc. Lond.* **96**, 271−299.

STEVENSON, I.P. & GAUNT, G.D. (1971) The Geology of the Country around Chapel en le Frith. *Mem. Geol. Surv. G. B.*, Sheet 99, p. 444.

TUCKER, M.E. (1985) Shallow marine carbonate facies and facies models. In: *Sedimentology: Recent Developments and Applied Aspects* (Eds Brenchley P.J. & Williams, B.J.P.) Spec. Publ. Geol. Soc. Lond. 18, pp. 147−169.

Spec. Publs int. Ass. Sediment. (1990) **9**, 55–78

Cyclic sedimentation in carbonate and mixed carbonate–clastic environments: four simulation programs for a desktop computer

G. M. WALKDEN* *and* G. D. WALKDEN[†]

*Department Geology and Petroleum Geology, University of Aberdeen, Aberdeen AB9 1AS, UK; and
[†]Department of Geological Sciences, University of Durham, Durham DH1 3LE, UK

ABSTRACT

Four programs are introduced for the BBC range of microcomputers that can produce simple simulations of tectonic, eustatic or other types of cyclothem in a variety of shallow- or deep-water carbonate and mixed carbonate–clastic environments. Cycle periods and magnitudes can be preset, randomized or expressed as the sum of up to three independent wave periods, and symmetric and asymmetric options are available. Where appropriate, subsidence is separated into a user-definable tectonic component and a facies-dependent compactional component. Compaction of the appropriate sediments takes place as the program is running. The programs generate synthetic sediment columns comprising a succession of depth- or lithology-defined facies, the accumulation rates of which are separately controllable.

'Cyclothem' simulates sedimentation on a shallow-water carbonate platform modelled from the late Dinantian of northern Britain. The program will reproduce tectonic or eustatic cycles and whilst shoaling will interrupt sedimentation, emergence will produce a subaerially-modified surface as commonly seen in the field. Marine facies are depth-defined, and where these are given uniform sedimentation rates the simulations assume a broadly chronostratigraphic character. The causes and significance of cyclothem asymmetry are examined. Cycle magnitude, sedimentation rates and subsidence rates can be separately defined and the effects of these on facies diversity and cyclothem thickness are discussed.

'Walther' uses an additional machine code routine to create an inset window that maps a delta. This grows, strands or drowns according to independent water depth controls and is used to introduce up to four non-carbonate facies. The program simulates British 'Yoredale'-type cyclothems and has parallels with Pennsylvanian mid-continent cycles of the USA. Both of these are considered to be eustatically controlled. The significance of an asymmetric wave and of compactional subsidence in maintaining conditions favourable for coal formation are also examined.

'Milankovich' returns to the shallow-water carbonate setting and simulates cyclothem distribution patterns arising from the simultaneous interaction of two or three secular controls on sea-level. The screen can be time- and thickness-scaled automatically to enable the investigation of fifth order (Milankovich-type) to third order (Vail-type) cycles. Modelling using this program demonstrates some of the dangers inherent in counting cycle numbers as a means of determining the lengths of Milankovich periods.

'Croll' models off-platform environments where sedimentation is not interrupted by emergence. Limestone–shale successions are an appropriate application, where controls on facies repetitions may be through factors other than sea-level fluctuations. Three facies can be simulated and independent accumulation rates set for each. Using the sum wave of up to three Milankovich periods to trigger facies changes it can be demonstrated once again that data derived from the geological record may provide misleading information as to the actual periods involved.

INTRODUCTION

Computer simulation provides an unique analytical tool in sedimentology that is close to becoming a basic research technique. Amongst other applications, it enables the analysis of sedimentary successions through the reversal of the usual interpretative route, that of working from the product (sediments and sedimentary rocks) back to the processes that created them. Instead, simulation can model processes themselves, perhaps exploring complex interactions of these over any convenient timescale, and it makes possible the direct testing and observation of how the product may have been arrived at.

The approach may not be fully objective since much of our understanding of geological processes is based upon rock interpretation. However, applied with caution and due regard to geological experience, simulation can provide unique insight into the importance and inter-relationships of specific processes of value to both research and teaching. Computer modelling should be regarded as potentially the best form of scientific experiment available to the earth scientist.

With increased interest in the possibility that cyclic processes may have dominated much of the stratigraphic record (e.g. Vail *et al.*, 1977; Crowell, 1978; Berger *et al.*, 1984; Ross & Ross, 1985; Blair & Bilodeau, 1988; Klein & Willard, 1989) several computer models of varying degrees of sophistication have appeared in recent years to test some of the concepts (e.g. Turcotte & Willemann, 1983; House, 1985; Schwarzacher & Schwarzacher, 1986; Read *et al.*, 1986; Goldhammer *et al.*, 1987; Read & Goldhammer, 1988; Spencer & Demicco, 1989). These generate cyclic patterns comparable with those seen in both shallow-water and deep-water successions, but in most cases the work has emphasized the results rather than the programs themselves which are not widely available.

Useful results are within the capabilities of standard microcomputers and in this paper a unified series of four programs is presented based upon the theme of cyclic sedimentation. The programs have been written for the BBC Model B and will also run on the B+ and the Master. They use BBC BASIC but an additional routine in one program uses machine code for speed and memory efficiency. It is hoped that the four programs will be found useful by research geologists and those involved in advanced teaching. They are being made generally available

on disk. Versions for use on other common microcomputers will be produced if there is sufficient demand.

The programs simulate a variety of controlling processes including orbital forcing, eustatic change, tectonic subsidence, delta progradation, sediment compaction and variable accumulation rates. The combinations available differ according to the version used. The rate, pattern or degree of influence of the more important variables can be user-defined within a wide range of values so that separate and interactive effects can be distinguished. A run is completed with the production of a simulated sedimentary column showing facies responses to the defined variables, and the screen displays a graphical record of changes in the principal parameters such as background subsidence and relative sea-level (Figs 1 & 2). Through rescaling time and sediment thick-

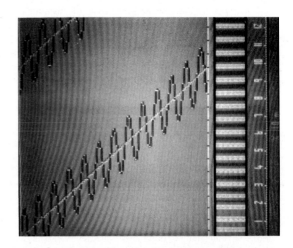

Fig. 1. Main screen mode of 'Cyclothem'. Screen width 3·9 Myr; column height 125 m; subsidence 2·5 cm kyr^{-1} cycles 30 m magnitude and 300 kyr period. Sedimentation rate is slow at 3·5 cm kyr^{-1} in order to demonstrate maximum facies diversity in the cyclothems. The menu screen (Fig. 3) shows the variables used in this run. An enlargement of the column is shown in Fig. 5. Cyclothem boundaries are emphasized by a tick to the left of the column. The deepest facies (wave-base set at 18 m) are red and show as dark grey, whilst the shallowest facies (peritidal base set at 2 m) are white and show as the lightest grey. Cyclothem boundaries are marked by subdued exposure surfaces (yellow, showing as mid-grey). Since sedimentation rate and subsidence rate are similar, the first few cyclothems are atypical to the 'basement effect' of the bottom of the screen.

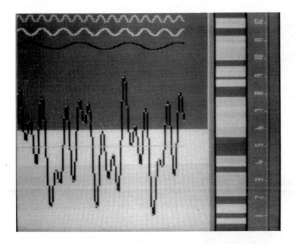

Fig. 2. Main screen mode of 'Croll'. Three Milankovich periods at 23, 41 and 100 kyr have been run simultaneously (top of screen) and are summed in the centre of the screen. The upper field of the screen is red (dark grey), the middle field is white (light grey) and the lower field is blue (slightly darker). Field thresholds are adjustable up and down and the sedimentation rate in each field can be separately set. The sediment column picks up the appropriate colour and rate as the sum wave builds across the screen. The sediment column is therefore an analogue of the peaks and troughs of the sum wave.

ness or by notionally reassigning the colour-coded facies, the programs can be used to model cyclicity and stratal complexity on different scales and in a variety of sedimentary settings. The programs may also be used to demonstrate some basic concepts relating to stratigraphic analysis and sequence stratigraphy.

The four programs, 'Cyclothem', 'Walther', 'Milankovich' and 'Croll' were successively developed using similar style and presentation. 'Cyclothem' simulates cyclicity in a marine environment in which facies type is expressed as a function of depth and time. Its most appropriate application is for autochthonous sediments such as carbonates. The more complex 'Walther' simulates a mixed carbonate–deltaic environment where facies type is determined by a combination of depth and delta progradation. 'Croll' generates sine waves simulating up to three astronomical forcing functions of any period, and sums these to control a simulated pelagic succession. 'Milankovich' incorporates this multiple independent sine wave facility into a simplified version of 'Cyclothem' to simulate orbitally forced or longer period Vail-type eustatic cycles in a shallow-water carbonate environment.

Throughout the description of the programs and the discussion of modelling and results, a distinction is made between a cyclothem and a cycle. A cyclothem is here defined as a sequence of sediments or sedimentary rocks created by a cyclic process. The use of the term is lithostratigraphic and a cyclothem may be symmetrical or asymmetrical. A cycle is defined as the process or sequence of environmental changes that creates a cyclothem. The use of this term is chronostratigraphic but it can be expressed in analogue form as a wave.

PROGRAM PRESENTATION

Much computer-simulated cyclicity has been based until now upon the format of the 'Fischer diagram' in which successive eustatic sea-level waves are drawn horizontally from left to right, and the associated sediment packets are drawn obliquely below these. Their angle is proportional to the rate of subsidence (e.g. Fischer, 1964; Goldhammer *et al.*, 1987). Such diagrams are an excellent means of analysing cyclicity, but are less appropriate for dynamic simulations where attention focuses on the sea-level/sediment-surface relationship. Our programs are based upon a format that expresses subsidence and eustacy in a less conceptual manner by graphing background subsidence as a rising curve modulated by an eustatic wave. Our sediment column builds alongside this curve when subsidence or eustatic rise are creating accomodation, and its surface becomes exposed when sea-level drops.

All programs use full colour main (Figs 1 & 2) and menu (Fig. 3) screens. The menu shows the current user-defined parameters and enables their resetting through an interactive routine. The main screen displays the real time simulation and is divided into two sectors. Most of the screen is occupied by the sea-level window in which separate graphs of tectonic subsidence and the sum of tectonic and eustatic sea-level change are drawn. The sediment column, comprising successively accreted, colour-coded facies selected from a pre-determined range, appears on the right. The width of the sea-level window is nominally 3·9 Myr, and the height of the fully developed sediment column is 120 m.

The range of user-defined variables differs in each of the programs but includes rate and pattern of background subsidence, magnitude and period of eustatic cycles, depth range of facies and rate of accumulation of each. In most cases variables can be

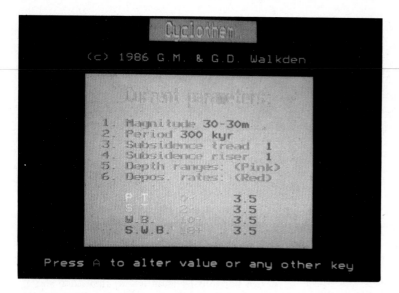

Fig. 3. Menu screen of 'Cyclothem'. The screen displays the current settings of all user-defined variables. 'Magnitude' and 'Period' refer to the cycle; 'Subsidence tread' and 'Subsidence riser' relate to the $x:y$ ratio of the subsidence line which is drawn in steps if required; 'Depth ranges' relates to the depth-defined facies, and 'Depos. rates' relates to their respective sedimentation rates.

set at actual time or thickness values and some can be allowed to vary randomly within preset limits. Other variables, such as sediment compaction and sedimentation lag times, are at preset values and must be changed by accessing the program. One or two more important modifications such as the form of the eustatic curve are featured in descendent versions of specific programs.

Reasonable default values are given to user-defined variables which enable the programs to be run in a demonstration mode directly following loading. At the conclusion of a run, determined by the completion to the top of the screen of the sediment column, the screen can be saved on disk for reloading later or the user can return direct to menu.

THE PROGRAM 'CYCLOTHEM'

Objectives

This program simulates shallow-water carbonate successions where a main controlling parameter is likely to be water depth. The underlying objectives are similar to those of Read *et al.* (1986) although presentation and scope are distinct. The original purpose of the program was to determine the influence of differing rates and patterns of background subsidence and eustatic cyclicity on the generation of 10 m-scale, 200–600 kyr marine cyclothems of the

sort seen in the late Lower Carboniferous (late Dinantian or late Viséan) of Britain and Europe (Walkden, 1987; Leeder, 1988). This periodicity is typical of late Palaeozoic cycles (e.g. Heckel, 1986) but is up to an order of magnitude longer than that modelled by previous authors such as Grotzinger (1986) for the Lower Proterozoic, Read & Goldhammer (1988) for the Ordovician and Goldhammer *et al.* (1987) for the Middle Triassic.

Late Dinantian cyclicity is most obvious in successions containing clastic–carbonate interactions of the Yoredale type (e.g. Johnson, 1972; Walkden & Davies, 1983), but where the clastic components fail the cyclothems are still observed in laterally equivalent carbonate successions (e.g. Mitchell *et al.*, 1978; Burgess & Mitchell, 1976; Walkden, 1987). They are characterized by shallowing-up facies sequences and well-developed exposure surfaces. (Ramsbottom, 1973; Walkden, 1974; Somerville, 1979; Walkden, 1987). On the Euramerican to worldwide scale, Carboniferous cyclicity is increasingly being ascribed to glacioeustacy (Goodwin & Anderson, 1985; Heckel, 1986; Veevers & Powell, 1987; Walkden, 1987) and recently Klein & Willard (1989) have made a distinction between eustatically-generated Kansas-type cyclothems and tectonically-generated Appalachian-type cyclothems. In the UK there is strong support for tectonic and autocyclic mechanisms for Carboniferous minor cyclicity (Bott & Johnson, 1967; Johnson, 1984; Bott, 1987; Leeder & Strudwick, 1987) although recently Leeder (1988) has accepted a eustatic model for late Dinantian carbonate cyclothems.

In 'Cyclothem' no clastic sediments are necessarily involved and the facies patterns created are a record of changing depth with time at the sediment surface caused by the simultaneous operation of three variables: subsidence, eustacy and sedimentation rate. Any sediment sequence in which a principal controlling factor on facies development has been water depth can therefore be simulated, or the program may simply be used to demonstrate depth as one controlling factor upon facies succession. As long as sedimentation rate is kept uniform during a run, each completed cyclothem can be regarded as a near but not perfect chrono-stratigraphic record of changing environments. Time occupied during emergence is not expressed chronostratigraphically but is shown proportionally by penetrative erosive effects. Once individual depth facies are assigned separate accumulation rates, the sediment column must be read throughout as a lithostratigraphic record.

Program variables

The depth of water at the sediment column is shown by a line directly tied to the sea-level curve. This rises and falls as the curve is drawn and the depth at the sediment column determines the 'facies' type selected. Facies are colour coded with reference to user-defined limits for tide base, normal wave base and storm wave base. The rate of accumulation for each facies can be independently set between 1 and $45 \, \mathrm{cm} \, \mathrm{kyr}^{-1}$. During emergence a single line of supratidal facies is deposited, and the surface becomes penetrated by stylized dissolution pits.

Magnitude and period of eustatic cycles are set directly in metres per thousands of years. Using a screen width of 3·9 Myr the minimum practical period is 100 kyr, but by scaling the screen by an order of magnitude, which also involves appropriate scaling of cycle magnitude and sediment depth ranges, cycles down to 10 kyr may be examined. In 'Milankovich' such scaling is done automatically.

Subsidence is not defined in absolute values but is given a vertical 'riser' component and an horizontal 'tread' in screen pixels. A ratio of 1:1 corresponds to a subsidence rate of $2.5 \, \mathrm{cm} \, \mathrm{kyr}^{-1}$, whilst a ratio of, say, 5:5 would achieve the same rate of subsidence but in larger steps. Subsidence, whether by steps or progressive, can be randomized within a given range to explore the effects of change whilst the program is running.

In order to obtain meaningful results from real time simulations all these input variables must be assigned realistic values. Bearing in mind uncertainties over the duration of Asbian and Brigantian stages (Leeder, 1988) average platform subsidence rates in the late Dinantian, a time of active basinal extension, were in the range $3–6 \, \mathrm{cm}$ kyr^{-1}. Leeder (1988) and Leeder & McMahon (1988) showed that basinal rates were a little faster, up to $8 \, \mathrm{cm} \, \mathrm{kyr}^{-1}$, but in general all rates slowed into the Namurian and Westphalian as thermal subsidence alone took over. These rates lie well within the normal range for subsidence in extensional basins and on thermally-subsiding passive continental margins which extend between 1 and $10 \, \mathrm{cm} \, \mathrm{kyr}^{-1}$ (e.g. Schlager, 1981; Mitchell & Reading, 1986). Cisne (1986) showed how subsidence might be stepwise, controlled by faulting, and Bott (1987) suggested that steps of up to 8 m might have affected Carboniferous basins. Blair & Bilodeau (1988) discussed tectonically-produced cyclothems in fault controlled regimes and Klein & Willard (1989) invoked episodic thrust loading as a factor in cyclothem development.

Since ancient subsidence rates are measured from sediment thicknesses, subsidence rates are also average sediment accumulation rates. It is well known that such figures can be more than an order of magnitude slower than rates measured or deduced from the Holocene (e.g. Wilson, 1974; Schlager, 1981; Sadler, 1981). Recent accumulation rates for carbonates range between 0·5 and $>1 \, \mathrm{m} \, \mathrm{kyr}^{-1}$, and averages vary according to facies type and location. Whilst it is possible that Holocene rates are atypical because of very rapid sea-level rises, the difference is more likely to reflect the importance of reworking, erosion and non-depositional hiatuses in the stratigraphical record, some of which 'Cyclothem' can model. Goldhammer *et al.* (1987) argued for lower overall accretion rates by making a distinction between sediment production rates and net platform accumulation rates, pointing out that much internally-produced sediment is eventually lost over the platform edge. This pattern was probably experienced on all platforms at some time in the late Dinantian in Britain (Walkden, 1987), and the importance of the distinction should be recognized in the choice of sedimentation rates.

Rates and patterns of eustatic change are far more difficult to model accurately, especially for ancient rocks. Holocene records infer glacioeustatic rises of up to $10 \, m \, kyr^{-1}$ (e.g. Donovan & Jones, 1979; Schlager, 1981) and Pleistocene deep-sea oxygen isotope curves, which mainly reflect the volume of water removed from the oceans (e.g. Imbrie & Imbrie, 1980; Imbrie, 1985), suggest similar rates. In terms of its geometry, therefore, a simple sine wave may not be a satisfactory analogue of glacioeustacy if the Quaternary is to be used as a guide.

In particular, the symmetry of a sine wave is not representative of the Quaternary situation in which major glaciations seem to have terminated faster than they began (Imbrie & Imbrie, 1980). There is no reason to suppose that smaller scale glacioeustatic fluctuations of the past should match Quaternary patterns, but Turcotte & Willemann (1983) noted during modelling that symmetrical oscillations in sea-level did not produce the strong asymmetry seen in many Pennysylvanian cycles, and Grotzinger (1986) and Read *et al.* (1986) opted for asymmetrical cycles using fast rises and slow falls. Goldhammer *et al.* (1987) managed to reproduce complex pentacyclic patterns observed in the field through super-imposing a sinusoidal oscillation on top of a longer period asymmetrical pulse.

Modelling using 'Cyclothem'

The problem of cyclothem asymmetry

Asymmetry is typical of shallow-water carbonate cyclothems (e.g. Wilson, 1974; James, 1979; Goodwin & Anderson, 1985) and many display upward-shallowing A−B−C sequences with a dis-conformable C−A boundary. In some cases this boundary is an emergent surface showing subaerial diagenetic modification (e.g. Walkden, 1987; Goldhammer *et al.*, 1987). Using 'Cyclothem 1', stepwise subsidence alone will produce strongly asymmetrical regressive or 'shallowing-up' patterns that match the geological record well, except that they lack significant emergence phenomena (Fig. 4). Furthermore, palaeokarsts and other forms of penetrative vadose diagenetic modification are commonly found directly capping subtidal facies (Walkden *et al.*, 1983; Walkden, 1987, p. 143; Goldhammer *et al.*, 1987, p. 868) indicating relatively sudden falls in base-level that precluded the develop-ment of intertidal facies. This cannot be modelled by stepwise subsidence.

In 'Cyclothem 2' a sine wave simulates $30 \, m \times 300 \, kyr$ eustatic cycles superimposed upon steady

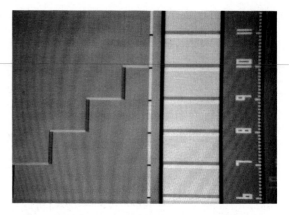

Fig. 4. 'Cyclothem 1' run using rapid subsidence of 10 m every 400 kyr. At the facies settings used here this sudden subsidence floods the platform and establishes sub wave-base facies for a short period (dark grey) followed by subtidal facies then peritidal (lightest grey). This pattern is produced regardless of the sedimentation rate used.

background subsidence. This produces a maximum rise rate during the upwave of c. $30 \, cm \, kyr^{-1}$ for cycles of 30 m magnitude, reducing to c. 10 cm at 10 m magnitude. These rises outpace low to average sedimentation rates and achieve deepening on inundation. The moderate magnitude is used to create patterns with high depth diversity. Emergence is marked, but the cyclothems produced are chrono-stratigraphic time−depth analogues which reflect the symmetrical deepening and shallowing of the input cycles and lack the asymmetry of the geological record (Fig. 5). To reproduce asymmetry requires either more complex facies dynamics or a change in the shape of the cyclic wave itself.

Ginsburg (1971) suggested that steady subsidence alone could result in small-scale asymmetrical cyclothems if sediment production was much slower following progradation of intertidal and supratidal facies. This would produce a 'lag' in sedimentation, permitting subsidence and marine inundation of supratidal flats before normal production rates were restored. Once again the presence in the field of penetrative meteoric vadose effects directly capping subtidal facies, indicative of sharp downward move-ment of sea-level, rules out universal application of the Ginsburg model, but his principle of a sedimen-tation lag remains. Subsequent modelling has incor-porated either a lag time of a few thousand years (Read *et al.*, 1986; Grotzinger, 1986) or a lag depth of 1 m or so (Goldhammer *et al.*, 1987) as a means of

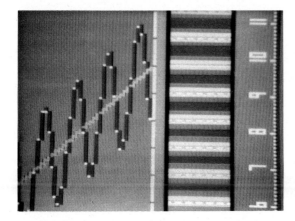

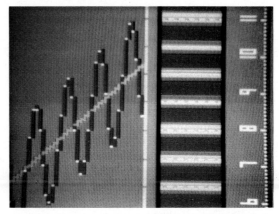

Fig. 5. 'Cyclothem 2' run with symmetrical wave-producing symmetrical cyclothems. Settings are as Fig. 1. Peritidal facies (white) and subtidal facies (light blue) are difficult to distinguish in monochrome. In general the darker the tone the deeper the facies. Cyclothem boundaries are marked by the tick to the left of the sediment column and by slight subaerial pitting at cyclothem boundaries. Scale in tens of metres.

Fig. 6. 'Cyclothem 3' showing the effect of an initial sedimentation lag. Variables are otherwise identical to Fig. 5. Peritidal facies are always absent at the base of cyclothems and subtidal facies are also much reduced. Cyclothems are otherwise not markedly asymmetrical with such high facies diversity, and cyclothem boundaries become generally more subdued towards the top of the column (numbered) owing to sedimentation + emergence failing to balance subsidence.

excluding intertidal deposits at the base of synthetic cyclothems.

However, time-defined or depth-defined lags of the scale used in these models only eliminate the intertidal window during transgression, leaving the deepest part of the cyclothem more or less symmetrically positioned. It is therefore only with cycles of small magnitude and consequent low depth diversity that obvious asymmetry can be achieved. For higher magnitude cycles, asymmetry is only simulated if the lag is given a long term influence, effective over a significant timespan of the sea-level upwave. Even in the extreme case modelled in 'Cyclothem 3', where the lag is an adjustable exponential function that decays over a few tens of thousands of years, the symmetry is not greatly reduced when cycles of moderate magnitude are being used (Fig. 6).

A more effective means of modelling asymmetry is by using a rapid upwave and slow downwave as shown by Pleistocene and Holocene eustacy (Imbrie & Imbrie, 1980). 'Cyclothem 4' has a preset asymmetry in which the upwave occupies approximately 25% of the cycle and the downwave the remainder. This pattern could preserve the chronostratigraphical property of the simulation but it has been combined with the sediment lag. The version produces cyclothems that have markedly asym-

metrical A−B−C, A−B−C patterns with intervening emergent surfaces (Fig. 7), even with relatively large cycle magnitudes.

Asymmetrical cycles increase the probability that regressive intertidal deposits would form at the top

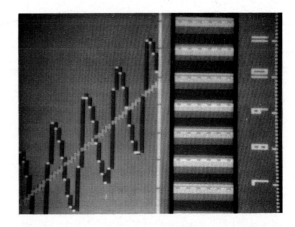

Fig. 7. 'Cyclothem 4' using an asymmetric 30 m × 300 kyr cycle creating well-developed asymmetric cyclothems. Settings as in Fig. 5. Peritidal facies and sometimes also subtidal facies are missing at the base of cyclothems owing to a fast transgression and a combined sedimentation lag. Peritidal facies are present at the tops of cyclothems as between 70 and 80 m.

of cyclothems and yet it has been noted that in many cases these are lacking, indicating sharp relative movements of sea-level. This problem was addressed by Read *et al.* (1986) who pointed out that the development of intertidal facies may greatly depend upon the rate of horizontal progradation of tidal flats across the platform. The presumed location of the modelled site (proximal or distal) in relation to a persistent shoreline is therefore of considerable importance. Since 'Cyclothem' is principally a time—depth simulation and not a real facies simulation, it is clear that although the intertidal 'window' might be well developed during regression, there is no guarantee that in the real situation it would be filled by prograding intertidal facies.

To reduce the opportunity for intertidal progradation, the size of this window must be reduced in the simulations. This could be done by moving to a 'square' wave form in which the fall is also relatively abrupt, for which there is some support from Pleistocene studies (e.g. Donovan & Jones, 1979). Alternatively, cycles of moderate magnitude (30 m × 250 kyr) should be used, the upwave of which significantly outpaces sedimentation so that during downwave the fall rate is at its fastest as sea-level approaches the sediment surface. Reducing the rate of accumulation of peritidal facies to less than that of subtidal facies, for which there is some support from Holocene carbonate sedimentation rates, further emphasizes the pattern.

Effects of changes in subsidence

More complex facies patterns are introduced if randomized background subsidence or stepwise background subsidence are reintroduced in addition to eustacy (Cyclothem 5). With small subsidence steps of a few metres, facies boundaries may be crossed and recrossed, especially during regression. The graph of background subsidence together with the response shown by relative sea-level can be matched directly to the facies responses in the sediment column (Fig. 8).

Using steady subsidence, any increase in the rate of subsidence increases the thickness of cyclothems (Fig. 10b) and marginally increases their facies diversity. The increased thickness occurs because the faster rate of subsidence 'stretches' the eustatic wave with less time taken up by exposure at the end of an eustatic rise and more in depositing sediments. The increased diversity appears because a faster overall subsidence rate enables deeper water to

Fig. 8. 'Cyclothem 5' combining the asymmetric 30 m wave with stepped subsidence of 10 m in 400 kyr as used in Fig. 4. Eustatic cycles remain clearly identifiable through the subsidence effect but the shallowing-up sequences are frequently briefly reversed owing to sudden subsidence. Note that exposure periods are also affected. Variables otherwise as in Fig. 5.

become established during the upwave. Exposure surfaces are thus poorly developed and if too much subsidence is allowed, alternatively modelled by reducing the sedimentation rate, sedimentation cannot keep pace and the platform finally drowns (*cf.* Schlager, 1981; Kendall & Schlager, 1981).

Effects of changes in cycle magnitude

Using steady subsidence and sedimentation rates, small cycle magnitudes produce cyclothems with low facies diversity and regular thickness, whilst larger cycle magnitudes produce cyclothems of greater facies diversity but identical thickness (Figs 9a, b). This observation contradicts the suggestion of Leeder (1988) that it should be possible to calculate the average magnitude of eustatic cycles from the resultant cyclothems. Only when it can be safely assumed that; (1) all accommodation created during the eustatic upwave was filled straight away as a result of fast sedimentation rates; and (2) there was no significant exposure of the sediment surface (i.e. there was no drop of sea-level below the platform surface), does cyclothem thickness less known or calculated subsidence give the magnitude of the eustatic cycle.

With fast sedimentation rates it is only the first cyclothem in any run that reflects to any degree the magnitude of the eustatic wave. Thereafter, the

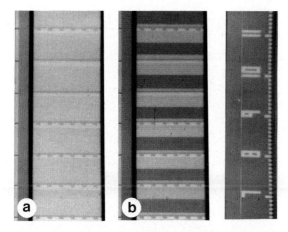 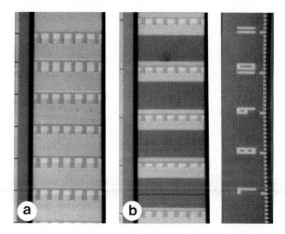

Fig. 9. Two 'Cyclothem 5' sediment columns for comparison of the effect of cycle magnitude. (a) 10 m magnitude, 300 kyr period, 2·5 cm kyr⁻¹ subsidence and 3·5 cm kyr⁻¹ sedimentation rate. (b) 15 m magnitude but other variables as in (a). Note identical thickness of cyclothems but contrasting facies diversity. Compare with Fig. 5 (30 m magnitude) which has otherwise identical variables, and with Fig. 10b which has faster subsidence and sedimentation rates.

Fig. 10. Two 'Cyclothem 5' sediment columns demonstrating the effect of increased sedimentation and subsidence. (a) Subsidence 2·5 cm kyr⁻¹, sedimentation 15 cm kyr⁻¹, cycles 30 m × 300 kyr; lengthy subaerial exposure and low facies diversity. (b) Subsidence 3·8 cm kyr⁻¹ sedimentation 15 cm kyr⁻¹, cycles 30 m × 300 kyr; moderate subaerial exposure and moderate facies diversity. A fast sedimentation rate increases the period of exposure, but not cyclothem thickness (a). It reduces facies diversity. Only faster subsidence will increase cyclothem thickness (b) and it increases facies diversity.

thickness of each cyclothem is equal only to the amount of subsidence during the period of the respective cycle. Such cycles lack asymmetry because sedimentation has kept pace with the rise, and all deposits are of shallow-water facies (Fig. 10). Similar calculations to those of Leeder (*op cit*) done by Read & Goldhammer (1988) probably assumed that sedimentation kept pace during the eustatic rise but will have omitted the portion of the eustatic cycle lost during emergence. Where cyclothems show upward-shallowing characteristics, only the diversity of facies can give any reliable indication of cycle magnitude.

THE PROGRAM 'WALTHER'

Objectives

Walther was principally developed as an experimental program that introduces simulated clastic sediments into carbonate cyclothems. Mixed clastic–carbonate cyclothems are widespread in the geological column but it is again the Carboniferous situation that forms the basis for modelling. Carboniferous cyclothemic patterns are extremely diverse (e.g. Francis, 1983), but in late Dinantian to early

Namurian sequences the Yoredale pattern is typical in northern Britain, in which limestone forms a basal transgressive unit followed by deltaic shale, sand and finally delta plain muds and coals (Johnson, 1959).

Comparable cyclicity starts at the same time in North America (Ramsbottom, 1973; Veevers & Powell, 1987) and extends into the Pennsylvanian and beyond. Equivalent Upper Carboniferous sequences in Europe are dominated by Coal Measures-type cyclothems (e.g. Ramsbottom *et al.*, 1974), in which the marine component is much reduced or absent. However, North American mid-continent cyclothems follow the Yoredale pattern but with emphasis on the limestone and little to no sand (e.g. Heckel, 1986). Sands become more important towards the east and west sides of the craton (Ferm, 1970; Driese & Dott, 1974; Klein & Willard, 1989).

Yoredale-type cyclothems are generally on the 5–20 m scale, although smaller repeats ('rhythms') do occur in their upper parts (e.g. Johnson, 1959; Heckel, 1986). These repeats are attributable to autocyclic processes, but it is worth noting that the pattern is comparable with the groups of cyclothems noted in carbonate successions by Goldhammer *et al.* (1987) attributed to eustacy. Full cycles had periods in the range 235–880 kyr (Driese & Dott, 1974; Heckel 1986; Walkden, 1987) and their probable glacioeustatic origin has already been discussed.

Program features

The style and format of Walther is close to that of 'Cyclothem'. The main differences lie in the use of a screen window in which 'clastic' facies are generated and an additional routine that enables 'compaction' of these to take place in the sediment column. A sine wave is used in the program to control eustacy, and in 'Walther 1' this is given a linear, almost 'castellated' expression and in 'Walther 2' it is asymmetric.

The introduction of specific clastic facies into the simulation necessitates some means of generating these stochastically so that the patterns produced are not merely an animation of the real situation. It also calls for feedback between facies type, water depth and sediment accumulation. The eventual choice, controlled by the need for clarity and simplicity, was to emulate some of the essential features of an actual growing delta lobe by generating a rudimentary facies map of a delta that develops with time (Fig. 11). The routine uses a 64 square grid, nominally representing an 8 × 8 km or 80 × 80 km sector of shelf, and five colour-coded facies, namely carbonate, prodelta mud, delta front sand–silt, delta-plain mud and peat. Facies transition within any individual square is mostly uni-directional but timing is random, and transition can only occur when an adjacent square is already occupied by the new facies. By seeding new facies at randomly determined points only along the northern margin of the window, a reasonably convincing pattern of southward prograding facies belts is created and separate runs produce quite different distributions and arrival times of facies. Sand advance commonly breaks into separate lobes but in general the patterns approach a generalized shallow-water wave-influenced or multi-distributary fluvial-dominated delta in which a broad irregular sand sheet advances seawards. This pattern is appropriate for modelling Yoredale sequences (e.g. Elliott, 1975, 1986).

By minor program alteration any one square in the window can be set as the reference point for the sediment column, distal or proximal, and it is the precise timing and rate of accumulation of specific facies appearing in this reference square that finds

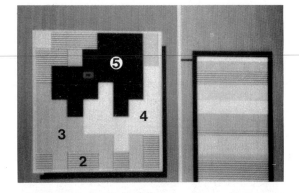
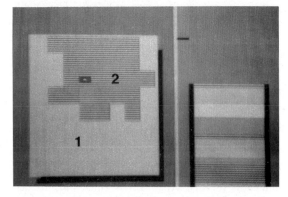

Fig. 11. Two successive stages in the development of the delta in 'Walther'. Facies are colour coded in the simulations but are numbered here to aid recognition in monochrome. (1) Carbonate; (2) pro-delta mud; (3) delta front sand–silt; (4) delta-plain mud; (5), delta-top peat. The black bar to the left of the sediment column represents sea-level and the oblong on the delta map is the reference point for the sediment column.

expression in the sediment column. This link between lateral facies distribution and vertical facies succession noted by Walther in 1884 (Middleton, 1973) gives the program its name.

Whilst the delta window provides a two dimensional facies mosaic, changing with time, the depth dimension is simulated at the sediment column using the rising and falling sea-level bar adopted in 'Cyclothem'. There are feedback routines that link water depth at the column and delta development in the window. Facies types can be allocated specific accumulation rates (e.g. fast sand and slow peat) and shallow-water facies are only permitted to prograde across the reference square when an appropriate water depth is developed above the sediment column. Emergence resulting from a fall of sea-level at the sediment column strands the delta locally, and shallow flooding promotes delta-top mud or peat. Rapid flooding, depending upon scale, causes either a reverse transition of facies (knockback routine) or the complete drowning of the delta and its replacement by transgressive facies (wipe routine).

Compaction is introduced first to demonstrate the difference between depositional thickness and compacted thickness in cyclothems, and second to show the effect of active compaction on overall subsidence rates. Ramsey & Moslow (1987) document compactional subsidence rates above unconsolidated Holocene deltaic sediments in Louisiana of >1 cm yr^{-1} which considerably outpaces normal tectonic subsidence, but some of this may be attributable to withdrawal of subsurface oil and brines. Because of the limited thickness of the sediment column, compaction modelling necessarily involves only interstitial water, but to be effective it has to go to completion in 'Walther' simulations more rapidly than is likely in the real situation. According to Chillingarian (1983), expulsion of interstitial water can be rapid down to 500 m but structural and adsorbed water in clays continues to be expelled beyond 1 km. The effect of more rapid compaction in our simulations is merely that of a thicker sediment column than our program is designed to accommodate, and therefore remains reasonably realistic. The routine starts after 20 m of sediment has been deposited and goes to completion at 20% of peat thickness and 40% of mud at around 100 m.

The basic sedimentological assumptions employed in 'Walther' are that during the initial phases of an eustatic rise, the delta top becomes rapidly flooded and clastic sedimentation rates are slowed to negligible. A wipe routine removes the delta from the window, simulating flooding of the abandoned surface, and delta development remains inhibited as the upwave continues, representing the effect of rising base levels in the hinterland and the backing up of the clastic supply systems. The sediment that is modelled during transgression is therefore nominally a carbonate rather than a mud, but since its rate of accumulation is user-defined its only intrinsic programmed characteristic is that it will not compact. A conscious limitation of this program, therefore, is that deltas cannot advance during a major transgression and marine conditions prevail at this stage. Coal Measures cyclothems or the highly proximal Yoredale situations of Ramsbottom (1973) are not well modelled, and large-scale delta abandonment, as in the shoal water deltas of the Mississippi (e.g. Elliott, 1986), cannot be reproduced within a single

eustatic cycle unless simulated by introducing stepwise subsidence. This activates the knockback routine and, depending upon the depth change involved, will usually trigger a reversal of facies followed by renewed progradation.

Once sea-level rise is complete, it is assumed that delta outbuilding would recommence, and the delta draw routine switches on. Clastic supply continues in abundance during fall as the notionally backed up sediment is rapidly reworked and cleared from the flanks of the hinterland (e.g. Walkden, 1987). Delta-top facies are given a depositional window between 2 m above and 1 m below sea-level. This simulates conditions in delta-top lagoons and ponds but can lead to unrealistic over-emphasis of these facies during fast falls of sea-level past the depositional surface. On emergence of the delta top, the delta draw routine switches off, although it is recognized that in distal areas beyond the site-modelled development would continue. Local reactivation of the top is permitted following renewed small-scale inundation, and steady subsidence under shallow conditions will promote delta-top facies.

Sediment accumulation rates in clastic systems are notoriously difficult to model and steady figures are quite meaningless for some facies. This problem is addressed by Sadler (1981) who noted that sedimentation rate is in approximate inverse proportion to the timespan over which it is measured. This reflects the difference between accumulation potential, determined by the dynamics of the sedimentary system, and preservation potential, ultimately determined by base level and subsidence. The controlling link between the two is sediment reworking and erosion. 'Walther' does not simulate sediment reworking but merely plots upward facies trends. As in 'Cyclothem', sedimentation rates are user-definable, but rates should be chosen with regard to preservation potential rather than accumulation potential.

Other menu options include the magnitude and period of the cycle. Subsidence can be set as linear or stepwise, and can be randomized. The overall rate of development of the delta (and the relative width of facies belts) is adjustable through the program and is not a menu option. The compaction routine can be disabled through the program.

Modelling using 'Walther'

Simulation of a complex interplay of processes such as subsidence, compaction, sea-level change and delta development makes great demands on basic programming and 'Walther' can provide only a very generalized synthesis. Nevertheless its simulations are at the very least thought provoking. The program runs slowly when the machine code delta routine is interfacing with the basic program so that it no longer runs accurately in real time.

Effects of tectonic and compactive subsidence

'Walther 1' is preset with a symmetrical 20 m magnitude flat-topped eustatic wave and is given uniform sedimentation rates for all facies (Fig. 12). Upwave initially outpaces sedimentation but as the wave levels out sedimentation begins to catch up. Shallowing is rapid as the sediment surface rises to meet falling sea-level and delta-top facies are consequently thin or possibly missing. During the first few cycles there is rarely sufficient opportunity for full progradation of facies, and coals are initially rarely developed. However, if these early cyclothems contain appreciable quantities of compactible prodelta mud, later cyclothems become thicker as a result of added compactive subsidence (Fig. 12). In particular the increased subsidence extends the depositional window for delta-top facies and thick peats occasionally appear during downwave or lowstand.

More rapid subsidence, or the equivalent effect through a slower overall rate of sedimentation,

therefore has the effect of increasing the importance of delta-top facies, as resetting these values demonstrates. Conversely, overall faster rates of sedimentation or slower subsidence increases the importance of the transgressive to early regressive delta-front portions of cyclothems.

Proximal and distal cyclothems

The above effects are similar to those achieved by setting the reference square to proximal and distal situations respectively. Carbonate-dominated distal cyclothems and delta-dominated proximal cyclothems, as noted in Yoredale sequences by Johnson (1972) and Ramsbottom (1973), are readily reproduced. Emergence of the distal cyclothems where the limestones are relatively thick would bring carbonates into the meteoric vadose zone with potential for karstification and subaerial diagenesis as in 'Cyclothem'. This would be unusual in proximal situations and demonstrates the reason for the normal lack of a diagenetic cap over limestones within fully developed Yoredale-type sequences.

Development of coals

The inclusion of peat in the facies succession has only limited significance, the main justification being that coal is a component of the 'ideal' Yoredale

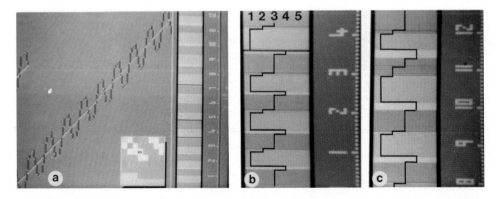

Fig. 12. A completed run of 'Walther 1' (a) together with enlargements of the sediment column showing the effect of compaction on cyclothem thickness (b and c). Variables as follows: Subsidence rate 3 cm kyr^{-1}, sedimentation rate 10 cm kyr^{-1}, cycles 20 m × 400 kyr. Facies are numbered as in Fig. 11, and a graphic log has been superimposed on the sediment column to aid facies recognition. Note how sand (3) and limestone (1) dominate in the lower part of the column where the muds (2, 4) and coal (5) have been compacted. This lost thickness has been transferred to the upper part of the column in the form of increased accommodation creating cyclothems that are 50% thicker. Sand and limestones are also thicker in upper cyclothems despite the fact that they do not compact. In the case of the limestone this is because compactional subsidence reduces the degree of emergence at the cycle boundary and flooding occurs earlier. In the case of the sand it is because of a feedback routine between delta development and water depth which will only permit the delta top to prograde across the reference square if the water depth is 2 m or less.

cyclothem. The only intrinsic features of peat in 'Walther' simulations are that it compacts more than other delta-top sediments and it can be assigned a different accumulation rate from the muds. Simulated peat is mainly an indication of maturity of the delta plain and of an extended opportunity for a coal-forming environment to develop.

In reality peat could equally appear during the window of the delta-top muds, or not at all. Nevertheless, it is notable that peat accumulation is normally maximized during the later parts of runs when compactive subsidence is active. The added effect of compaction reduces the length of emergence during the downwave because both the sediment surface and sea-level are falling during this part of a cycle. This permits sedimentation to continue for longer before emergence terminates it, providing more opportunity for the delta to mature and for delta-top sediments to accumulate.

Using 'Walther 1', however, this situation is meta-stable and although extensive peats can be produced at lowstand the crucial conditions of either relatively fast subsidence or relatively slow sedimentation leave the system vulnerable. Once these additional delta-top sediments are themselves undergoing burial, their own compactive contribution to overall subsidence can quickly lead to sedimentation being outpaced which results in permanent drowning of the sediment column. Clearly, since our compaction is modelled as a faster process than in the real situation (but who can be sure?), the drowning problem is magnified, but the effect is clearly equivalent to that produced by an appropriately thicker sediment sequence than modelled within a single screen.

Lowstand with a symmetrical eustatic cycle of moderate magnitude could therefore be an unlikely scenario for peat formation in sequences that contain a lot of coal. More stable would be the highstand coal built on top of rapidly accumulated clastic facies prior to sea-level fall. This situation is difficult to model, though, because it requires either a rapid burst of delta development at highstand or a longer period of highstand for full development of the delta, including peat accumulation, to take place. Thick delta-top sequences are in any case unlikely in these circumstances as fast overall sedimentation rates reduce their importance. The scenario also involves dramatic elevation of the delta top during downwave and lowstand, with (in the real situation) consequent dramatic draining, erosion and oxidation of delta-plain facies. For these reasons extensive and repeated delta-plain coals may not be highstand effects either.

A better scenario may be the asymmetrical eustatic wave with a fast rise and slow fall, as developed in 'Cyclothem 4'. This accords with the conclusions of Heckel (1986) regarding the origin of mixed clastic carbonate mid-continent cycles. In 'Walther 2' the upwave–highstand period is 25% of the cycle, and the downwave–lowstand period is the remaining 75%. This wave–form produces fast inundation of the delta top, and there follows much time during highstand and downwave for the re-establishment of the delta and the progradation of the delta plain. Coals may be produced anywhere between highstand and lowstand. Depending upon the magnitude of the cycle, the slow downwave enables background and compactive subsidence to counteract at least part of the sea-level fall (Fig. 13). This has the effect of extending the sedimentation window, increasing the stability of the pattern under a range of sedimentation rates, and reducing the deleterious elevation of the delta during emergence.

When cycle magnitudes are small these effects are enhanced and sea-level rise effectively proceeds in a series of steps. Because emergence is never marked, sedimentation is possible for most of the time and the delta regularly matures with the appearance of thick delta-top facies including peat. If delta-top facies alone are assigned slow accumulation rates with this type of cycle, a limited degree of delta lobe switching can be simulated as subsidence outpaces delta-top accumulation rates and facies transition is reversed to sand through the knockback routine in order to fill the accommodation created.

Quantum subsidence

Stepwise or quantum subsidence has its greatest effect on 'Walther 2' simulations when a moderate to low cycle magnitude is employed (Fig. 14). Sudden subsidence of >2 m will activate the knockback routine putting the normal facies transition into reverse. Resumed progradation will normally eventually reinstate the interrupted facies and the result is the repetition of prodelta or delta-front facies in the lower or central parts of cyclothems and of delta-top facies in the upper parts. Of particular appeal is the appearance of seam splits in coals (e.g. Broadhurst & Simpson, 1983) occurring where a coal swamp was suddenly drowned. Columns produced by such interplay between eustatically-initiated progradational sequences and tectonically-

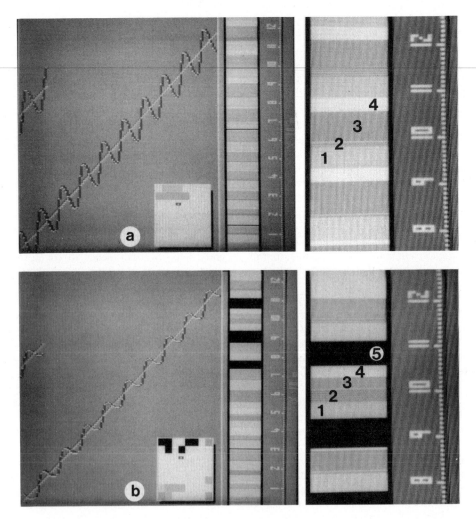

Fig. 13. Two runs of 'Walther 2' (asymmetric wave) together with enlargements of the sediment column to show the effects of different cycle magnitudes on facies development. Facies are numbered as in Figs 11 & 12. Variables as follows: subsidence rate 3 cm kyr^{-1}, sedimentation rate 10 cm kyr^{-1}, cycle period 400 kyr. (a) A cycle magnitude of 20 m and delta-top facies are poorly developed. (b) A magnitude of 10 m and delta-top facies are well developed. Delta development rate in both runs is the same. Note that the best development of delta-top facies occurs where there is added compactive subsidence.

triggered reversals can be an impressive simulation of the facies diversity seen within actual cyclothemic sequences.

In general the program 'Walther' demonstrates simply and for the first time, how delta−marine interactions of the Yoredale type can result from an eustatic regime. The synthetic cyclothems are distinct from the predicted eustatic pattern of Leeder & Strudwick (1987, fig 7.2) and show that in all but distal situations a limestone−shale boundary should

normally be diachronous rather than erosive and coals are a usual component. These features were believed by Leeder & Strudwick (1987) to be characteristic only of the patterns produced by avulsion or by rapid tectonic subsidence alone followed by introduction of a delta lobe. Evidently the facies sequences produced by the three processes remain difficult to distinguish and the real situation is likely to contain elements of each.

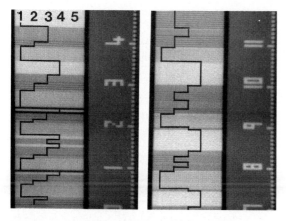

Fig. 14. 'Walther 2' run with stepped subsidence comprising 4 m every 14 kyr. The graphic log, facies numbers and preset variables are as in Figs 11 & 12. The sudden subsidence causes a reversed facies transition that can affect any facies with the exception of that at the base of the cyclothem. This means that all lithologies including coals and limestones are commonly split.

PROGRAMS 'MILANKOVICH' AND 'CROLL'

Objectives

Unlike 'Cyclothem' and 'Walther' which are principally for simulating facies diversity within cyclothems, the programs 'Milankovich' and 'Croll' are intended for modelling cyclothem distributions. The recognition of repetitive patterns in certain types of sedimentary sequence forms an important part of the evidence that Milankovich-band periodicities (e.g. Imbrie, 1985; Chappell & Shackleton, 1986; Weedon, 1989) may be recognizable in the geological record (Grotzinger, 1986; Goldhammer *et al.*, 1987; Read & Goldhammer, 1988; Schwarzacher, 1975, 1987; Schwarzacher & Fischer, 1982; Olsen, 1984; House, 1985). Although there have been recent warnings that the Milankovich periodicities may not be as predictable as hitherto assumed, especially through the geological past (Winograd *et al.*, 1988; Laskar, 1989), some convincing patterns comprising cyclic units grouped into clusters (megacycles) emerge from studies of shallow- to deep-water sediments from a wide range of ages and environments (e.g. Schwarzacher & Fischer, 1982; Goldhammer *et al.*, 1987).

As its name suggests, the program 'Milankovich' simulates astronomically-forced Milankovich-type functions. These are modelled as waves, and adopt-

ing the basic assumption that it is these signals that control glacioeustacy the sum of these is used as a direct control on sea-level. The program shares most features of style and presentation with 'Cyclothem' from which it is directly derived, including the shallow-water carbonate setting. Three waves replace the single wave in 'Cyclothem' including one that is asymmetrical, with an upwave that is 25% of the total period. Each can be assigned an independent period and magnitude through the menu, and periods well beyond the Milankovich band are acceptable so that the program can also simulate 'Vail'-type cycles (Vail *et al.*, 1977). In order to accommodate this range, the screen and all user-defined variables based upon absolute values can be scaled as required at 390 780 or 3900 kyr. Subsidence can again be set in a steady or stepwise mode but there is no longer independent control on the depth range and deposition rate of each facies.

The program 'Croll' is designed to simulate a deeper-water environment in which emergence of the sediment pile is not possible, but where orbitally-forced factors might nevertheless affect sediment type. This may be more directly through temperature or climate rather than through eustacy (e.g. Arthur *et al.*, 1984; House, 1985; Arthur *et al.*, 1986; Weedon, 1986; Raiswell, 1988). Up to three independent waves of user-definable period and magnitude are simultaneously drawn and the sum of these controls the simulation. By defining appropriate upper and lower thresholds the peaks and troughs of the sum wave can be used to trigger changes between up to three facies, and each can be assigned a different sedimentation rate.

The name of the program acknowledges the self-made scientist James Croll LLD FRS, who was the first to recognize a connection between orbitally-driven variations and climatic change (Croll, 1864). Croll was born and brought up in the Scottish Deeside village of Banchory, where the present authors have lived for the last 20 years and where the programs described in this paper were developed.

Modelling using 'Milankovich'

In certain shallow-water carbonate sequences such as the Middle Triassic of northern Italy (Goldhammer *et al.*, 1987) and the Upper Triassic Lofer sequence of the northern Alps (Schwarzacher, 1975), a key feature is the presence of 1 m-scale cyclothems grouped into megacyclic packets of around five. Goldhammer *et al.* noted that mega-cyclothems showed a marked upward-thinning of the component minor cyclothems and modelled the pattern using a computer simulation. A single mega-cycle of this type is easily synthesized using a relatively short period of eustatic rise and fall (e.g. 20 kyr) which has a rise rate that significantly outpaces sedimentation and a static position in relation to an absolute datum line so that sediments can only build as far as the top of the upwave (Fig. 15). As the wave repeatedly floods then exposes the platform surface, separate increments of sediment are added, but the amount deposited with each cycle decreases as a result of the successive up-building of the sediment column which extends the period of exposure. The result is a perfect upward-thinning packet of minor cyclothems, the total number of which is dependent upon the balance between sedimentation rate, cycle period and cycle magnitude, and will vary with a change in any one of these. In order to make this pattern repeat itself some means of achieving sudden subsidence is necessary. This can be achieved tectonically (e.g. Cisne, 1986) by introducing step-wise subsidence (Fig. 16), or eustatically through an additional longer period asymmetric wave (Goldhammer *et al.*, 1987) that will rapidly drown the surface (Fig. 17). These simulations provide

useful predictive observations for use in the field where one should expect to see the greatest depth diversity within minor cyclothems at the bases of megacyclic packets and the greatest degree of sub-aerial alteration above the minor cyclothems towards the tops of the packets.

Goldhammer *et al.* (1987) found that they could reproduce their five-fold thinning-up clusters by set-

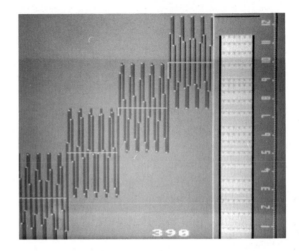

Fig. 16. 'Milankovich' run using a 5 m × 20 kyr symmetrical cycle and 2·5 m steps at 100 kyr intervals. The result is a repeated pattern of five thinning-upward minor cyclothems arranged in 2·5 m packets or clusters.

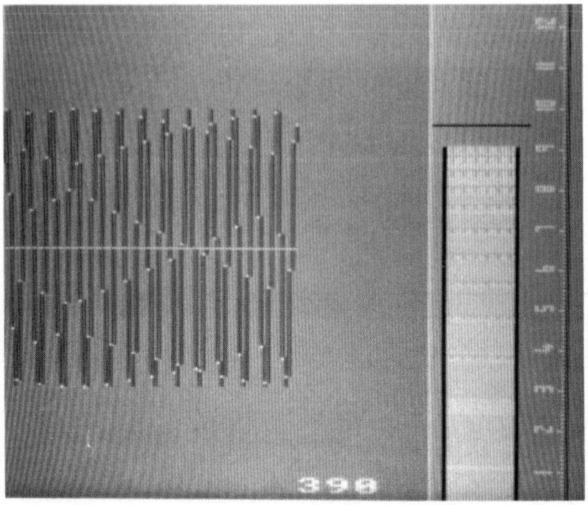

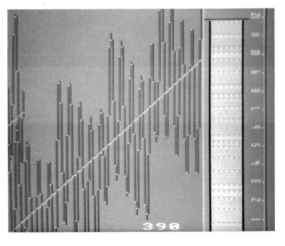

Fig. 17. 'Milankovich' run using an 8 m × 20 kyr symmetric wave and a 4 m × 100 kyr asymmetric wave to produce five-fold thinning-upward clusters. The run is almost indistinguishable from Fig. 16. Sedimentation lag depth 1 m.

Fig. 15. A 'Milankovich' simulation using a 390 kyr screen and a stationary 7 m × 20 kyr wave. Provided that the rise rate of the cycle always significantly outpaces the sedimentation rate (here set at 7 cm kyr^{-1}) then a thinning succession of minor cyclothems will result. Note how the degree of exposure at minor cyclothem boundaries increases as the time spent during deposition decreases.

ting a fifth-order 20 000 year period sine wave against a fourth-order 100 000 year asymmetric wave comprising a rapid rise and slow fall. This relationship, with only a small departure was used successfully throughout their experiments and supported the conclusion that the Middle Triassic was dominated by Milankovich periodicities closely comparable with the Pleistocene.

Using 'Milankovich' and the same orders of sedimentation and subsidence rates employed by Goldhammer *et al.*, together with their asymmetric fourth-order cycle, we have found that it is possible to reproduce their asymmetric pentacyclic clusters using fifth-order waves right down to 10 000 years. The results of two series of these simulations are presented in Fig. 18 where, for a given magnitude of the long-period wave, period and magnitude of the short-period wave are plotted against each other. This permits the recognition of a wedge-shaped 'zone of pentacycles' that extends from high magnitude 20 000 year cycles with a wide range in possible magnitude down to 10 000 years or less with a range of possible magnitude diminishing to zero. Examples are shown in Fig. 19.

A key to the preservation of fewer cyclothems than cycles in the stratigraphical record lies in the loss of cycles below the platform surface (e.g. Read & Goldhammer, 1988). 'Milankovich' can be used to demonstrate that if the magnitude of the fifth-order (20 000 year) cycle is small enough its upwaves are insufficient to flood the platform effectively during the fourth-order (100 000 year) sea-level low and as many as one third of the cycles go unrecorded because they are lost beyond the platform edge. In addition to this, during the fourth-order high when the platform is flooded, such low-magnitude short-period cycles may not fall sufficiently to produce emergence of the platform surface so that the number of evident cyclothems can be further reduced to less than half the total number of cycles.

It is also important to note from modelling that although a true cycle ratio of around 5:1 lies within a reasonably wide range of possible variations in cycle magnitude and sedimentation rate (Fig. 18), high-magnitude short-period cycles fail to develop the strongly asymmetric pattern recorded in the field by Goldhammer *et al.* (1987). In order to achieve marked thinning of cyclothems towards the top of a megacyclic packet, deposition must be confined to the brief period of emergence that occurs when sedimentation has nearly filled the accommodation created by the successive highs of the fifth-order

(a)

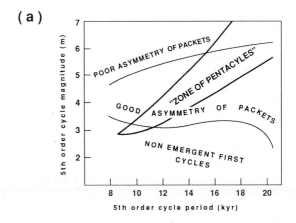

(b)

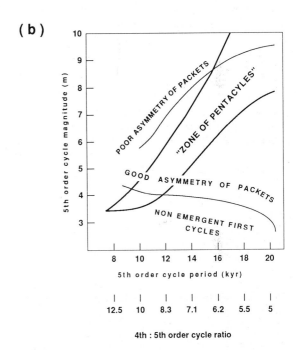

Fig. 18. Graphs constructed using empirical data from 'Milankovich' runs to show the range of settings under which a series of pentacyclic packets comprising five or so thinning-up minor cyclothems can be produced. Examples of runs are seen in Figs 17 & 19.

(a) No sedimentation lag on inundation, fourth-order asymmetric cycle set at 4 m × 100 kyr, fifth-order settings as shown on 'x' and 'y' coordinates.

(b) Sedimentation lag depth of 1 m; fourth-order asymmetric cycle set at 4 m × 100 kyr, fifth-order settings as shown on 'x' and 'y' coordinates.

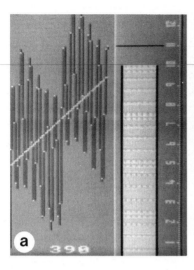 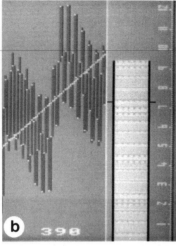

Fig. 19. Two 'Milankovich' runs compared.

(a) Fifth-order cycle 7 m × 16 kyr; fourth-order cycle 4 m × 100 kyr. Note that at least one cycle is regularly lost because it fails to produce emergence of the platform.

(b) Fifth-order cycle 5 m × 13 kyr; fourth-order cycle 4 m × 100 kyr. Two to three cycles are regularly lost and the first or second cycle in each packet is only briefly emergent. The degree of exposure at each subsequent minor boundary progressively increases.

Sedimentation lag depth in each case 1 m.

cycles, as in Fig. 15. This situation cannot be achieved with large magnitude fifth-order cycles because the maximum accommodation they create is never filled before the fourth-order rise takes place. This therefore considerably reduces the area in Fig. 15 in which good asymmetric 'pentacycles' can form.

The implication of this modelling is that there is no exclusive reason for assuming that the 5:1 cyclothem ratio seen in platform sediments by Goldhammer *et al.* (1987) was in fact the effect of known Milankovich periods. A cycle ratio of up to 10:1 is possible, although from modelling using 'Milankovich' it can be seen that a ratio of ≈5·5−6·5 is the most likely.

A further application of 'Milankovich' is in modelling some of the general concepts surrounding sequence stratigraphy by using an interplay of two symmetrical cycle periods. For example, combining appropriate fourth-order (e.g. 300 kyr) and third-order (e.g. 3 Myr) cycles of moderate magnitude will simulate parasequences and sequences (e.g. Van Wagoner *et al.*, 1988), together with a type 1 (erosional) sequence boundary. The typical thinning of parasequences accompanying increasing exposure periods towards a sequence boundary is readily reproduced, and depending upon magnitude of cycles and rate of sedimentation chosen, several cycles may be lost at the sequence boundary itself. Since the program is designed for modelling shallow-water successions, type 2 sequence boundaries, where emergence and erosion are not characteristic, are less easily modelled.

Modelling using 'Croll'

The program 'Croll' is primarily intended to demonstrate the effects of combining Milankovich-type forcing functions of different periods to model pelagic facies such as basinal limestone−shale sequences. The relationship between orbitally-forced climatic change and sea-level is extremely complex and is influenced by a number of factors such as oceanic conditions and volume and location of ice as well as total insolation (e.g. Imbrie, 1985; Crowley *et al.*, 1987). Since 'Croll' is not a sea-level simulation but an orbital-forcing simulation we have not included the asymmetric wave facility. Emergence of the sediment pile cannot occur so that sedimentation is continuous. Since the three facies can be given independent accumulation rates, it is possible to simulate both changes in carbonate productivity and changes in sediment influx (productivity cycles and dilution cycles of Einsele, 1982).

The effect of summing symmetrical waves of different periods is to produce 'beats' at specific multiples of the input frequencies. Using the sum wave to trigger facies changes ensures that the sediment response also has a broadly symmetrical pattern on a number of different scales. This is demonstrated where standard Milankovich frequencies representing 100 000 (eccentricity), 41 000 (obliquity) and 23 000 (precession) years are used with a uniform sedimentation rate (Fig. 20a). However, by adjusting the sedimentation rates of the three facies, the simulations assume radically different appearances (Fig. 20b−d). These transformations could be used

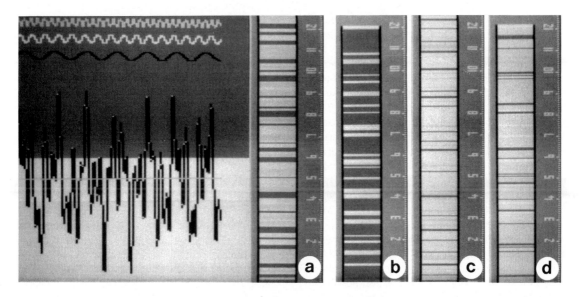

Fig. 20. Four runs of 'Croll' (a–d) compared. Screen presentation is as described in Fig. 2, but here the screen width is doubled to 480 kyr. Three colours are used for the three background fields and in the sediment column, but it is sufficient that only two show clearly in monochrome. In each run the Milankovich periods are held at 23, 41 and 100 kyr and the sum waves are identical, but the deposition rates for each field corresponding to peak, centre and trough positions of the sum wave are changed. (a) Sedimentation rates for each screen field are identical and the sediment column is an obvious match to the peak and trough positions. (b) Sedimentation rates for the peak field are fast, the rest slow. (c) Sedimentation rates for the central field are fast, the rest slow. (d) Sedimentation rates for the trough sector are fast, the rest slow.

It will be noted that although the four columns have been produced from an identical sum wave pattern as shown in (a), they are almost impossible to correlate.

to model quite different responses to forcing functions in an off-platform environment such as temperature-controlled changes in carbonate productivity (e.g. Fig. 20b), clastic influx-controlled variations resulting from climatic changes in source hinterlands (Fig. 20c), or more complex switching mechanisms such as depth- or circulation-determined oscillations between anoxia and turbulence or perhaps the combined effect of warm humid extremes inhibiting carbonate through sediment influx and cold dry extremes inhibiting carbonate through temperature (Fig. 20d).

Amongst these simulations, there are superficial similarities to limestone–shale sequences such as the Lower Lias of Dorset (e.g. House, 1985; Weedon, 1986; Hallam 1986). This succession consists of <1 m scale thin limestones and bituminous shales separated by beds of mudstone. Sellwood (1972) considered that the alternations represent variations in clastic influx; Weedon (1986) and House (1985) thought that climatic change was responsible

and Hallam (1986) reiterated warnings about diagenetic overprinting. It should be noted from Fig. 20 that the total number of alternations recorded will depend upon the sensitivity of the thresholds set. For example, in the case of Fig. 20c, the higher the mud threshold is placed the fewer and thinner the muds will be. When more than one Milankovich frequency is in operation some peaks generated by the lowest frequency can be lost through destructive interference so that pure counts of bed alternations over a known chronological interval (e.g. House, 1985) could give an unrepresentative picture of the frequencies in operation. Furthermore, because thresholds could be different across a basin, any evidence based on comparison of bed numbers (e.g. Hallam, 1986) may be unreliable.

The significant characteristic of Milankovich sequences is the presence of recurrent patterns or signals. These have been detected in the Lias using power spectral analysis (Weedon, 1986, 1989) and Schwarzacher & Fisher (1982) drew attention to limestone—shale sequences of Carboniferous and Cretaceous age that display regular 'bundling' of bed thicknesses into cycles on a scale between <1 m and 2.5 m. By analysing bedding plane distribution within a large number of cycles, Schwarzacher & Fisher demonstrated a five-fold grouping pattern which they attributed to the operation of a 20 000 year forcing function within a longer period of 100 000 years. This was modelled by Schwarzacher (1981) using a simple mathematically-generated simulation containing a stochastic means of proliferating the shale horizons.

Using 'Croll' an idealized 5 : 1 pattern of 'bunched' cycles is easily reproduced using a 100 kyr and 20 kyr period. This is achieved by adjusting the sedimentation rate as above so that facies triggered by wave peaks (representing muds) are deposited several times faster than facies triggered by troughs (Fig. 21). As with 'Milankovich' it is also possible to achieve bunching using wave frequencies with a ratio of >5:1 if the mud threshold is high enough to ensure that only the highest peaks trigger the muds. Regular bunching is also produced by running a 19 kyr period against 23 kyr, and a convincing 5:1 bunching pattern is produced running a 23 kyr period against a 41 kyr period (Fig. 22). More complex bunched patterns can be reproduced using a three-period model involving standard Milankovich frequencies. Such simulations confirm that in

environments that enjoy long term stability, bunching or other forms of repetition are a diagnostic character of orbitally-forced sedimentation patterns. What they also show, however, is that sophisticated statistical methods may be required to extract the Milankovich signals from the sedimentary record, especially if the patterns have become masked by independent changes in sedimentation dynamics or by diagenesis.

DISCUSSION AND CONCLUSIONS

Used properly, computer modelling is one of the best forms of experiment available to the earth scientist. The programs 'Cyclothem', 'Walther', 'Milankovich' and 'Croll' demonstrate that useful results, with applications both in research and teaching, are well within the scope of a widely available microcomputer.

Asymmetrical shallowing-up cyclothems capped by clear emergence features such as palaeokarsts are a characteristic product of an eustatic cycle with an asymmetric wave. Such asymmetry is the general rule and reflects a sea-level rise rate that significantly outpaced sedimentation rate, but in some successions such as the Asbian (Dinatian) of northern England asymmetry is not always obvious. This may mean that rise rates were slow, perhaps because cycle magnitudes were not great, or that the cycles were not asymmetrical.

Where there is also a tendency for peritidal deposits to be missing at the tops of cyclothems, again noted in the Asbian, a relatively rapid sea-level fall is indicated which may be further evidence of an eustatic wave that was not asymmetric and fell about as fast as it rose. In such cases cycle magnitude clearly exceeded the thickness of the cyclothem because the rate of sea-level fall was already fairly rapid as it passed the sediment surface, but it is not possible to arrive at a close estimate of actual cycle magnitude. Assuming constant sedimentation rates, cyclothem thickness in these circumstances is entirely determined by the subsidence rate and the period of time between emergent events.

Brigantian (late Dinantian) cyclothems have greater facies diversity and are thicker and more obviously asymmetrical than Asbian ones (Walkden, 1987). These differences might be accounted for by increased cycle magnitude and asymmetry leading to the greater facies diversity, or a faster overall rate of subsidence. A longer cycle period alone (e.g.

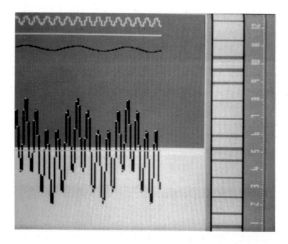

Fig. 21. 'Croll' run using 20 and 100 kyr waves and their sum to show how 'bunched' limestone—shale (light/dark) couplets could be produced. In this run the central field of the screen is reduced, the upper (dark) field is given a slow sedimentation rate whilst the lower field is fast. Screen width 390 kyr.

Walkden, 1987) would lead to an increase in cyclothem thickness but would not increase the facies diversity. If faster subsidence were responsible, some smaller cycles may have failed to produce emergence, decreasing the number of recorded cyclothems over a given period. Changes in cycle magnitude and asymmetry prelude the onset of the main Gondwanan glaciation timed at the beginning of the Namurian by Powell & Veevers (1987), and would be responses to changes in distribution and massing of a growing ice body. Veevers & Powell (1987) acknowledged Yoredale cyclicity from the beginning of the Brigantian, but miss the evidently smaller cycles of the Asbian that start about 5–9 kyr earlier.

Mixed carbonate–clastic Yoredale cyclothems are naturally asymmetric and do not necessarily imply an asymmetric cycle. However, modelling confirms that the basal limestones seen in Yoredale successions mainly represent the upwave phase of the cycle when clastic influx was reduced owing to flooding of sourcelands, and the faster this took place the more effective would be the inhibition of clastics. In proximal situations the tops of the limestones should be marked by a diachronous transition to shale created after the cycle peaked and clastic sediments were once again released from the source areas. Emergence features should only be a typical feature of distal Yoredale cyclothems where limestone deposition continued well into the downwave stage of the cycle. These generalizations are in accord with field observations.

Thick coals are not a normal feature of Yoredale cyclothems. Using a symmetrical wave, peat preservation seems less likely if it accumulated only during highstand, and the sedimentation dynamics of the system as a whole are unstable if it formed only during lowstand. Peat accumulation is best encouraged where an asymmetric cycle with a fast upwave and slow downwave stretches the delta-top window. This situation is enhanced where compactional subsidence is taking place and when cycle magnitude is relatively small. The ideal pattern is where there is little relative sea-level fall on the downwave as a result of a slow downwave being balanced by subsidence. In these circumstances delta-top facies are encouraged anywhere between highstand and lowstand and thick coals could form. It follows that Yoredale cycles could have had larger magnitudes than late Upper Carboniferous cycles and that coals formed where the combination of cycle magnitude and subsidence rate, including compactional subsidence, created the balance necessary to stretch the

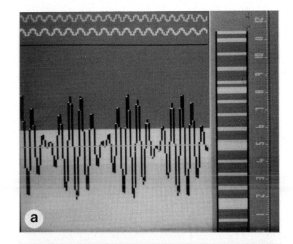

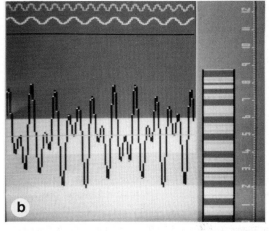

Fig. 22. Two runs of 'Croll' to show that bunching can be produced using periods other than 20 and 100 kyr. (a) The two periods are 19 and 23 kyr. (b) The two periods are 23 and 41 kyr. Screen width 390 kyr. By adjusting the thresholds or the sum wave magnitude, bunches of any number of beds up to seven can be produced from these wave frequencies.

delta-top window. Sedimentation rates also have an effect on the relative importance of transgressive and regressive portions of cyclothems. In general, slow sedimentation leads to regressive-dominated cyclothems whilst fast sedimentation leads to transgressive-dominated cyclothems.

Grouping and bunching of emergent surfaces in platform environments, or of facies such as limestone–shale couplets in off-platform environments, may be an indication that more than one secular control has been in operation, perhaps including stepwise tectonic subsidence. Correct identification of signals in Milankovich successions depends upon a sufficiency of data and an adequate statistical approach to its handling. Symmetry and bed repetitions are characteristic of Milankovich sequences but mere counting of bed numbers or bed thicknesses against time, or of bed ratios, cannot be relied upon to reveal the actual frequencies in operation. Bed or cyclothem clustering with apparent Milankovich ratios can be arrived at through wave frequencies that are very unlike the familiar Milankovich periods.

The programs used in this study are available on disk with a brief explanatory and operating manual. Enquiries should be addressed to the senior author at the Department of Geology and Petroleum Geology, Marischal College, University of Aberdeen, Aberdeen AB9 1AS, UK.

REFERENCES

ARTHUR, M.A., DEAN, W.E., BOTTJER, D.J. & SCHOLLE, P.A. (1984) Rhythmic bedding in Mesozoic−Cenozoic pelagic carbonate sequences: The primary and diagenetic origin of Milankovich-like cycles. In: *Milankovich and Climate* (Eds Berger, A., Imbrie, J., Hays, J., Kukla, G. & Saltzman, B.) pp. 191−222. Reidel, Boston.

ARTHUR, M.A., BOTTJER, D.J., DEAN, W.E., FISCHER, A.G., HATTIN, D.E., KAUFFMANN, E.G., PRATT, L.M. & SCHOLLE, P.A. (1986) Rhythmic bedding in Upper Cretaceous pelagic carbonate sequences: varying sedimentary response to climatic forcing. *Geology* 14, 153−156.

BERGER, A., IMBRIE, J., HAYS, J., KUKLA, G. & SALZMAN, B. (Eds) (1984) *Milankovich and Climate*. Plates 1&2, p. 895. Reidel, Boston, USA.

BLAIR, T.C. & BILODEAU, W.L. (1988) Development of tectonic cyclothems in rift, pull apart and foreland basins: sedimentary response to episodic tectonism. *Geology* 16, 517−520.

BOTT, M.H.P. (1987) Subsidence mechanisms of Carboniferous basins in northern England. In: *European Dinantian Environments* (Eds Miller, J., Adams, A.E. & Wright, V.P.) pp. 21−31. John Wiley, Chichester, UK.

BOTT, M.H.P. & JOHNSON, G.A.L. (1967) The controlling mechanism of Carboniferous cyclic sedimentation. *Quart. J. Geol. Soc. Lond.* 122, 421−441.

BROADHURST, F.M. & SIMPSON, I.M. (1983) Syntectonic sedimentation, rifts and faults reactivation in the Coal Measures of Britain. *J. Geol. Soc. Chicago* 91, 330−337.

BURGESS, I.C. & MITCHELL, M. (1976) Visean lower Yoredale limestones on the Alston and Askrigg Blocks, and the base of the D2 subzone in northern Britain. *Proc. Yorks. Geol. Soc.* 40, 613−630.

CHAPPELL, J. & SHACKLETON, N.J. (1986) Oxygen isotopes and sea level. *Nature* 324, 137−140.

CHILLINGARIAN, G.V. (1983) Compactional diagenesis. In:

Sediment Diagenesis (Eds Parker, A. & Sellwood, B.W.) pp. 57−169. Reidel, Dordrecht.

CISNE, J.L. (1986) Earthquakes recorded stratigraphically on carbonate platforms. *Nature* 323, 320−322.

CROLL, J. (1864) On the physical cause of the change of climate during geological epochs. *Phil. Mag.* 28, 121−137.

CROWELL, J.C. (1978) Gondwanan glaciation, cyclothems, continental positioning and climatic change. *Am. J. Sci.* 278, 1345−1372.

CROWLEY, T.J., MENGEL, J.G. & SHORT, D.A. (1987) Gondwanaland's seasonal cycle. *Nature* 329, 803−807.

DONOVAN, D.T. & JONES, E.J.W. (1979) Causes of worldwide changes in sea-level. *J. Geol. Soc. Lond.* 136, 187−192.

DRIESE, S.G. & DOTT, R.H. (1974) Model for sandstone-carbonate 'cyclothems' based on Upper Member of Morgan Formation (Middle Pennsylvanian) of northern Utah and Colorado. *Bull. Am. Assoc. Petrol. Geol.* 68, 574−597.

EINSELE, G. (1982) Limestone−marl cycles (Peridoites): Diagnosis, significance, causes—a review, In: *Cyclic and Event Stratification* (Eds Einsele, G. & Seilacher, A. pp. 8−53. Springer Verlag, Berlin.

ELLIOTT, T. (1975) The sedimentary history of a delta lobe from a Yoredale (Carboniferous) cyclothem. *Proc. Yorks. Geol. Soc.*, 40, 505−536.

ELLIOTT, T. (1986) Deltas, In: *Sedimentary Environments and Facies* (Ed. Reading, H.G.) pp. 111−154. Blackwell Scientific Publications, Oxford.

FERM, J.C. (1970) Allegheny deltaic deposits. In: *Deltas* (Eds Morgan, J.P. & Shaver, R.H.) pp. 246−255. Spec. Publ. Soc. Econ. Palaeontol. Mineral 15, Tulsa.

FISCHER, A.G. (1964) The Lofer cyclothems of the Alpine Triassic. In: *Symposium on Cyclic Sedimentation* (Ed. Merriam, D.F.) pp. 107−150. Kansas State Geol. Surv. Bull. 169.

FRANCIS, E.H. (1983) Carboniferous. In: *The Geology of Scotland* (Ed. Craig, G.Y.) pp. 253−296. Scottish Academic Press, Edinburgh, UK.

GINSBURG, R.N. (1971) Landward movement of carbonate mud: new model for regressive cycles in carbonates (Abstract), *Bull. Am. Assoc. Petrol. Geol.* 55, 340.

GOLDHAMMER, R.K., DUNN, P.A. & HARDIE, L.A. (1987) High frequency glacio eustatic sea-level oscillations with Milankovich characteristics recorded in middle Triassic platform carbonates in northern Italy. *Am. J. Sci.* 287, 853−892.

GOODWIN, P.W. & ANDERSON, E.J. (1985) Punctuated aggradational cycles: a general hypothesis of episodic stratigraphic accumulation. *J. Geol.* 93, 515−533.

GROTZINGER, J.P. (1986) Cyclicity and palaeoenvironmental dynamics, Rocknest platform, north west Canada. *Bull. Geol. Soc. Am.* 97, 1208−1231.

HALLAM, A. (1986) Origin of minor limestone−shale cycles: Climatically induced or diagenetic? *Geology* 14, 609−612.

HECKEL, P.H. (1986) Sea level curve for Pennsylvanian eustatic marine transgressive−regressive depositional cycles along mid-continent outcrop belt, North America. *Geology* 14, 330−334.

HOUSE, M.R. (1985) A new approach to an absolute timescale from measurements of orbital cycles and sedimentary microrhythms. *Nature* 315, 721–725.

IMBRIE, J. (1985) A theoretical framework for the Pleistocene ice ages. *J. Geol. Soc. Lond.* 142, 417–432.

IMBRIE, J. & IMBRIE, J.Z. (1980) Modelling the climatic response to orbital variations. *Science* 207, 943–953.

JAMES, N.P. (1979) Shallowing upward sequences in carbonates. *Geosci Can* 4, 126–136.

JOHNSON, G.A.L. (1959) The Carboniferous stratigraphy of the Roman Wall district, western Northumberland. *Proc. Yorks. Geol. Soc.* 32, 83–130.

JOHNSON, G.A.L. (1972) Lateral variation of marine and deltaic sediments in cyclothem deposits with particular reference to Visean and Namurian of northern England. *C.R. 4eme Congr. Etud. Stratigr. Geol. Carbonif.* 1, 323–329.

JOHNSON, G.A.L. (1984) Carboniferous sedimentary cycles in Britain controlled by plate movements. *C.R. 9eme Congr. Etud. Stratigr. Geol. Carbonif.* 3, 367–371.

KENDALL, C.G. ST. C. & SCHLAGER, W. (1981) Carbonates and relative changes in sea level. *Marine Geology* 44, 181–212.

KLEIN, G. deV. & WILLARD, D.A. (1989) Origin of the Pennsylvanian coal-bearing cyclothems of North America. *Geology* 17, 152–155.

LASKAR, J. (1989) A numerical experiment on the chaotic behaviour of the Solar System. *Nature* 338, 237–238.

LEEDER, M.R. (1988) Recent developments in Carboniferous geology: a critical review with implications for the British Isles and N.W. Europe. *Proc. Geol. Assoc.* 78, 73–100.

LEEDER, M.R. & MCMAHON A.H. (1988) Upper Carboniferous (Silesian) basin subsidence in northern Britain. In: *Sedimentation in a Synorogenic Basin Complex; the Upper Carboniferous of NE Europe* (Eds Besley, B. & Kelling, G.) pp. 43–52. Blackie, Oxford, UK.

LEEDER, M.R. & STRUDWICK, A.E. (1987) Delta-marine interactions: a discussion of sedimentary models for Yoredale-type cyclicity in the Dinantian of northern England. In: *European Dinantian Environments* (Eds Miller, J., Adams, A.E. & Wright, V.P.) pp. 93–114. John Wiley, Chichester, UK.

MIDDLETON, G.V. (1973) Johannes Walther's law of the correlation of facies. *Bull. Geol. Soc. Am.* 84, 979–988.

MITCHELL, A.H.G. & READING, H.G. (1986) Sedimentation and tectonics. In: *Sedimentary Environments and Facies* (Ed. Reading H.G.), pp. 471–519. Blackwell, Oxford, UK.

MITCHELL, M., TAYLOR, B.J. & RAMSBOTTOM, W.H.C. (1978) Carboniferous. In: *The Geology of the Lake District* (Ed. Moseley, F.) pp. 168–188. Yorkshire Geological Society.

OLSEN, P.E. (1984) Periodicity of lake level cycles in the Late Triassic Lockatong Formation of the Newark Basin. In: *Milankovich and Climate* (Eds Berger, A., Imbrie, J., Hays, J., Kukla, G. & Saltzman, B.) pp. 129–146. Reidel, Boston.

POWELL, C. McA. & VEEVERS, J.J. (1987) Namurian uplift in Australia and South America triggered the main Gondwanan glaciation. *Nature* 326, 177–179.

RAISWELL, R. (1988) Chemical model for the origin of minor limestone–shale cycles by anaerobic methane oxidation. *Geology* 16, 641–644.

RAMSBOTTOM, W.H.C. (1973) Transgressions and re-

gressions in the Dinantian: a new synthesis of British Dinantian stratigraphy. *Proc. Yorks. Geol. Soc.* 39, 567–607.

RAMSBOTTOM, W.H.C., GOOSENS, R.F., SMITH, E.G. & CALVER, M.A. (1974) Carboniferous. In: *The Geology and Mineral Resources of Yorkshire* (Eds Rayner, D.H. & Hemingway, J.E.) pp. 30–45. Yorkshire Geological Society.

RAMSEY, K.E. & MOSLOW, T.F. (1987) A numerical analysis of subsidence and sea level rise in Louisiana. *Coastal Sediments '87*, ASCE New Orleans.

READ, J.F. & GOLDHAMMER, R.K. (1988) Use of Fischer plots to define third order sea-level curves in Ordovician peritidal cyclic carbonates, Appalachians. *Geology* 16, 895–899.

READ, J.F., GROTZINGER, J.P., BOVA, J.A. & KOERSCHNER, W.F. (1986) Models for generation of carbonate cycles. *Geology* 14, 107–110.

ROSS, C.A. & ROSS, J.R.P. (1985) Late Paleozoic depositional sequences are synchronous and worldwide. *Geology* 13, 194–197.

SADLER, P.M. (1981) Sediment accumulation rates and the completeness of stratigraphic sections. *J. Geol.* 89, 569–584.

SCHLAGER, W. (1981) The paradox of drowned reefs and carbonate platforms. *Bull. Geol. Soc. Am.* 92, 197–211.

SCHWARZACHER, W. (1975) *Sedimentation Models and Quantitative Stratigraphy*, p. 382. Elsevier, Amsterdam.

SCHWARZACHER, W. (1987) The analysis and interpretation of stratification cycles. *Palaeo-oceanography* 2, 79–95.

SCHWARZACHER, W. (1981) Quantitative correlation of a cyclic limestone-shale formation. In: *Quantitative Stratigraphic Correlation* (Eds Cubitt, J.M. & Reyent, R.A.) pp. 275–286. John Wiley, Chichester, UK.

SCHWARZACHER, W. & FISCHER, A.G. (1982) Limestone-shale bedding and perturbations of the Earth's orbit. In: *Cyclic and Event Stratification* (Eds Einsele, G. & Seilacher, A.) pp. 72–95. Springer Verlag, Berlin.

SCHWARZACHER, W. & SCHWARZACHER, W. (1986) The effect of sealevel fluctuations in subsiding basins. *Computers Geosci* 12, 225–227.

SELLWOOD, B.W. (1972) Regional environmental change across a Lower Jurassic stage boundary in Britain. *Palaeontology* 15, 125–157.

SOMERVILLE, I.D. (1979) Minor cyclicity in late Asbian limestones in the Llangollen district of North Wales. *Proc. Yorks. Geol. Soc.* 42, 317–341.

SPENCER, R.J. & DEMICCO, R.V. (1989) Computer models of carbonate platform cycles driven by subsidence and eustacy. *Geology* 17, 165–168.

TURCOTTE, D.L. & WILLEMANN, J.H. (1983) Synthetic cyclic stratigraphy. *Earth Planet. Sci. Lett.* 63, 89–96.

VAIL, P.R., MITCHUM, R.M. & THOMPSON, S. (1977) Global cycles of relative changes of sea level. In: *Seismic stratigraphy — Applications to Hydrocarbon Exploration*, pp. 83–97. Am. Assoc. Petrol. Geol., Memoir 26.

VAN WAGONER, H.W., POSAMENTIER, H.W., MITCHUM, R.M.JR., VAIL, P.R., SARG, J.F., LOUTIT, T.S. & HARDENBOL, J. (1988) An overview of the fundamentals of sequence stratigraphy and key definititions. In: *Sea Level Changes: an Integrated Approach*, pp. 39–46. Spec. Publ. Soc. Econ. Paleontol. Mineral.

VEEVERS, J.J. & POWELL, C. McA. (1987) Late Paleozoic

glacial episodes in Gondwanaland reflected in transgressive−regressive depositional sequences in Euramerica. *Bull. Geol. Soc. Am.* **98**, 475−487.

WALKDEN, G.M. (1974) Palaeokarstic surfaces in Upper Visean (Carboniferous) limestones of the Derbyshire block, England. *J. Sed. Petrol.* **44**, 1232−1247.

WALKDEN, G.M. (1987) Sedimentary and diagenetic styles in late Dinantian carbonates of Britain. In: *European Dinantian Environments* (Eds Miller, J., Adams, A.E. & Wright, V.P.), pp. 131−156. John Wiley, Chichester, UK.

WALKDEN, G.M. & DAVIES, J. (1983) Polyphase erosion of subaerial omission surfaces in the late Dinantian of Anglesey, North Wales. *Sedimentology* **30**, 861−878.

WEEDON, G.P. (1986) Hemi-pelagic shelf sedimentation and climatic cycles: the basal Jurassic of S. Britain. *Earth Planet. Sci. Lett.* **76**, 321−335.

WEEDON, G.P. (1989) The defection and illustration of regular sedimentary cycles using Walsh power spectra and filtering, with examples from the Lias of Switzerland. *J. Geol. Soc. Lond.* **146**, 133−144.

WILSON, J.L. (1974) *Carbonate Facies in Geologic History.* Springer Verlag, Berlin.

WINOGRAD, I.J., SZABO, T.B., COPLEN, A.C. & RIGGS, A.C. (1988) A 250000 year climatic record from Great Basin vein calcite−implications for Milankovich theories. *Science* **242**, 1275−1279.

Spec. Publs int. Ass. Sediment. (1990) **9**, 79–108

Middle Triassic carbonate ramp systems in the Catalan Basin, northeast Spain: facies, systems tracts, sequences and controls

F. CALVET*, M. E. TUCKER[†] *and* J. M. HENTON[†]

* *Dept. G.P.P.G., Facultat de Geologia, Universitat de Barcelona, Zona Universitaria de Pedralbes, 08028 Barcelona, Spain; and* [†] *Department of Geological Sciences, University of Durham, Durham DH1 3LE, UK*

ABSTRACT

The Triassic of the Catalan Basin, eastern Spain, is 'Germanic' in character, with Buntsandstein, Muschelkalk and Keuper Units, succeeded by the Norian Imon Formation. Carbonate platforms developed in the Lower Muschelkalk (Anisian) and Upper Muschelkalk (Ladinian) and in the Imon. The Muschelkalk platform carbonates, the subject of this paper, are separated by the Middle Muschelkalk evaporites and marls. These platforms were broadly of the carbonate ramp type, with the facies very persistent laterally in the older platform sequence, but much more varied in the younger platform. Small-scale shallowing-upward cycles of various types are a feature of many units in both the Lower and Upper Muschelkalk. The Lower Muschelkalk has prominent palaeokarst horizons and in the Upper Muschelkalk, mud mound-reef complexes are a feature of one stratigraphic domain. Many of the Muschelkalk carbonates are dolomitized, with several types of dolomite distinguishable.

The Mid-Triassic strata of the Catalan Basin can be divided into two depositional sequences, and the systems tracts philosophy can be applied. Depositional sequence 1 consists of the uppermost part of the Buntsandstein (a lowstand systems tract of lutites, carbonates and evaporites, stratigraphically equivalent to the Röt of the German and North Sea basins) and the Lower Muschelkalk. The latter can be divided into a transgressive systems tract (of onlapping shallowing-upward cycles) and a highstand systems tract of a broadly aggrading-regressive peritidal unit. Depositional sequence 2 consists of the Middle Muschelkalk (a lowstand systems tract of gypsum-anhydrite and marl) and the Upper Muschelkalk. The latter can be divided into a transgressive systems tract of onlapping oolites, lagoonal facies, outer ramp cycles, mud mound–reef complexes and stromatolitic grainstone shoals, succeeded by a highstand systems tract package of laminated 'basinal' dolomicrites to peritidal dolomites. The two depositional sequences are third-order cycles which can be interpreted *largely* in terms of crustal extension followed by regional subsidence. The carbonate platforms developed at times of tectonic quiescence in the Catalan Basin. The metre-scale shallowing-upward cycles (parasequences) *may* be the result of orbital forcing in the Milankovitch band (they are fourth- and fifth-order cycles).

INTRODUCTION

In western and central Europe the Triassic was a period of crustal extension that resulted in the formation of a complex network of grabens and troughs (Vegas & Banda, 1982; Ziegler, 1982). In the northeastern part of the Iberian Peninsula, the extension resulted in the reactivation of major NW–SE and NE–SW trending Hercynian faults in the Palaeozoic basement (Vegas, 1975) and these controlled the graben development in this region. The main Triassic

basins located in the north, northeastern and eastern part of the Iberian Peninsula area: the Pyrenees Basin, the Catalan Basin, the Ebro Basin and the complex Valencia–Cuenca Basin. These basins are of intracratonic type, similar to the central European German Basin and the North Sea Basin.

The Triassic Catalan Basin is about 300 km in length, NE–SW, and approximately 200 km across. It is separated from the Ebro Basin to the south by the Soria—Montalban—Orpesa High (Fig. 1). The northern boundary of the Catalan Basin, separating it from the Pyrenean rift—wrench system, is the hypothetical Girona High. At times, the Catalan Basin was divided into 'sub-basins' through syn-sedimentary faulting and differential subsidence, and three tectono-sedimentary domains are recognized within the basin with different thicknesses and/or stratigraphies. The Catalan Basin opened towards the southeast into the Tethys sea. There may well have been a basement high along the southeast side of the basin, in the region of the present-day Balearic islands.

Triassic sedimentation in western and central Europe is broadly cyclic, and this has been related to changes in sea-level, rates of subsidence and/or rates of siliciclastic input (Busson, 1982; Ziegler, 1982). The cyclicity is expressed through an alternation of 'transgressive', dominantly carbonate sequences and 'regressive' siliciclastic and/or evaporite sequences. In the northeastern Iberian Peninsula, the Lower Muschelkalk, Upper Muschelkalk and Imon Formation, of Anisian, Ladinian and probably Norian ages respectively, represent the major transgressive episodes and mostly consist of carbonate ramp facies. In the Catalan Basin and eastern Valencia Basin, deposits of all three transgressive periods are de-

veloped. However, in the Iberian Range and in Pyrenees Basin, to the west and north, only the deposits of the last two transgressions are present (Virgili *et al.*, 1977; Garrido-Megias & Villena, 1977; Gandin *et al.*, 1982; Virgili *et al.*, 1983; Virgili, 1987). In the Meseta highlands of central Spain, only the deposits of the last transgression (Imon) are present. On a broad scale, therefore, the extent of the transgression increased successively and resulted in an onlapping arrangement of the carbonate facies (Marzo & Calvet, 1985). This is basically a reflection of the crustal extensional and regional subsidence and/or eustatic sea-level rises in western and central Europe during the Triassic.

This paper: (1) presents a general outline of the stratigraphy and sedimentology of the Lower Muschelkalk (Anisian) and Upper Muschelkalk (Ladinian) in the Catalan Basin; (2) analyses the Middle Triassic carbonate ramp systems in the Catalan Basin in terms of the depositional sequences and systems tracts philosophies, which have been developed from the study of seismic reflection profiles (Hubbard *et al.*, 1985; Haq *et al.*, 1987; Vail *et al.*, 1987; papers in Wilgus *et al.*, 1988); (3) compares the regional Triassic sedimentary cycles with the third-order cycles of the global sea-level curve of Haq *et al.* (1987); and (4) discusses the origin of third- and fourth/fifth-order cycles in the Middle Triassic in terms of tectonic and eustatic mechanisms.

GEOLOGICAL AND STRATIGRAPHICAL SETTING

In the northeastern part of the Iberian Peninsula there are three mountain chains: the Pyrenees, the Iberian Range and the Catalan Coastal Range. The Ebro Basin, Tertiary in age, was formed during the Alpine orogeny as the foreland basin of the three mountain chains. Between the Catalan Coastal Range, dominated by major NE—SW basement faults, and the Iberian Range, dominated by major NW—SE basement faults, and the Iberian Range, dominated by NW—SE basement faults, there is the Linking Zone where the dominant structural direction is E—W (Guimera, 1984, 1988a, 1988b; Anadon *et al.*, 1985).

The structure of the Coastal Range (Figs 2 & 3) is dominated by longitudinal, near-vertical basement faults which trend from NE—SW to ENE—WSW and form a right-stepping, *en echelon* array

Fig. 1. Triassic basins of the Iberian Peninsula and France.

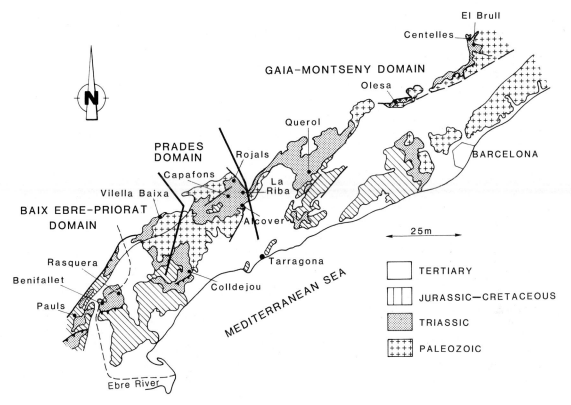

Fig. 2. Geological map of the Catalan Mountain Range. Adapted from Salvany (1986).

(Guimera, 1984, 1988a, 1988b; Anadon *et al.*, 1985). During the Alpine compressive phase, these faults moved sinistrally with local transpression. During the Neogene extension, some of these faults were reactivated as normal faults. Another set of strike−slip faults, trending NW−SE, also occur in the Catalan Coastal Range and in some cases these displace the NE−SW faults (Guimera, 1984, 1988a, 1988b).

The Triassic rocks in the Catalan Range have been divided into three tectono-sedimentary domains on the basis of regional variations in thickness, facies and stratigraphy (Marzo, 1980; Marzo & Calvet, 1985; Salvany & Orti, 1987). The domains for the Muschelkalk are the Gaia−Montseny in the northeast, Prades in the central part, and Baix Ebre−Priorat in the southwest (Fig. 2). The thickness of the Triassic sequence in the Catalan Basin as a whole is very variable, from 500 to > 800 m, through the strong tectonic control on deposition, particularly of the clastic facies.

The Triassic in the Catalan Range has been divided into six lithostratigraphic units (Virgili, 1958; Marzo & Calvet, 1985) which from the base to the top are:
1 Buntsandstein. Conglomerates, sandstones, mudstones and rare evaporites, 60−310 m thick. The upper part is a distinctive lutite−carbonate−evaporite unit, stratigraphically equivalent to the Röt of northern Europe.
2 Lower Muschelkalk. Limestones and dolomites, 70−120 m thick.
3 Middle Muschelkalk. Sandstones, mudstones and evaporites, and rarely volcanics, 50−130 m thick.
4 Upper Muschelkalk. Limestones, dolomites and shales, 100−140 m thick.
5 Keuper. Evaporites, marls, carbonates and locally volcaniclastic sediments and lavas, 50−150 m thick.
6 Imon Formation. Mostly dolomites, 40−70 m thick.

Figure 4 summarizes the main stratigraphical, chronostratigraphical, palaeontological and sequential features of the Triassic in the Catalan Basin.

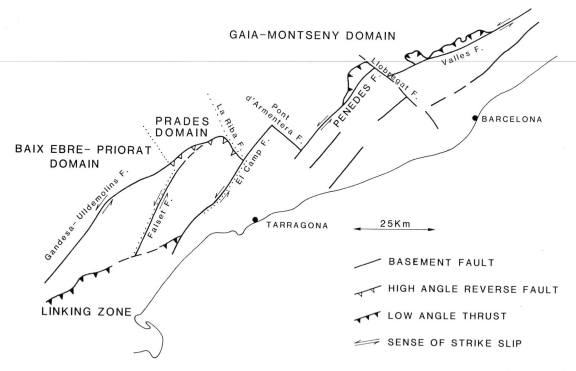

Fig. 3. Tectonic scheme of the Catalan Mountain Range. Adapted from Anadon *et al.* (1985) and Guimera, 1988a.

LOWER MUSCHELKALK

The Lower Muschelkalk in the Catalan Basin varies in thickness from 70 m in the more northeastern area to 120 m towards the southwest. The lower boundary is transitional down into the lutite−carbonate−evaporite unit of the uppermost part of the Buntsandstein (Marzo, 1980). By contrast, there is a sharp contact at the top of the Lower Muschelkalk between the Colldejou Unit, represented by supra-tidal dolomites, and the Middle Muschelkalk clastic−evaporitic facies above.

The Lower Muschelkalk has been divided into four units (Calvet & Ramon, 1987), from base to top: (1) El Brull laminated limestones and dolomites; (2) Olesa bioclastic limestones; (3) Vilella Baixa bioturbated limestones; and (4) Colldejou white dolomites. The El Brull, Olesa and Vilella Baixa Units are partially to totally dolomitized to form grey dolomites. A regional discontinuity separates these lower grey dolomites from the overlying, white Colldejou dolomites (Figs 5 & 7).

The age of the Lower Muschelkalk is Anisian. The El Brull Unit probably corresponds to the upper part of the Lower Anisian (Solé *et al.*, 1987) from the presence of *Stellapollenites theirgartii* and

Fig. 4. [*Opposite*] Stratigraphical, chronostratigraphical and depositional sequences of the Lower and Middle Triassic of the Catalan Basin. (1) *Stellapollenites thiergartii* (Madler) CLEMENT-WESTERHOF *et al.*, 1974; *Voltziaceaesporites heteromorpha* KLAUS, 1964 (Solé *et al.*, 1987). (2) *Paraceratites hispanicum* Kutassy, *Paraceratites evolutospinosus* Tornsquist, *Paraceratites flexuosiformis* Tornsquist (Virgili, 1958). (3) *Mentzelia mentzeli* DUNKER, 1851 (Calzada & Gaetani, 1977). (4) *Stellapollenites thiergartii* (Madler) CLEMENT-WESTERHOF *et al.*, 1974; *Praecirculina granifer* KLAUS, 1964 (Solé *et al.*, 1987). (5) *Daonella lommeli* (Virgili, 1958; Marquez, 1983). (6) *Metapolygnathus mungoensis* DIEBEL (March, 1986). (7) *Prothachyceras steinman, Prothachyceras hispanicum, Hungarites pradoi* (Virgili, 1958). (8) *Metapolygnathus mungoensis* DIEBEL, *Pseudofurnishius murcianus* VAN DEN BOOGAARD (March, 1986) (9) *Camerosporites secatus* KESCKIK, 1985, *Praecirculina granifer* (Leschik) KLAUS, 1960, *Duplicisporites granulatus* (Leschik) SCHEURING, 1970 (Solé *et al.*, 1987). (10) *Camerosporites secatus* KESCKIK, 1955, *Patinasporites densus* LESCHIK, 1955 (Solé *et al.*, 1987).

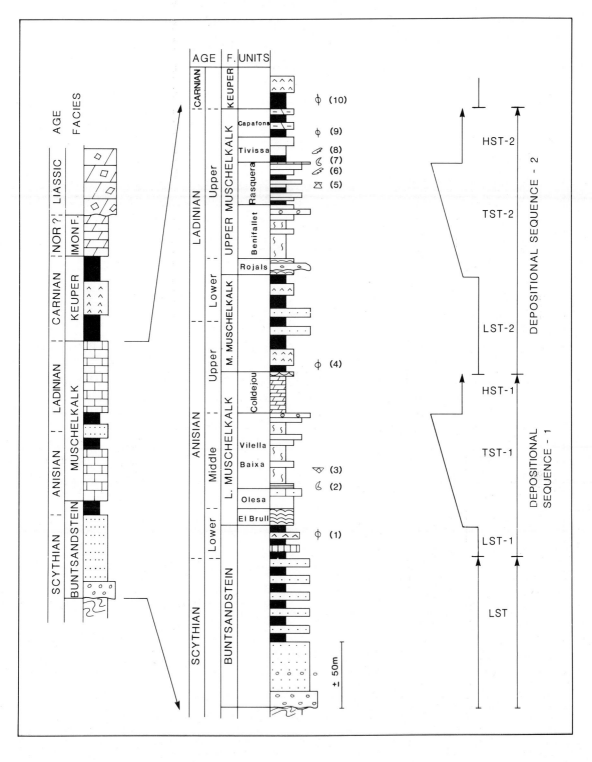

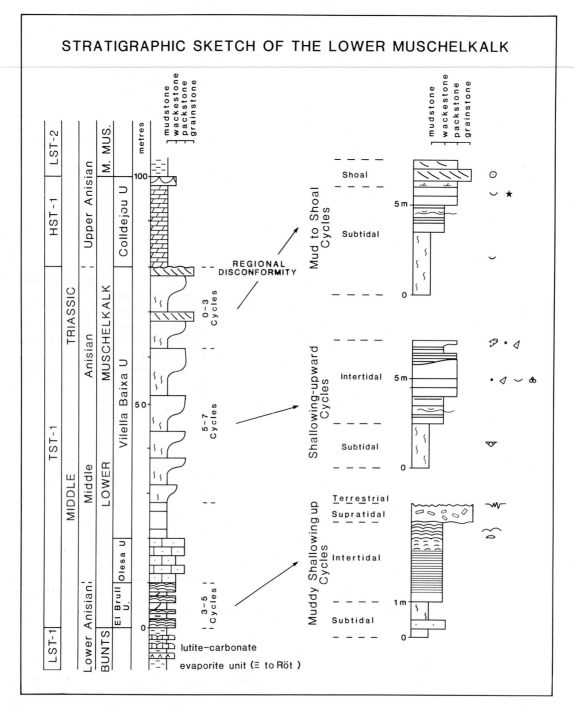

Fig. 5. Stratigraphy, chronostratigraphy and depositional sequences of the Lower Muschelkalk.

Voltziaceaesporites heteromorpha in the lutite—carbonate—evaporite unit of the uppermost Buntsandstein (Röt equivalent). The Olesa Unit is Middle Anisian on the basis of *Paraceratites* (Schmidt, 1932; Virgili, 1958). The Vilella Baixa Unit is also Middle Anisian, dated by the presence of the brachiopod *Mentzelia mentzeli* (Calzada & Gaetani, 1977). The Colldejou Unit probably corresponds to the lower part of the Upper Anisian, as deduced from the occurrence of *Stellapollenites theirgartii* and *Praecirculina granifer* in the lower part of the Middle Muschelkalk (Solé *et al.*, 1987).

El Brull laminated carbonates unit

This unit, 6—14 m thick, basically consists of well-laminated limestones and dolomites. There is a rapid transition up from the Buntsandstein into this unit and its upper boundary is marked by an erosion surface. The lower part of the El Brull Unit, 1—6 m thick, consists of two or three dolomite—marlstone cycles. The ochre-brown dolomites (0·2—2 m thick) have flat, smooth millimetre-scale lamination in the lower part passing up into undulating and domal stromatolites. Intraclasts, mud cracks and wave-rippled oolites occur towards the top of the dolomites, and may be covered by thin stromatolites. The grey marlstone, 0·1—0.5 m thick, is massive with no sedimentary structures. The dolomite—marlstone cycles are interpreted as intertidal—supratidal to terrestrial deposits with relatively hypersaline waters and a strong clastic input.

The upper part of the El Brull Unit varies from 3 to 13 m, and contains two to three cycles. Each cycle (Fig. 5) consists of three lithofacies, in ascending order: (1) grey, bioturbated lime mudstone—wackestone with sparse fauna; (2) grey lime mudstone—wackestone with planar millimetre-scale lamination to centimetre-scale domal stromatolites towards the top, where there are also mudcracks, intraclasts, ripples and some millimetre-size lensoid evaporite mold casts; and (3) local breccias, composed of intraclasts. These cycles are interpreted as shallowing-upward cycles of muddy, shallow-subtidal to supratidal deposits.

Olesa bioclastic limestones unit

The Olesa Unit, 4—12 m thick, consists of grey bioclastic limestones, from lime mudstone to grainstone. The upper boundary is a sharp lithofacies change from these bioclastic limestones to the bio-

turbated limestones of the Vilella Baixa Unit. The two main Olesa facies are: (1) lime mudstones to wackestones with scarce fauna (bivalves and ostracods). The beds are 0·4—1 m thick, with a generally massive base passing up into parallel, millimetre-scale laminated and rare wave-rippled carbonates; (2) packstones—wackestones in 0·2—0.3 m thick beds, with a marine fauna (small gastropods, bivalves, echinoderms, peloids, foraminifera and ostracods) and locally oncoids (10—20 mm in size). With the absence of emergence indicators, these facies are interpreted as lagoonal deposits.

Vilella Baixa bioturbated limestones unit

This unit is 20—90 m thick (increasing towards the southeast) and varies from lime mudstones to oolitic grainstones. The upper boundary is a sharp and important regional disconformity. The Vilella Baixa Unit mostly consists of six lithofacies: thin-bedded wackestones, bioturbated lime mudstones, massive lime mudstones, skeletal wackestones and packstones, oolitic—skeletal grainstones and packstones, and dasyclad packstones. These occur in four types of sequence: (1) muddy sequences; (2) shallowing-upward sequences; (3) mud—shoal sequences; and (4) dasyclad-capped sequences.

1 Muddy sequences. These are located in the lower part of the Vilella Baixa Unit and consist of massive lime mudstone, with subordinate wackestone and packstone, in 1—3 m thick lenticular bodies. They reach 200 m across and have no discernible internal structure. They are interpreted as mud banks, perhaps similar to those of the inner Florida Shelf (e.g. Bosence *et al.*, 1985). The mud banks were located in a very shallow, inner lagoon with restricted water movement.

2 Shallowing-upward sequences. These sequences (Figs 5 & 6) are located in the intermediate part of the Vilella Baixa Unit. The lower part of each sequence, 1—10 m thick, consists of grey to beige bioturbated pseudonodular lime mudstones, interpreted as shallow-subtidal deposits. The brachiopod *Mentzelia mentzeli* occurs as whole unabraded specimens. The upper part is composed of grey wackestone—packstone beds, 0·5—1·0 m thick, which are massive, parallel-laminated (centimetre-scale) and wave-rippled. This upper part contains bivalve fragments, echinoderm debris, ostracods, dasyclads, foraminifera, peloids and quartz grains. The upper part is very shallow-subtidal to intertidal in origin. The thickness of each shallowing-upward sequence

Fig. 6. Detail of shallowing-upward sequence in Vilella Baixa Unit, Lower Muschelkalk. The lower part is made of bioturbated lime mudstone (b), and the upper part is composed of wackestone–packstone.

varies from 2 to 25 m, and there are five to seven cycles at each locality. The lateral persistence of each sequence is at least a kilometre. The shallowing-upward sequences were deposited in a shallow-ramp setting, probably in the outer more seaward part of the lagoon.

3 Mud–shoal sequences. These sequences (Fig. 5) are located in the upper part of the Vilella Baixa Unit. In general they consist of bioturbated lime mudstone in the lower part, wackestone–packstone in the intermediate part, as described for the shallowing-upward sequences, but in addition grainstones are present at the top. The grainstones are commonly oolitic and have cross-bedding and current ripples. They also contain echinoderm and bivalve fragments, corals, intraclasts and peloids. These mud–shoal sequences are 4–8 m thick, and there are up to three cycles developed. The oolitic grainstone is interpreted as a shoal deposit of the sand belt of a shallow ramp.

4 Dasyclad-capped units. Dasyclad packstones in 0·1–0·3 m thick beds are located at the top of some shallowing-upward and mud–shoal sequences. This restricted facies is interpreted as the deposit of small lagoons and lakes upon a coastal carbonate plain developed when the shallowing-upward and mud–shoal sequences became subaerial.

Overall, the Vilella Baixa Unit represents a general transgressive sedimentary package from the lagoonal mud banks of the lower part to the shallowing-upward sequences of the middle part and finally to the mud–shoal sequences of the upper part.

Colldejou white dolomites unit

The Colldejou Unit is 20–40 m thick (increasing towards the southwest) and consists of white to beige dolomicrites in 0·1–0·5 m thick beds. The upper boundary is a regional disconformity related to a laminated crust, 0·1–0·4 m thick with tepees and locally a paleokarst, supratidal stromatolites and collapse breccias. The laminated crust has been interpreted as a caliche crust by Esteban *et al.* (1977a). The disconformity marks the boundary between the Lower Muschelkalk carbonate facies and the Middle Muschelkalk clastics and evaporites.

The main facies of the unit are: (1) white–beige–dolomicrites–dolomicrosparites with a parallel millimetre-scale lamination, small wave-ripples, mud cracks, and lensoids and rosettes of calcite pseudo-morphs after gypsum; (2) stromatolites; and (3) local dolowackestones and some dolopackstones, with bivalves, dasyclads, peloids and ooids.

The following features of the Colldejou Unit suggest depositions in an intertidal–supratidal and sabkha-type environment: (1) scarce fauna; (2) pseudomorphs after gypsum; (3) parallel-lamination, the consequence of the lack of bioturbation; (4) local breccias and intraclasts; and (5) planar and desiccated stromatolites.

Dolomitization

The Lower Muschelkalk contains two principal dolomite types. Above the regional disconformity, 'white dolomites' (Calvet & Ramon, 1987) form the Colldejou Unit, as described above. 'Grey dolomites' occur below the disconformity and extend down to a dolomitization front which is discordant to bedding on an outcrop and regional scale (Fig. 7). The sedimentological features of the Colldejou Unit suggest that it was deposited in an inter- to supratidal and sabkha-type setting. Dolomitization of the white dolomites is pervasive and there is a fine preservation of small-scale sedimentary structures.

The grey dolomites below the regional disconformity vary in thickness from 5 to 80 m. In the Gaia–Montseny Domain grey dolomites replace the upper and locally the intermediate part of the Vilella Baixa Unit (Fig. 7). In the Prades Domain the

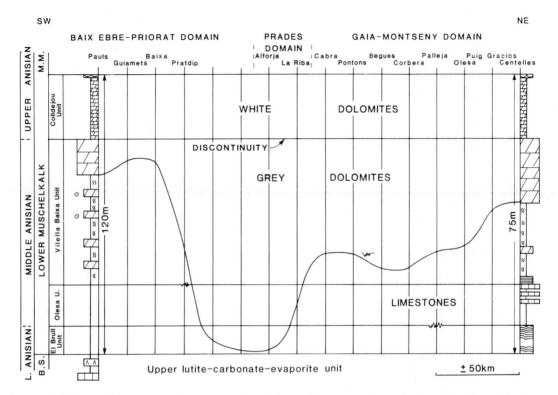

Fig. 7. Distribution of dolomite types in the Lower Muschelkalk with respect to the regional stratigraphy and facies.

Vilella Baixa, Olesa and El Brull Units are all affected, whereas in the Baix Ebre−Priorat Domain dolomite only replaces the top few metres of the Vilella Baixa Unit. Thus the grey dolomites have replaced a variety of carbonate facies, some of which have a normal marine fauna. The grey dolomites contrast with the white dolomites by being typically massive (Fig. 8), dark grey dolosparites, with euhedral to subhedral crystals, $30-200\,\mu m$ in diameter. The style of dolomitization is much more variable than the white dolomites with locally only partial dolomitization and a variable degree of fabric preservation. Chert nodules ($0 \cdot 01-0 \cdot 3\,m$ in size) occur at some horizons within the grey dolomites. The white and grey dolomites formed during early diagenesis and shallow burial, and resulted from processes relating to the sequence boundary at the top of the Lower Muschelkalk and/or the regressive sequence above it. Similar relationships to that of the grey and white dolomites have been described by Nichols & Silberling (1980) and Theriault & Hutcheon (1987).

Fig. 8. Massive aspect of 'grey dolomites' of Olesa Unit, overlain by bedded, cyctic Vilella Baixa Units, Lower Muschelkalk.

Paleokarsts

Paleokarsts are a feature of the Lower Muschelkalk and they have been described by Gottis & Kromm (1967), Ramon (1985) and Ramon & Calvet (1987). They are developed in the limestones of the El Brull, Olesa and Vilella Baixa Units, and their occurrence and degree of development vary in different domains. The eastern part of the Gaia–Montseny Domain has the most prominent paleokarsts, and they are associated with Pb–Zn–Ba mineralization (Andreu *et al.*, 1987).

The main paleokarst level has a morphology of horizontal cavities from a few centimetres to more than 3 m in height and from a few metres to more than 100 m in length. The cavities are filled by clays and silts, red to brown in colour, with centimetre–decimetre angular blocks from cave-roof collapse (Fig. 9). The surface morphology is smooth to angular. A vuggy porosity occurs in the carbonates close to the main paleokarst horizons. These cavities are interpreted as caves developed in the meteoric phreatic environment (Fig. 10).

Sedimentary model

The El Brull laminated carbonates are interpreted as a low-energy broadly tidal-flat deposit and the Olesa bioclastic limestones are lagoonal deposits. The Vilella Baixa Unit represents a general transgressive sequence from the lagoonal mud banks of the lower part of this unit to the shallowing-upward

cycles of the middle part to the mud–shoal sequences of the upper part. On the scale of the whole Lower Muschelkalk, these three units also form a broad, transgressive sequence from peritidal deposits (El Brull Unit) to lagoonal facies (Olesa) to the sand shoals at the top of the Vilella Baixa Unit (Fig. 11). The Colldejou Unit, above the regional disconformity, is an intertidal–supratidal/sabkha deposit of a broadly regressive nature.

On a large scale, the Lower Muschelkalk carbonates belong to the ramp sedimentary model of Ahr (1973), Read (1985) and Tucker (1985), with all the various facies deposited in the protected inner ramp and high-energy sand belt, shallow-ramp environments. The model proposed is similar to the low relief shelf (Brady & Rowell, 1976) and the homoclinal-ramp barrier-bank type of Read (1985).

UPPER MUSCHELKALK

In the Catalan Ranges, the thickness of the Upper Muschelkalk varies from around 100 m in the north to more than 140 m in the south. There is a rapid transition from the Middle Muschelkalk fluvial clastics and evaporites into the Upper Muschelkalk carbonates and a gradual transition (over some metres) from the carbonates into overlying Keuper evaporitic, mostly sulphate facies. Like the Lower Muschelkalk, the Upper Muschelkalk occurs in three Domains: Gaia–Montseny, Prades and Baix Ebre–Priorat, bounded by major faults (Fig. 2), but

Fig. 9. Paleokarst cavern (b) in the Lower Muschelkalk filled by angular blocks. The line (arrow) corresponds to the 'dolomitization front' of the massive 'grey dolomites' (d).

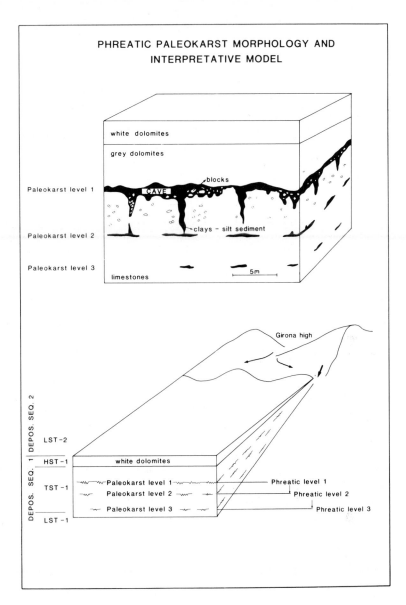

PHREATIC PALEOKARST MORPHOLOGY AND INTERPRETATIVE MODEL

Fig. 10. Phreatic paleokarst morphology in the Lower Muschelkalk and interpretative model related to the depositional systems tracts.

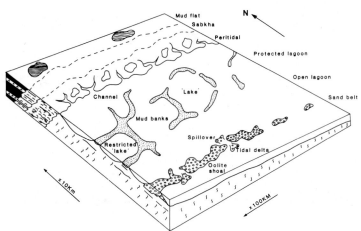

Fig. 11. Interpretative environmental model for El Brull, Olesa and Vilella Baixa Units in the Lower Muschelkalk.

by way of contrast, the carbonate facies of the regions show important facies differences in the middle part of the Upper Muschelkalk (Fig. 12).

The Upper Muschelkalk has been divided into various units (Figs 12 & 13) in the different domains (Calvet *et al.*, 1987). Four units have been recognized in the Gaia−Montseny Domain which, from the base to the top are: (1) Rojals oolitic limestones−dolomites; (2) Benifallet bioturbated limestones−dolomites; (3) Querol stromatolites; and (4) Capafons dolomites, marls, shales and breccias. In the Prades Domain the Upper Muschelkalk has been divided into five units: (1) Rojals Unit; (2) Benifallet Unit; (3) La Riba reefs; (4) Alcover laminated dolomites; and (5) Capafons Unit. In the Baix Ebre−Priorat Domain the Upper Muschelkalk can be divided into five units: (1) Rojals Unit; (2) Benifallet Unit; (3) Rasquera limestones−dolomites and shales with *Daonella*; (4) Tivissa limestones−dolomites and shales; and (5) Capafons Unit. The stratigraphically lowermost units (Rojals and Benifallet) and the uppermost unit (Capafons) are present in all the Domains.

The age of the Upper Muschelkalk is Ladinian. The Rojals and Benifallet Units are probably Lower Ladinian. The Rasquera Unit is of Upper Ladinian age, as determined by the presence of the bivalve *Daonella lommeli* (Virgili, 1958; Marquez, 1983), the conodonts *Sephardiella mungoensis* Diebel and *Pseudofurnishius murcianus* Van den Boogaard (Hirsch, 1976; March, 1986), and the ammonoids

Prothachyceras steinmann, *Prothachyceras hispanicum*, *Hungarites pradoi*, etc. (Virgili, 1958). The Capafons Unit corresponds to the Upper Ladinian, as dated by the presence of *Camerosporites secatus*, *Praecirculina granifer* and *Duplicisporites granulates* (Solé *et al.*, 1987).

Rojals oolitic limestones−dolomites unit

The Rojals Unit, 6−17 m thick, varies from lime mudstone to oolitic grainstone and locally contains sandstones. The upper boundary is a sharp contact, marked by a 0·1−0·2 m thick marl level.

Two different sequences are present in the Rojals Unit. In some areas a bed thinning-upward sequence consists of: (1) oolitic grainstone, 0·2−2 m thick, with a lenticular morphology and erosional base, and herringbone and large-scale cross-bedding decreasing in set size towards the top. This facies is interpreted as a shallow-marine high-energy deposit, and it is overlain by (2) oolitic−bioclastic packstone with small-scale cross-stratification and flaser bedding, and wackestone with lenticular bedding. This facies is interpreted as intertidal, and passes up into (3) planar-laminated stromatolites. This thinning-upward sequence (Figs 13 & 14) is interpreted as a tidal-channel fill and sand shoal deposit (1) and mixed mud−sand flat deposits (2 and 3). In other areas, a more muddy shallowing-upward unit is present.

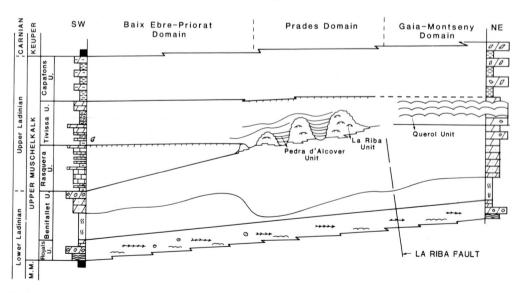

Fig. 12. Stratigraphical cross-section through the Catalan Basin during Upper Muschelkalk times.

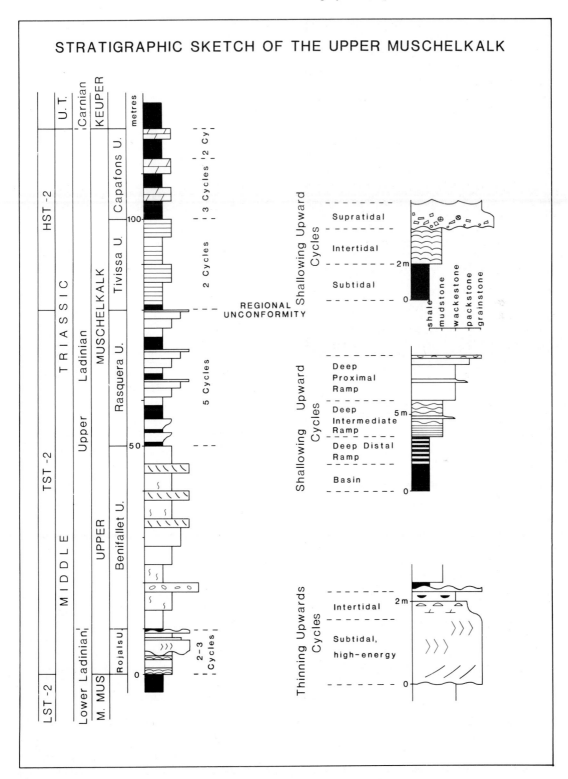

Fig. 13. Stratigraphy, chronostratigraphy and depositional sequences of the Upper Muschelkalk in the Baix Ebre–Priorat Domain.

Fig. 14. Lower and middle part of a tidal channel sequence in the Rojals Unit, Upper Muschelkalk. Erosive base (line) and oolitic grainstones with cross-bedding (t).

higher energy deposits displaying shoaling sequences. The oolitic grainstones are interpreted as sand shoals deposited in the sand belt of a shallow ramp.

Querol stromatolites unit

This unit, 2–12 m thick, is present in the Gaia–Montseny Domain only, and it is represented by metre-scale domal stromatolites. The main facies are: (1) dolomitic domal stromatolites of LLH-C type, 1–8 m in wavelength and up to 2 m in height; (2) packstones–grainstones with lenticular morphology and bipolar cross-bedding; (3) lime mudstones–wackestones. These facies are interpreted as tidal deposits, with tidal channels (packstone–grainstone facies) located between the huge stromatolite mounds.

La Riba reefs unit

The main feature of the Upper Muschelkalk in the Prades Domain is the presence of reefs (Calvet & Tucker, 1988a) which are completely dolomitized (Tucker *et al.*, 1989). The reefs are only present in this domain and they occur predominantly in the eastern part. The reefs are 5–60 m in height and several hundred metres across (Fig. 15). Some reefs were apparently linear structures, 1–2 km across, with a NE–SW orientation, parallel to the palaeo-strike of the basin. Others may have been more discrete, isolated reefs, circular in plan view. The number and size of the reefs decrease towards the western part of the domain.

The reefs have a distinct facies zonation: mound core, reef framework, reef-top bedded facies, flank facies and inter-reef facies (Fig. 16, Calvet & Tucker, 1988a). The mound core facies consists of massive lime mudstone, mostly micritic and peloidal, with scattered skeletal grains such as molluscs and algae (*Tubiphytes*). No framework is apparent, although the core mudstone is commonly mottled; this could be the result of bioturbation. Lenses of coarser, more sandy skeletal–peloidal sediment occur in the mound core. There is an increase in faunal diversity towards the top of the mound core facies and within the dipping flank beds.

The framework facies consists of coral-*Tubiphytes* framestone, coral framestone (*Thecosmillia* up to 0·4 m high) with sponges and platy algae such as *Archaeolithoporella* (Esteban *et al.*, 1977a; Calvet &

Benifallet bioturbated limestones–dolomites unit

This unit, 20–50 m thick, is variable in lithofacies, with lime mudstones to oolitic grainstones. In the Gaia–Montseny Domain the upper boundary is a sharp contact with the Querol stromatolites unit. In the Prades Domain the upper boundary is a sharp lithological change to the La Riba Reef Unit and Alcover Unit, with ferruginization and silicification in the latter case. In the Baix Ebre–Priorat Domain there is also ferruginization and silicification close to the contact with the Rasquera Unit.

The main facies of the Benifallet Unit are: (1) bioturbated grey, lime mudstones and wackestones, in 0.4–1 m thick beds, partially to totally dolomitized. Locally this facies grades to intraformational conglomerate; (2) massive wackestones, usually dolomitized, and (3) oolitic–bioclastic packstones to grainstones, with cross-bedding.

The lower part of the Benifallet Unit is interpreted as a lagoonal facies and this grades upwards to

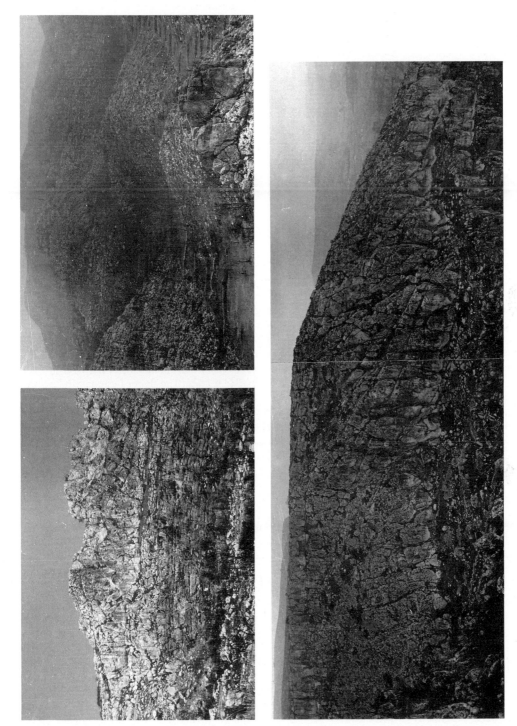

Fig. 15. La Riba reefs, Upper Muschelkalk. Upper left: massive mud-reef mound with dipping flank beds, resting upon well-bedded Benifallet dolomites. Upper right: view down on to reefs showing their elongate nature. Lower: reef with central massive core and well-developed, prograding flank beds.

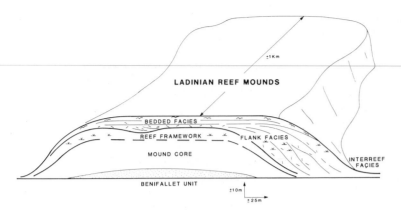

Fig. 16. Sedimentological model for the La Riba Reef Unit.

Tucker, 1988a), peloidal cementstone and skeletal—peloidal grainstone. Marine cements are prominent in the flank and framework facies and at some localities include former aragonite botryoids and isopachous fibrous calcite crusts.

Reef-top facies consist of planar and cross-bedded dasyclad grainstones with *Diploporia annulata* (Hemleben & Feels, 1977). Grainstones also accumulated on the margins of the buildups as thin slope sediments, but locally there are thick slope beds with clinoforms and metre-scale patch reefs. Above the reef-top grainstone facies are several metres of laminated, stromatolitic, fenestral dolomites, locally with columnar stromatolites, floe calcite (now dolomite), intraclast horizons and vadose pisoids associated with tepees structures. These rocks are a flat-bedded cap to the reef—mound complexes, and at the margins of the buildups they form a thin, brecciated and slumped facies. The inter-reef facies consists solely of a 0·1 m-thick, bioclastic—peloidal wackestone—packstone, equivalent to up to 60 m of reef! The reef complexes are onlapped by the Alcover laminated dolomites Unit above a sharp, undulating, locally erosive, unconformity, interpreted as a paleokarstic surface (Calvet & Tucker, 1988a). Some ferruginization and silicification occur at this surface.

The La Riba Unit probably passes laterally into the Querol stromatolites unit towards the north and into distal ramp facies (Rasquera Unit) towards the southwest. A similar arrangement of reefal complexes in the outer shallow ramp, decreasing in size offshore, has been described by Mesolella *et al.* (1974) from the Silurian Michigan Basin, and by Brandner & Resch (1981) in the Ladinian—Carnian reef complexes of the Austrian Alps.

Alcover laminated dolomites unit

This unit consists of thin-bedded, finely-laminated dolomicrites in the lower part, passing into a variety of bedded facies in the upper part. The Alcover Unit is principally located in the eastern part of the Prades Domain, filling the inter-reef depressions. This unit directly overlies the Benifallet Unit in the inter-reef areas, and onlaps and then overlaps the La Riba Reef Unit. The upper boundary below the Capafons Unit is represented by a centimetre-thick ferruginous level.

The lower part of the Alcover Unit, 15—20 m thick, consists of thin-bedded dolomicrites. The dolomicrite beds are 0·04—0·1 m thick and they are separated by millimetre-thick dolomitic marlstones. The colour of the dolomicrite facies is ochre-yellow at outcrop but grey on fresh surfaces. This facies contains millimetre-thick laminae, convoluted beds and intraformational folds and slides. Hemleben & Feels (1977) and Esteban *et al.* (1977b) have described shear-wrinkles, grooves, tool marks and chevron structures. Silicified evaporite nodules occur locally. The thin-bedded dolomites contain exquisitely preserved fish, reptiles, soft-bodied organisms and insects at several levels (Via & Villalta, 1966, 1975; Beltan, 1972, 1975; Hemleben & Feels, 1977; Via *et al.*, 1977; Via, 1987; Via & Calzada, 1987). According to Esteban *et al.* (1977b), this facies was deposited in quiet and anoxic conditions with rare density currents.

The upper part of the Alcover Unit consists of thin-bedded to massive dolowackestones, dolomitized oolitic—bioclastic packstone—grainstone and stromatolites. These facies represent the development of extensive shallow water over the region after the progressive filling of the inter-reef depressions by the lower part of the Alcover Unit.

Rasquera limestones−dolomites and shales with Daonella Unit

The Rasquera Unit, 23−40 m thick, is located only in the Baix Ebre−Priorat Domain. The upper boundary is defined by a hardground surface. The Rasquera Unit mostly consists of five lithofacies (Calvet & Tucker, 1988b). These facies occur in a distinct order forming cycles, 1·5−12 m thick, repeated five times in the 23−40 m thick Rasquera Unit.

The five facies are: (1) Marlstone−shale facies up to 7 m thick. The colour is generally grey-green to pale brown. Bedding is poorly defined, but locally there is a millimetre-scale paper lamination. Bioturbation is generally absent and fossils are rare. This facies is interpreted as a basinal facies. (2) Marlstone with thin-interbedded limestones facies. The thickness of this facies varies from 0·5 to 7 m. The limestone beds are 0·01−0·05 m thick and grey in colour, and the marlstone beds are generally a few centimetres thick. The thin-bedded limestones are, in general, laminated lime mudstone, with scarce thin-shelled bivalves (*Daonella*) concentrated along some laminae. This facies is interpreted as distal deep ramp. (3) Thin-bedded limestone facies. The thickness of this facies varies from 0·3 to 6 m. Thin-bedded limestones dominate this facies but they are separated by thin marlstones and millimetre-thick dolomitic partings. Three types of limestone bed can be distinguished: planar beds, nodular beds and less commonly, graded beds. The bioturbation increases towards the top of this facies package. This facies has been intepreted as intermediate deep ramp deposits. (4) Thick-bedded wackestone facies. The thickness of this facies varies from 0·3 to 4 m and the grey wackestone beds are 0·2−1·0 m thick. The wackestones contain thin-shelled bivalves, echinoderms and ostracods. This facies has been interpreted as proximal deep ramp. (5) Packstone−coquina facies. This facies occurs at the top of the cycles and does not reach more than 0·5 m in thickness. In general, the top surface has an irregular, almost scalloped morphology. The macrofaunal assemblage varies from a bivalve−ammonoid coquina to an assemblage dominated by millimetre-scale *Tubiphytes* microbuildups. The facies is interpreted as the shallowest-water facies of the cycles.

The Rasquera Unit facies constitute five principal cycles. The ideal vertical sequence (Fig. 13) consists of all five facies: marlstone−shale, marlstone with thin-bedded limestones, thin-bedded limestones, thick-bedded wackestones and packstone−coquina. In general, each cycle shows the following upward trends: (1) increasing grain-size; (2) increasing carbonate content; (3) increasing bioturbation intensity; (4) increasing bed thickness; and (5) increasing shallow-water fauna.

These facies and cycles are interpreted as distal, intermediate and proximal deep ramp (Calvet & Tucker, 1988b). There is little evidence of intense storm activity compared with other ramp models (Markello & Read, 1981; Aigner, 1984; Wright, 1986).

Tivissa limestones−dolomites and shales unit

The Tivissa Unit is located only in the Baix Ebre−Priorat Domain. The thickness of this unit varies from 25 to 50 m, and the main facies are: (1) marlstone−shale, 1−8 m thick; (2) marlstone with thin-interbedded limestones, up to 8 m thick. The limestone beds are 0·02−0·1 m thick and the marlstone beds are 0·01−0·08 m thick, decreasing towards the top. Locally the limestone beds show erosive bases and graded bedding; they are interpreted as storm beds; (3) thick-bedded lime mudstones and wackestones, 0·2−2 m thick, which are generally dolomitized; and (4) bioclastic packstone facies. The Tivissa Unit consists of these facies arranged into two cycles. The ideal vertical sequence consists of marlstone−shale, marlstone with thin-interbedded limestones (this facies is poorly developed in the second cycle), and thick-bedded facies. These facies and their sequences are interpreted as outer ramp shallowing to inner ramp deposits.

Capafons dolomites, marls, shales and breccia unit

The thickness of this unit varies from 25 to 45 m and it is present in all three sedimentological domains. The Capafons Unit grades into the overlying Keuper evaporitic facies and has a variety of lithofacies: (1) shale; (2) marlstone interbedded with thin-bedded limestones−dolomites; (3) thin-bedded marly dolomite with millimetre-scale lamination, ripples, evaporite mold casts and possible roots; (4) thick-bedded dolosparite; (5) oolitic grainstone, mostly dolomitized, with cross-bedding including herringbone type; (6) thick-bedded lime mudstone and wackestone; (7) massive dolomitic breccias with centimetre-scale evaporite mold casts; (8) breccias with chert and celestine nodules; and (9) bioclastic–oncolitic grainstone–packstone. These facies are ordered into shallowing-up sequences (Figs 13 & 17) and locally some thinning-upward sequences (tidal channel deposits) are present. Broadly, the Capafons Unit is interpreted as agitated shoal through lagoonal-marsh to hypersaline supratidal deposits. Santisteban & Taberner (1987) studied the sedimentology and diagenetic processes of this unit in the Prades Mountains.

Sedimentary model

The Rojals oolitic limestone–dolomite facies is interpreted as a high-energy tidal deposit, and the Benifallet Unit consists of a lagoonal deposit in the lower part and shoal sequences in the upper part, deposited on a shallow ramp. These two units are developed in the three domains and together represent a general transgressive cycle. However, after the deposition of the Benifallet Unit, some control may have led to the development of different sedimentary sequences in the various domains.

In the Baix Ebre–Priorat Domain the initial transgressive sequence continued with the Rasquera Unit in deep ramp facies. The Tivissa Unit (deep to shallow ramp deposits) and the Capafons Unit (lagoonal-marsh-mud flat deposits) together represent a general regressive sequence.

Fig. 17. Shallowing-upward sequence in the Capafons Unit of the Upper Muschelkalk composed of subtidal shales (sb), intertidal planar stromatolites (i) and supratidal breccias (sp).

In the Prades Domain the transgressive sequence represented by the Rojals and Benifallet Units is followed by the La Riba reef Unit. These reefal complexes were terminated by subaerial exposure, and then the Alcover Unit was deposited in the inter-reef areas, onlapping and eventually overlapping the reefs. The regressive sequence which started with the Alcover Unit continues into the Capafons Unit. In the Gaia–Montseny Domains, the transgressive sequence is represented by the Rojals and Benifallet Units, and the succeeding regressive sequence is represented by the Querol and Capafons Units.

Overall these Upper Muschelkalk carbonates, like those of the Lower Muschelkalk, belong to the ramp sedimentary model of Ahr (1973), Read (1985) and Tucker (1985). The model proposed started with a homoclinal ramp, barrier-bank type and evolved into a homoclinal ramp with buildups.

DEPOSITIONAL SYSTEMS TRACTS IN THE CATALAN BASIN

Several attempts have been made to define depositional sequences in the Triassic of the Iberian Peninsula (e.g. Garrido-Megias & Villena, 1977; Orti, 1987). Esteban & Robles (in Anadon et al., 1979) and Marzo & Calvet (1985) established sequences for the Catalan Basin according to the depositional sequence concept of Vail et al. (1977). In this paper, two depositional sequences are proposed for the Mid-Triassic strata of the Catalan Basin following the depositional systems tracts philosophy of Haq et al. (1987), Vail et al. (1987), Bally (1987) and papers in Wilgus et al. (1988). This has been derived largely from the study of seismic reflection profiles and a consideration of how depositional systems respond to major changes in sea-level (second- and third-order, operating on a timescale of several to several tens of millions of years). Although with the seismic stratigraphic approach there has been much emphasis on global eustatic changes of sea-level as a fundamental control, with little attention, at least in the early days, to regional tectonic effects, the depositional systems tracts philosophy can be applied to sedimentary sequences regardless of how the relative rises and falls of sea-level were brought about. The separation of global eustatic from regional tectonic mechanisms of sea-level change remains one of the major difficulties in interpreting sedimentary sequences (compare Hubbard *et al.*, 1985 and Vail *et al.*, 1984, for example and the review of Hubbard, 1988). Sarg (1988) has described the systems tracts of a carbonate rimmed shelf and Fig. 18 illustrates the systems tracts for a carbonate ramps.

The Mid-Triassic strata of the Catalan Basin can be divided into two depositional sequences (Figs 4, 19 & 20), here termed Depositional sequence 1 (which includes the uppermost part of the Buntsandstein and Lower Muschelkalk) and Depositional sequence 2 (which includes the Middle Muschelkalk and Upper Muschelkalk). The Buntsandstein below the lutite−carbonate−evaporite unit is a separate depositional sequence, of the lowstand type.

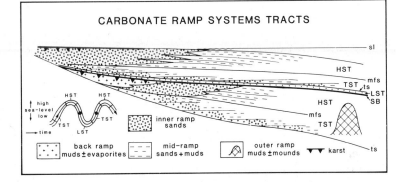

Fig. 18. Depositional systems tracts for a carbonate ramp. The rationale is that stratal patterns are controlled by relative changes of sea-level (determined by eustasy and/or tectonics; see also Fig. 25) and these also control the sediment accommodation potential. During specific time-intervals of the relative sea-level curve, particular depositional systems tracts are established: lowstand (LST), transgressive (TST), highstand (HST) and shelf-margin wedge (SMW, not appropriate here). On a carbonate ramp, deposits of the TST and HST generally form the major part of the sequence. TST strata typically consist of an onlapping, retrogradational package, which may consist of parasequences (each of which will generally be a shallowing-upward cycle). The overlying HST strata will form an aggradational−progradational package, which again may consist of a parasequence set of shallowing−upward cycles. In this figure, the TST and HST of the lower sequence are shown as the facies of a ramp with a barrier shoreline and a back-ramp lagoon, whereas in the upper sequence the TST and HST are shown as the facies of a ramp with a strandplain (see Tucker & Wright, 1990, for a review of carbonate ramp types). The lowstand systems tract may be represented by a paleokarstic surface and calcareous soils, sabkha evaporites, or fluvial-lacustrine facies, depending on the magnitude of the relative sea-level fall, hinterland relief and the climate. SB is the sequence boundary, ts is the transgressive surface, mfs is the maximum flooding surface and sl is sea-level.

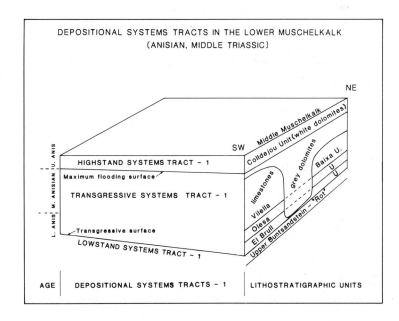

Fig. 19. Depositional systems tracts and lithostratigraphical units in the Lower Muschelkalk (Anisian).

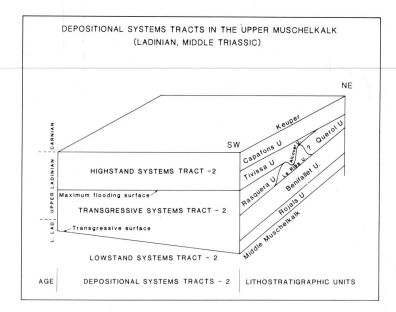

Fig. 20. Depositional systems tracts and lithostratigraphical units in the Upper Muschelkalk (Ladinian).

Depositional sequence 1 (Uppermost Buntsandstein and Lower Muschelkalk)

Depositional sequence 1 (Figs 5 & 19) consists of the lutite−carbonate−evaporite unit of the uppermost Buntsandstein which represents the lowstand systems tract, here named LST-1 succeeded by the El Brull Unit (peritidal deposits), the Olesa Unit (lagoonal deposits) and the Vilella Baixa Unit (from lagoonal to shoal deposits) which are the result of a broadly transgressive cycle of deposition and are thus the transgressive systems tract, termed TST-1. The lutite−carbonate−evaporite unit represents the first marine incursion into the Catalan Basin. This sequence reflects a minor transgression, before the major sea-level rise depositing the TST-1 of the Lower Muschelkalk. The Colldejou Unit (intertidal−sabkha deposits) of the upper part of the Lower Muschelkalk is a highstand systems tract (HST-1). The major transgressive surface (Ts-1) is located at the boundary between the Buntsandstein lutite−carbonate−evaporite unit and the El Brull laminated limestone−dolomite Unit (Lower Muschelkalk). The regional paraconformity located at the top of the transgressive cycle, represents the maximum flooding surface (Mfs-1).

Depositional sequence 2 (Middle and Upper Muschelkalk)

Depositional sequence 2 (Figs 13 & 20) in-cludes the Middle Muschelkalk and the Upper Muschelkalk. The continental clastic and evaporitic facies of the Middle Muschelkalk represent the lowstand systems tract (LST-2) of depositional sequence 2. The Rojals Unit (high-energy tidal deposits), the Benifallet Unit (from lagoonal to shoal deposits) and the Rasquera Unit (deep ramp deposits) in the Baix Ebre−Priorat Domain constitute a broadly transgressive cycle of sedimentation. The Rojals Unit, the Benifallet Unit and the La Riba reefs Unit in the Prades Domain, likewise, are the result of a general transgressive cycle, as is the Rojals Unit and Benifallet Unit in the Gaia−Montseny Domain. These units of the different domains thus belong to the transgressive systems tract, TST-2.

The Tivissa Unit (deep to shallow ramp deposits) and the Capafons Unit (lagoonal-marsh-mud flat facies) in the Baix Ebre−Priorat Domain; the Alcover Unit (anoxic to shallow-water facies) and the Capafons Unit in the Prades Domain, and the Querol Unit (stromatolites−oolites) and the Capafons Unit in the Gaia−Montseny Domain, all have the same regressive character and constitute a highstand systems tract of Depositional sequence 2, i.e. HST-2. The transgressive surface (Ts-2) is located at the boundary between the Middle Muschelkalk fine clastic facies and the Rojals oolitic limestones unit (Upper Muschelkalk). The condensed faunal level at the top of the Rasquera Unit in the Baix Ebre−Priorat Domain and the erosive surface on the La Riba Reef that is onlapped by the Alcover Unit beds is the maximum flooding surface, Mfs-2.

In the Iberian Range, there is a lateral passage towards the east from lower Keuper facies into marine limestones of the Upper Muschelkalk (A. Arche, pers. comm.). This lateral change, using time lines, suggests a downlap geometry for the upper part of the Upper Muschelkalk, interpreted above as a highstand systems tract.

NATURE OF SEQUENCE BOUNDARIES IN THE TRIASSIC

The lower boundary (Fig. 4) of the Triassic system in the Catalan Basin is an angular unconformity with the Palaeozoic, whereas the Triassic upper boundary is a disconformity with Lower Liassic breccias, possibly of Sinemurian−Carixian age (Giner, 1980; Salas, 1987). The Palaeozoic strata below the angular unconformity (a first-order unconformity) were deformed during the Hercynian orogeny and then extensively weathered and eroded (Virgili *et al.*, 1974; Marzo, 1980). The time represented by this unconformity is post-Stephanian to Lower Triassic or even Upper Permian (Virgili *et al.*, 1974; Marzo, 1980). Towards the Iberian Ranges and in association with palaeohighs, Buntsandstein sedimentation started during the Anisian or even during the Ladinian (Sopeña, 1979; Perez-Arlucea & Sopeña, 1985; Virgili, 1987).

Boundaries of Depositional sequence 1

The lower boundary of depositional sequence 1 corresponds to the disconformity−unconformity between the lutite−carbonate−evaporite unit and the underlying dominantly siliciclastic Buntsandstein. In the Catalan Basin, this is a paleosol horizon, termed the Prades paleosol by Marzo (1980). In the Pyrenees (J. Gisbert, pers. comm.) and in the Cantabrian Mountains (Garcia-Mondejar *et al.*, 1986), there is a marked influx of siliciclastics at this level. In the Iberian Range, the lower boundary is an angular unconformity (Lopez-Gomez, 1987; Perez-Arlucea, 1987). The upper boundary (Figs 4, 5 & 19) is a regional disconformity−unconformity represented by an abrupt lithological contact between the carbonates of the Lower Muschelkalk and the overlying siliciclastics and evaporites of the Middle Muschelkalk. Related to the unconformity

there is a laminated crust with tepees, erosion surfaces, supratidal stromatolites and collapse breccias. Possibly, the paleokarst described previously in the Lower Muschelkalk and interpreted as phreatic in origin is related to this boundary which could be the subaerial exposure surface of the paleokarst complex.

The transgressive surface (Ts-1) of Depositional sequence 1, situated at the contact between the lutite−carbonate−evaporite unit in the upper Buntsandstein and the Lower Muschelkalk is gradational in the Catalan Range. However, in some areas close to palaeohighs in the Iberian Range, the Lower Muschelkalk is unconformable upon the Buntsandstein with an onlap geometry (Lopez-Gomez, 1987; Perez-Arlucea, 1987). The maximum flooding surface (Mfs-1) of Depositional sequence 1 is a regional paraconformity. This surface, probably very extensive (>100 km) in an E−W direction, was buried by downlap deposition of the highstand systems tract. The Mfs-1 in the Catalan Basin separates the grey dolomites below from the white dolomites above.

Boundaries of Depositional sequence 2

The lower boundary of Depositional sequence 2 naturally corresponds to the upper boundary (Fig. 4) of Depositional sequence 1. However, the upper boundary (Figs 4, 13 & 20) of Depositional sequence 2 is difficult to define precisely in the Catalan Range because the Capafons Unit (the last unit of the Upper Muschelkalk) is transitional into the Keuper facies.

The transgressive surface (Ts-2) of Depositional sequence 2 is located at the boundary between the fine clastic Middle Muschelkalk facies and the limestone−dolomite facies of the Upper Muschelkalk. The TST-2 in the western part of the Iberian Range displays an onlap geometry with regard to the underlying clastics (Garcia-Gil, 1989). The maximum flooding surface (Mfs-2) of Depositional sequence 2 varies in nature according to the particular domain. In the Baix Ebre−Priorat Domain, where deepwater facies occur it is located at the top of the Rasquera Unit at a centimetre-thick condensed bed rich in ammonoids. In the Prades Domain, Mfs-2 is located between the La Riba−Benifallet Units and the overlying Alcover Unit. The latter has an onlapping arrangement with respect to the La Riba reefs.

DEPOSITIONAL CYCLES: TECTONICS VERSUS EUSTASY

The pattern of middle Triassic sedimentation described in this paper from the Catalan Basin is of two depositional sequences, each beginning with lowstand clastics and/or evaporites, overlain by a broadly transgressive sequence of onlapping shallow-water carbonates, which pass up into highstand aggradational−progradational carbonates. This pattern is similar to that in the Triassic of the German Basin and southern North Sea. It is instructive to compare the depositional systems tracts of the Catalan Basin with the global sea-level curve of Haq et al. (1987) (see Fig. 21), which is derived from data from Svalbard, the Dolomites of Italy and the Salt Range of Pakistan. The third-order cycles 1.2, 1.3 and 1.4 of Haq et al. cannot be recognized in the fluvial Buntsandstein, but cycle 1.5 could correspond to the first marine transgression into the Catalan Basin which deposited the lutite−carbonate−evaporite unit in the upper part of the Buntsandstein

and marked the beginning of Depositional sequence 1. In fact, this evaporitic unit corresponds to the Solling−Röt Halite of the German Basin−southern North Sea, which was deposited following the first transgression from Tethys. The lower boundary of Depositional sequence 1 is thus at least of regional extent and significance. Cycle 2.1 of Haq et al. (1987; Fig. 21) would appear to correspond to the major transgression which deposited the Lower Muschelkalk, that is the TST-1 and HST-1 of depositional sequence 1 in the Catalan Basin. The succeeding Middle and Upper Muschelkalk, with LST-2, TST-2 and HST-2, would correspond to third-order cycle 2.2, although there is a small difference in age with respect to the upper boundary.

The underlying control on base-level changes within the Catalan Basin is of considerable interest in view of the discussions of eustatic versus tectonic controls on relative sea-level changes. In the Catalan Basin, three phases of crustal extension can be distinguished: early Triassic (prior to and during deposition of the Buntsandstein), Anisian−Ladinian

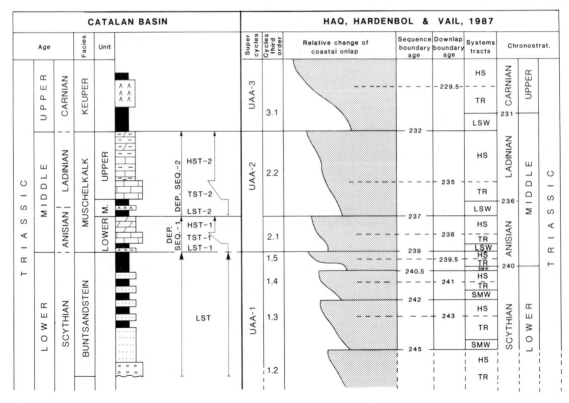

Fig. 21. Comparison between the Triassic of the Catalan Basin and the proposed sequence stratigraphy of Haq et al. (1987).

(prior to and during deposition of the Middle Muschelkalk) and Carnian (during deposition of the Keuper facies). The evidence for rifting is best provided by the considerable changes in sediment thickness (see Figs 22 & 23). Movements on basement faults within the Catalan Basin resulted in great thickness variations, especially in clastic and evaporite units. In the Ebro Basin, lateral thickness changes are even more marked (Jurado, 1989). The late Permian–Scythian extension phase is recognized throughout western Europe and in the northern North Sea resulted in the formation of the Viking Graben. This rifting initiated coarse clastic deposition (the Buntsandstein) with sequences reaching several kilometre thickness in some graben structures. The cessation of this extension is marked by the Hardegsen Unconformity, prominent in the southern North Sea and German Basin, and this is overlain by marls and then the Röt Halite. This unconformity separates synrift sediments below from postrift sediments above.

Extension in Middle Muschelkalk and Keuper times in the Catalan Basin was accompanied by extrusion of basaltic lavas. Sedimentation took place close to sea-level, and the differential subsidence controlled by basement faults led to thick sequences of evaporites in the sub-basin centres (as in the Baix Ebre–Priorat Domain for example, Fig. 22; Marzo & Calvet, 1985). By way of contrast, thicknesses of the Lower Muschelkalk and Upper Muschelkalk are much more uniform across the Catalan Basin, and in the case of the Lower Muschelkalk the facies

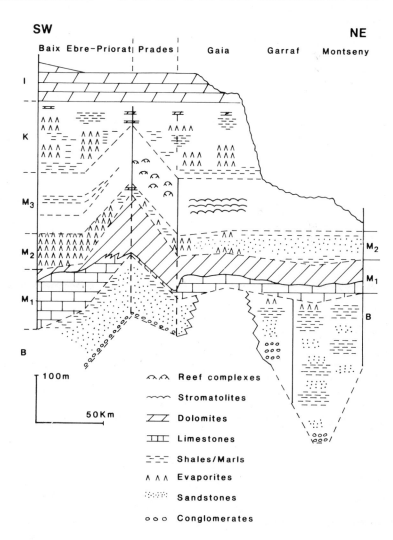

Fig. 22. Schematic stratigraphy of the Catalan Basin Triassic showing the thickness variations in the Buntsandstein, Middle Muschelkalk and Keuper. From Marzo & Calvet (1985).

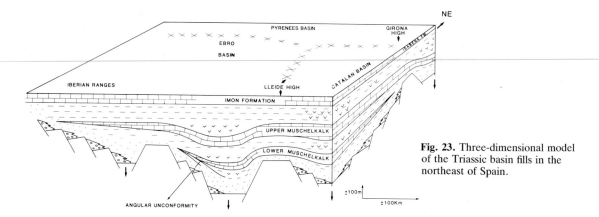

Fig. 23. Three-dimensional model of the Triassic basin fills in the northeast of Spain.

are very similar too in all domains. The dramatic facies variations in the central part of the Upper Muschelkalk are a consequence of the great range of depositional environments that existed, the water circulation patterns and nutrient supply, all of which were a function of ramp geometry, some structural control, and at times a more open connection with Tethys to the east. The Imon Formation, which has not yet been studied by the authors, is also a carbonate sequence of very uniform facies and thickness (Figs 22 & 23).

The formation of sedimentary basins in extensional tectonic regimes has received much attention in recent years, and the nature of their sedimentary fills and subsidence histories are now well documented (e.g. Sclater & Christie, 1980; Dewey, 1982; Hubbard, 1988; Thorne & Watts, 1989). The rifting phase is characterized by rapid subsidence and the deposition of coarse clastics, especially close to basin margin faults. Rates of subsidence depend on the amount of stretching and width of the basin, but they are typically the order of a metre or more per 1000 years. For narrow basins (<50 km), heat loss during rifting can lead to even higher rates (e.g. Pitman & Andrews, 1985). The rapid subsidence during rifting is generally followed by a much slower, regional thermal subsidence as a result of lithospheric cooling. Again rates vary depending on the extent and duration of the rifting, but typical values are the order of $10-20$ cm per 1000 years immediately after rifting, decreasing to a few cm $1000\,yr^{-1}$ over $20-30$ Ma.

Now with regard to the Catalan Basin, the early Triassic extensional event was a major rifting episode, accounting for the onset of deposition after a long period of subaerial erosion and for the nature of the sediment: fanglomerates, braidplain sandstones and floodplain mudrocks (Marzo, 1980). The upward passage from fluvial clastics into the lutite—carbonate—evaporite unit (Röt equivalent) in the upper part of the Buntsandstein, deposited close to sea-level, and the succeeding transgressive systems

Table 1. Duration of depositional systems tracts, sedimentation rates and cycle periodicities in the Catalan Basin Triassic. LST is the major part of the Buntsandstein. LST-1, the first systems tract of Depositional sequence 1, is the uppermost part of the Buntsandstein, the lutite—carbonate—evaporite unit, equivalent to the Röt of northwestern Europe. TST-1 and HST-1 form the Lower Muschelkalk. In Depositional sequence 2, LST-2 is the Middle Muschelkalk, and TST-2 and HST-2 form the Upper Muschelkalk.

Depositional systems tracts	Duration (Ma)	Thickness (m)	Max. sed. rate (cm $1000\,yr^{-1}$)	Max. no. of cycles	Cycle periodicity
LST	7	50−300	4	—	—
LST-1	1	10−25	2.5	—	—
TST-1	0.6	30−116	20	16	37 000
HST-1	1	20−40	4	—	—
LST-2	1	50−130	13	30	33 000
TST-2	1	49−107	11	15	70 000
HST-2	3	50−95	3	7	430 000

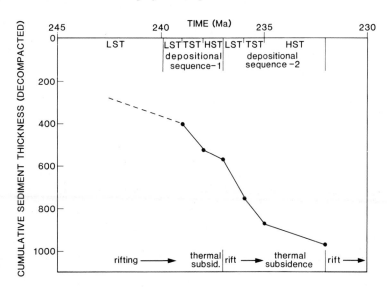

Fig. 24. Schematic graph of cumulative sediment thickness (decompacted) against time, with the extensional phases and depositional sequences also shown.

tract of the Lower Muschelkalk can be explained in terms of regional subsidence and/or eustatic sea-level rise.

It is instructive here to consider sedimentation rates (see Table 1, Fig. 24). Since much of the upper Buntsandstein through to the lower part of the Upper Muschelkalk (Benifallet Unit) was deposited close to sea-level, the sediment thickness very accurately reflects the accommodation, i.e. the space available for sediment. The accommodation depends on subsidence and eustacy. Although there are many errors in simply calculating sedimentation rates (notably the problems of non-deposition, compaction and uncertainty of the amount of time involved), it is useful to compare these figures with the typical subsidence rates mentioned above for extensional basins. If the figures are similar, then the sequence could be explained simply in terms of subsidence. If the sedimentation rates are much higher than the subsidence rates, then a eustatic sea-level rise is implicated.

The sedimentation rate for the Buntsandstein is really quite low (4 cm $1000\,\mathrm{yr}^{-1}$), but this probably reflects the time required to fill the basin. Rifts are commonly starved of sediment in the early stages as a result of the rapid subsidence (Pitman & Andrews, 1985). The sedimentation rates for the TST-1 and HST-1 of the Lower Muschelkalk are around 20 cm $1000\,\mathrm{yr}^{-1}$ and 4 cm $1000\,\mathrm{yr}^{-1}$, respectively. These figures, minimum values, are the same order of magnitude as those for regional subsidence, and it is particularly interesting to note the apparent decline in sedimentation rate through the Lower Muschelkalk (see Fig. 24). This could reflect a decreasing rate of thermal subsidence after the extension. In Depositional sequence 2, Middle through Upper Muschelkalk, it is noteworthy that there is a high sedimentation rate for the Middle Muschelkalk, a time of extension. The reason for this is the deposition close to sea-level and the ability of evaporites to be precipitated quickly and so keep an area of rapid subsidence at base level. Again, the sedimentation rates of the TST-2 and HST-2 in the Upper Muschelkalk, 6 cm $1000\,\mathrm{yr}^{-1}$ and 2·5 cm $1000\,\mathrm{yr}^{-1}$, respectively, are similar to regional subsidence rates, and they also decrease with time.

Thus, bearing in mind the limitations of the information available, it is possible to explain the pattern of sedimentation in the two depositional sequences as simply a response to the tectonic regime: two phases of extension, each followed by regional subsidence, permitting deposition of the synrift Buntsandstein and postrift Lower Muschelkalk, and then the synrift Middle Muschelkalk followed by the postrift Upper Muschelkalk (Figs 24 & 25). The fact that these Lower and Middle Triassic depositional cycles can be correlated over the whole of western Europe in the Germanic-type Triassic is simply a result of the whole region being subjected to the same extensional tectonic regime. It does not require global eustatic sea-level changes. Considering the facies sequence in the Upper Muschelkalk, a somewhat more rapid rise due to some other mechanism is perhaps required

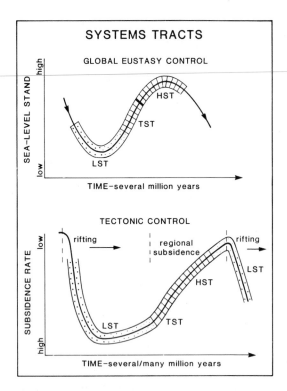

Fig. 25. Simplistic sketch suggesting that a similar sequence of systems tracts will be produced through tectonic rifting and regional subsidence, as through global eustasy. Superimposed on these third-order (10^6 yr) relative changes of sea-level, which give rise to depositional sequences there may well be shorter time-scale, fourth and fifth order (10^5 and 10^4 yr) sea-level changes, perhaps resulting from the effects of Milankovitch rhythms, which give rise to parasequences and parasequence sets. In the tectonic model, rifting leads to the creation of sources and sinks, and siliciclastics are deposited in fans, passing in both space and time into alluvial plains and lakes, and evaporites (if arid), of a lowstand systems tract. With denudation and gradual loss of relief, and deposition, clastics give way to carbonates as relative sea-level rises through regional subsidence, establishing a transgressive systems tract. Stratal onlap of the TST facies is followed by carbonate aggradation and progradation of the highstand systems tract as the magnitude of regional subsidence declines and relative sea-level begins to fall. The latter results in the formation of the sequence boundary. A major paleokarst, evaporite horizon or fluvial sandstone may develop as the lowstand systems tract is established again.

for the development of the La Riba mud mounds upon the Benifallet lagoonal unit. And then the Catalan Basin apparently suffered a short-lived emergence episode after the La Riba Unit, with

paleokarstification, and this was followed by rapid reflooding of the basin. The effect of some barrier, probably structural, to the east/southeast may account for the rapid sea-level changes before and after the La Riba Reef Unit. The lower and upper boundaries of the La Riba Reef Unit are parasequence (or subsequence) boundaries. They represent much shorter time-scale sea-level changes responsible for the depositional sequences themselves.

On a much larger scale than western Europe, the Triassic was also a time of crustal extension before the main phase of continental rifting and separation and new ocean formation in the Jurassic (Tethys and Atlantic). An increase in plate tectonic activity and seafloor spreading rates causing ocean basin volume changes would lead to eustatic changes in sea-level of global extent. Both rises and falls can be expected in Triassic times. It is a problem to distinguish these eustatic effects from relative sea-level rises due to regional subsidence, but one important feature of these plate-tectonic induced eustatic sea-level changes is that the rate is much lower (<1 cm $1000 \, \text{yr}^{-1}$) than that of the regional thermal subsidence following a phase of rifting.

FOURTH–FIFTH-ORDER SHALLOWING-UPWARD CYCLES

The carbonate ramps of the Lower and Upper Muschelkalk were formed during the transgressive and highstand phases of two major periods of relative sea-level rise, when terrigenous clastic influx was greatly reduced, and at a time of relative tectonic quiescence. As discussed earlier, many stratigraphic units in the Muschelkalk consist of metre-scale, shallowing-upward cycles (see Figs 5 & 13). These parasequences are apparently the result of relatively rapid sea-level rises, to initiate the cycle, followed by a stillstand or slow fall, during which time tidal flats or sand shoals prograde or aggrade. The exposure horizons at the tops of cycles may represent long periods of time.

The origin of small-scale shallowing-upward carbonate cycles is a contentious issue, with sedimentary, glacioeustatic and tectonic mechanisms all available, with their adherents (see reviews of Hardie, 1986 and Tucker & Wright, 1990). An orbital forcing mechanism is gaining popularity, it has to be admitted, but to demonstrate Milankovitch-type cyclicities in sedimentary strata requires a sequence with a sufficiently large number of cycles so that statistical techniques can be employed. Unfortunately, such an approach cannot be used in the Catalan Basin where the number of cycles is relatively small (and this is the case with many other sequences in the geological record consisting of repeated shallowing-upward cycles). However, it is pertinent to recall that in the Ladinian strata of the Latemar Platform of the Dolomites in northern Italy, rocks of equivalent age to the Upper Muschelkalk, as well as in the Carnian Dürrenstein Formation and Norian Dolomia Principale, Milankovitch-band cyclicities have been recognized in the patterns of thickness variations of small-scale cycles, suggesting an orbital-forcing control (Hardie *et al.*, 1986; Goldhammer *et al.* 1987). The maximum duration of the Muschelkalk cycles, obtained from a simple division of time by cycle number, ranges from 30 000 to 600 000 years. This is comparable to, and much longer than, the duration of Latemar cycles. One major factor leading to gross over-estimates of cycle duration is the complete absence of cycles when sea-level drops below the platform margin or does not reach the inner region of a carbonate ramp. Although it cannot be proved, and persuaded by the evidence from Italy it is considered most likely that the shallowing-upward cycles in the Lower and Upper Muschelkalk are the result of orbital forcing.

The Middle Muschelkalk has many metre-scale repetitions of red marl−nodular gypsum (shallow sub-littoral to sabkha) and laminated gypsum−nodular gypsum (lagoonal to sabkha) which could also be responses to glacioeustatic fourth−fifth-order cycles of sea-level change.

It is worth pointing out that the pattern of rapid sea-level rise and slow fall or stillstand which appears to explain the fourth−fifth-order shallowing-upward cycles of probable glacioeustatic/orbital forcing origin is different from that of the much thicker second−third-order cycles, which typically are depicted as slow sea-level rises and apparently rapid sea-level falls (see Fig. 21; Vail *et al.*, 1977; Haq *et al.*, 1987). There has been much discussion over the apparent rapid falls (e.g. Miall, 1986), and in many people's minds now (e.g. Hubbard, 1988, and this paper), the slow rises are largely the result of regional tectonic subsidence following rifting, rather than some global eustatic phenomenon, for which at present there is no known mechanism. The apparent slow falls and rapid rises of relative sea-level of many fourth−fifth-order cycles is consistent with a glacioeustatic origin, where the buildup of polar ice caps is slow, but their melting is fast. Another problem is whether there

were Triassic ice caps to wax and wane, but that is not an issue for this paper.

CONCLUSIONS

1 In the Triassic of the Catalan Basin, three carbonate platforms occur above the Buntsandstein: the Lower Muschelkalk, the Upper Muschelkalk and the Imon Formation, of Anisian, Ladinian and Norian ages, respectively. Between these carbonate units are wedged regressive clastics and evaporites of the Middle Muschelkalk and Keuper, of Upper Anisian−Lower Ladinian and Carnian age.

2 The three carbonate units in the Catalan Basin are the result of deposition in different ramp sedimentary models. The model proposed for the Lower Muschelkalk is similar to the homoclinal-ramp, barrier-bank type of Read (1985), and the Upper Muschelkalk is similar to the homoclinal-ramp, barrier-type that evolves into a homoclinal ramp with buildups.

3 The mid-Triassic strata of the Catalan Basin can be divided into two depositional sequences: Depositional sequence 1 includes the uppermost Buntsandstein and Lower Muschelkalk and Depositional sequence 2 includes the Middle Muschelkalk and Upper Muschelkalk. The lowstand systems tracts are represented by the lutite−carbonate−evaporite unit at the top of the Buntsandstein and the Middle Muschelkalk (both continental to marine evaporitic facies). Deposits of the transgressive systems tracts are composed of peritidal deposits in the lower part of the transgressive section passing up into sand belt deposits, reef complexes and deep ramp facies in the upper part. The two highstand systems tracts show a general aggrading to regressive character and the carbonate facies are mostly of a restricted nature.

4 On a broad scale the Triassic carbonate deposits in the northeastern Iberian Peninsula display an onlapping, retrogressive and back-stepping arrangement. The Buntsandstein, Middle Muschelkalk and Keuper sediments were deposited during phases of crustal extension and rifting, resulting in major thickness changes across the Catalan Basin, and locally basalt eruption. Deposition of the Muschelkalk carbonates took place during the succeeding period of regional subsidence and tectonic quiescence. Although it is notoriously difficult to separate the roles of tectonics and eustasy in controlling sea-level changes in sedimentary basins, in this case the pattern of sedimentation in the two depositional sequences can be explained as primarily a function of the extensional tectonic regime, against a background of first−second-order tectono-eustatic sea-level changes through seafloor spreading rate variations. See Fig. 25.

5 Metre-scale shallowing-upward cycles (parasequences) in the Muschelkalk are the result of fourth or fifth-order sea-level changes, of rapid sea-level rise and slow sea-level fall or stillstand. They could well be the result of orbital forcing in the Milankovitch band.

ACKNOWLEDGEMENTS

The authors are grateful to Carole Blair from Durham for typing the manuscript and Karen Atkinson from Durham and B. Andres from Barcelona for drafting the figures. The authors are particularly grateful to their colleagues Alfredo Arche, Mateo Esteban, Rob Gawthorpe, Dave Hunt, M. Marc, A. Marquez, Mariano Marzo, Federico Orti, Ramon Salas, N. Sole, X. Ramon and C. Virgili for discussion and suggestions during the preparation of this paper and reviews of the manuscript. The work was in part supported by the Natural Environment Research Council and the Acciones Integradas Hispano−Britanica/British Council and the CICYT Project number PB 0322.

REFERENCES

AHR, W.M. (1973) The carbonate ramp: an alternative to the shelf model. *Trans. Gulf Coast Assoc. Geol. Soc.* **23**, 221−225.

AIGNER, T. (1984) Dynamic stratigraphy of epicontinental carbonates, Upper Muschelkalk (M. Triassic), South German Basin. *N. J. Geol. Paläont. Abh.* **169**, 127−159.

ANADON, P., CABRERA, L., GUIMERA, J. & SANTANACH, P. (1985) Paleogene strike−slip deformation and sedimentation in the southeastern margin of the Ebro Basin. In: *Strike−slip Deformation, Basin Formation and Sedimentation* (Eds Biddle, K.T. & Christie-Blick, N.) Spec. Publ. Soc. Econ. Paleont. Mineral. 37, pp. 303−318.

ANADON, P., COLOMBO, F., ESTEBAN, M., MARZO, M., ROBLES, S., SANTANACH, P. & SOLE-SUGRANES, L.L. (1979) Evolución tectonoestratigráfica de los Catalánides. *Acta Geol. Hispánica* **14**, 242−270.

ANDREU, A., CALVET, F., FONT, X. & VILADEVALL, M. (1987) Las mineralizaciones de Pb-Zn-Ba en el Muschelkalk inferior de los Catalánides. *Cuad. Geol. Ibérica* **11**, 779−795.

BALLY, A.W. (1987) *Atlas of Seismic Stratigraphy*, Vol. **1**, Am. Assoc. Petrol. Geol. Studies in Geology, No. 27.

BELTAN, L. (1972) La faune ichthyologique du Muschelkalk de la Catalogne. *Mem. Roy. Acad. CC y AA*, **XLI**, 279−325.

BELTAN, L. (1975) A propos de l'Ichthyofaune Triasique de la Catalogne Espagnole. *Coll. Int. CNRS*, **218**, 273−280.

BOSENCE, D.W.J., ROWLANDS, R.J. & QUINE, M.L. (1985) Sedimentology and budget of a Recent carbonate mound, Florida Keys. *Sedimentology* **32**, 317−343.

BRADY, M.J. & ROWELL, A.J. (1976) An Upper Cambrian subtidal blanket carbonate, eastern Great Basin. *Brigham Young Univ., Geol. Stud.* **23**, 153−163.

BRANDNER, R. & RESCH, W. (1981) Reef development in the Middle Triassic (Ladinian and Cordevolian) of the northern Limestone Alps near Innsbruck, Austria. In: *European Fossil Reef Models* (Ed. Toomey, D.F.) Spec. Publ. Soc. Econ. Paleont. Mineral. 30, pp. 203−231.

BUSSON, G. (1982) Le Trias comme pēriode salifere. *Geol. Rundschau* **71**, 857−880.

CALVET, F. & RAMON, X. (1987) Estratigrafia, sedimentología y diagénesis del Muschelkalk inferior de los Catalánides. *Cuad. Geol. Ibérica* **11**, 141−169.

CALVET, F. & TUCKER, M.E. (1988a) Triassic (Upper Muschelkalk) mud mounds and reefal complexes, Catalan Basin, Spain (Abstracts). 9th IAS Regional Meeting, Leuven, Belgium, pp. 36−37.

CALVET, F. & TUCKER, M. (1988b) Outer ramp cycles in the Upper Muschelkalk of the Catalan Basin, northeast Spain. *Sed. Geol.* **57**, 185−198.

CALVET, F., MARCH, M. & PEDROSA, A. (1987) Estratigrafia, sedimentología y diagénesis del Muschelkalk superior de los Catalánides. *Cuad. Geol. Ibérica* **11**, 171−197.

CALZADA, S. & GAETANI, M. (1977) Nota paleoecológica sobre *M. Mentzelli* (Brachiopoda, Anisiense, Catalánides). *Cuad. Geol. Ibérica* **4**, 157−168.

DEWEY, J.F. (1982) Plate tectonics and the evolution of the British Isles. *J. Geol. Soc.* **139**, 371−412.

ESTEBAN, M., POMAR, L., MARZO, M. & ANADON, P. (1977a) Naturaleza del contacto entre Muschelkalk inferior y Muschelkalk medio de la zona de Aiguafreda (Provincia de Barcelona). *Cuad. Geol. Ibérica* **4**, 201−210.

ESTEBAN, M., CALZADA, S. & VIA, L. (1977b) Ambiente deposicional de los yacimientos fosiliferos del Muschelkalk Superior de Alcover−Montral. *Cuad. Geol. Ibérica* **4**, 189−200.

FONTBOTE, J.M. (1954) Las relaciones tectónicas de la Depresión del Vallés-Penedés con la Cordillera Prelitoral Catalana y con la Depresión del Ebro. *Bol. Real. Soc. Esq. Hist, Nat.* (tomo homenaje Profesor E. Pacheco) 281−310.

GANDIN, A., TONGIORGI, M., RAU, A. & VIRGILI, C. (1982) Some examples of the Middle-Triassic marine transgression in southwestern Mediterranean Europe. *Geol. Runds.* **71**, 881−894.

GARCIA-GIL, S. (1989) *Estudio sedimentologico y paleogeografico del Triasico en el Tercio Noroccidental de la Cordillera Iberica (Provincias de Guadalajara y Soria).* PhD Thesis, Madrid University, Vol. 1, p. 406.

GARCIA-MONDEJAR, J., PUJALTE, V. & ROBLES, S. (1986) Caracteristicas sedimentologicas, secuenciales y tectonoestratigraficas del Triasico de Cantabria y Norte de Palencia. *Cuad. Geol. Iberica* **10**, 151−172.

GARRIDO-MEGIAS, A. & VILLENA, J. (1977) El Trias Germánico en España: Paleogeografia y análisis secuencial. *Cuad. Geol. Ibérica* **4**, 37–56.

GINER, J. (1980) *Estudio sedimentológico y diagenético de las formaciones carbonatadas del Jurásico de los Catalánides, Maestrazgo y rama aragonesa de la Cordillera Ibérica (sector oriental).* PhD Thesis, Barcelona University, p. 316.

GOTTIS, M. & KROMM, F. (1967) Sur l'existence d'un épisode régressif au sein du Muschelkalk inférieur sur la bordure occidentale du Massif Catalan. *Actes Soc. Linnéenne Bordeaux* **104**, 3–4.

GUIMERA, J. (1984) Palaeogene evolution of deformation in the northeastern Iberian Peninsula. *Geol. Mag.* **121**, 413–420.

GUIMERA, J. (1988a) *Estudi estructural de l'enllaç entre la Serralada Ibérica i la Serralada Costenera Catalana.* PhD Thesis, Barcelona University, p. 600.

GUIMERA, J. (1988b) Rasgos principales de las estructuras compresivas y distensivas alpinas de los "Iberides" orientales. *II Congreso Geológico de España*, Granada, Comunicaciones, Vol. 2, pp. 149–152.

HAQ, B.L., HARDENBOL, J. & VAIL, P.R. (1987) Chronology of fluctuating sea levels since the Triassic. *Science* **235**, 1156–1167.

HARDIE, L.A. (1986) Stratigraphic models for carbonate tidal-flat deposition. *Q. J. Colorado Sch. Mines* **81**, 59–74.

HARDIE, L.A., BOSELLINI, A. & GOLDHAMMER, R.K. (1986) Repeated subaerial exposure of subtidal carbonate platforms, Triassic, Northern Italy: Evidence for high frequency sea-level oscillations on a 10^4-year scale. *Paleoceanography* **1**, 447–457.

HEMLEBEN, Ch. & FEELS, D. (1977) Fossilführende dolomitisierte Plattenkalke aus dem "Muschelkalk superior" bei Montral (prov. Tarragona, Spanien). *N. J. Geol. Paläont. Abh.* **154**, 186–212.

HIRSCH, F. (1976) Sur l'origine des particularismes de la faune du Trias et du Jurassique de la plate-forme africano-arabe. *Bull. Soc. Geol. France* **XVIII**, 543–552.

HUBBARD, R.J. (1988) Age and significance of sequence boundaries on Jurassic and early Cretaceous rifted continental margins. *Bull. Am. Assoc. Petrol. Geol.* **72**, 49–72.

HUBBARD, R.J., PAPE, J. & ROBERTS, D.G. (1985) Depositional sequence mapping to illustrate the evolution of a passive continental margin. In: *Seismic Stratigraphy II* (Eds Berg, O.R. & Woolverton, D.). Mem. Am. Assoc. Petrol. Geol. 39, pp. 93–115.

JURADO, H.J. (1989) *El Triasico del subsuelo de la Cuenca del Ebro.* PhD Thesis, Barcelona University, Vol. 1, p. 259.

LOPEZ-GOMEZ, J. (1987) Aspectos sedimentologicas y estratigraficos de las facies Buntsandstein y Muschelkalk entre Cueva del Hierra y Chelva (provincias de Cuenca y Valencia), Serrania de Cuenca, España. *Cuad. Geol. Iberica* **11**, 647–664.

MARKELLO, J.R. & READ, J.F. (1981) Carbonate ramp-to-deeper shale shelf transitions of an Upper Cambrian intrashelf basin, Nolichucky Formation, southwest Virginia Appalachians. *Sedimentology* **28**, 573–597.

MARQUEZ, A. (1983) *Bivalvos del Triásico medio del Sector Meridional de la Cordillera Ibérica y de los Catalánides.*

PhD Thesis, University of Valencia, p. 429.

MARCH, M. (1986) *Conodontos del Triásico Medio de los sectores meridionales de la Cordillera Ibérica y de los Catalánides.* MSc Thesis, Valencia University, p. 136.

MARZO, M. (1980) *El Buntsandstein de los Catalánides: Estratigrafia y procesos de sedimentación.* PhD Thesis, Barcelona University, p. 317.

MARZO, M. & CALVET, F. (1985) *Guia de la excursión al Triásico de los Catalánides.* II Coloquio de Estratigrafia y Paleogeografia del Pérmico y Triásico de España. La Seu D'Urgell, p. 175.

MESOLELLA, K.J., ROBINSON, J.D., McCORMICK, L.M. & ORMISTON, A.R. (1974) Cyclic deposition of Silurian carbonates and evaporites in Michigan basin. *Bull. Am. Assoc. Petrol. Geol.* **58**, 34–62.

MIALL, A.D. (1986) Eustatic sea-level changes interpreted from seismic stratigraphy: a critique of the methodology with particular reference to the North Sea Jurassic record. *Bull. Am. Assoc. Petrol. Geol.* **70**, 131–137.

NICHOLS, K.M. & SILBERLING, N.J. (1980) Eogenetic dolomitization in the pre-Tertiary of the Great Basin. In: *Concepts and Models of Dolomitization* (Eds Zenger, D.H., Dunham, J.B. & Ethington, R.L.). Spec. Publ. Soc. Econ. Paleont. Mineral. 28, pp. 237–246.

PEREZ-ARLUCEA, M. (1987) Distribución paleogeográfica de las unidades del Pérmico y del Triásico en el sector Molina de Aragón-Albarracín. *Cuad. Geol. Ibérica* **11**, 607–622.

PEREZ-ARLUCEA, M. & SOPEÑA, A. (1985) Estratigrafia del Pérmico y Triásico en el sector central de la Rama Castellana de la Cordillera Ibérica. *Estudios Geológicos* **41**, 207–222.

PITMAN, W.C. & ANDREWS, J.A. (1985) Subsidence and thermal history of small pull-apart basins. In: *Strike–slip Deformation, Basin Formation and Sedimentation* (Eds Biddle, K.T. & Christie-Blick, N.) Spec. Publ. Soc. Econ. Paleont. Mineral. 37, pp. 45–49.

ORTI, F. (1987) Aspectos sedimentólogicos de las evaporitas del Triásico y del Liásico inferior en el E de la Peninsula Ibérica. *Cuad. Geol. Ibérica* **11**, 837–858.

RAMON, X. (1985) *Estratigrafía y Sedimentología del Muschelkalk inferior del Dominio Montseny-Llobregat.* MSc Thesis, University of Barcelona, p. 100.

RAMON, X. & CALVET, F. (1987) Estratigrafía y sedimentología del Muschelkalk inferior del Dominio Montseny–Llobregat (Catalánides). *Estudios Geológicos* **43**, 471–487.

READ, J.F. (1985) Carbonate platform facies models. *Bull. Am. Assoc. Petrol. Geol.* **69**, 1–21.

SALAS, R. (1987) *El Malm i el Cretaci inferior entre el Massís de Garraf i la Serra D'Espadà. Anàlisi de conca.* PhD Thesis, Barcelona University, p. 345.

SALVANY, J.M. (1986) *El Keuper dels Catalànids: sedimentologia i petrologia.* MSc Thesis, Barcelona University, p. 128.

SALVANY, J.M. & ORTI, F. (1987) El Keuper de los Catalánides. *Cuad. Geol. Ibérica* **11**, 215–236.

SANTISTEBAN, C. & TABERNER, C. (1987) Depósitos evaporíticos de ambiente sabkha preservados como pseudomorfos en dolomita, en los materiales superiores de la facies Muschelkalk de la Sierra de Prades (Tarragona). *Cuad. Geol. Ibérica* **11**, 199–214.

SARG, J.F. (1988) Carbonate sequences stratigraphy. In:

Sea-level Changes: an Integrated Approach (Eds Wilgus et al.) Spec. Publ. Sec. Econ. Paleont. Mineral. 42, 155–181.

SCHMIDT, M. (1932) Uber die Ceratiten von Olesa bei Barcelona. *Bull. Inst. Cat. Hist. Nat.* **XXXII**, 195–222.

SCLATER, J.G. & CHRISTIE, P.A.F. (1980) Continental stretching: an explanation of the post–mid-Cretaceous subsidence of the central North Sea basin. *J. Geophys. Res.* **85**, 3711–3739.

SOLE, N., CALVET, F. & TORRENTO, L. (1987) Análisis palinológico del Triásico de los Catalánides (NE España). *Cuad. Geol. Ibérica* **11**, 237–254.

SOPEÑA, A. (1979) Estratigrafia del Pérmico y Triásico del nordeste de la provincia de Guadalajara. *Seminarios de Estratigrafia, serie Monográfica* **5**, 1–329.

THERIAULT, F. & HUTCHEON, I. (1987) Dolomitization and calcitization of the Devonian Grosmont Formation, Northern Alberta. *J. Sed. Petrol.* **57**, 955–966.

THORNE, J.A. & WATTS, A.B. (1989) Quantitative analysis of North Sea subsidence. *Bull. Am. Assoc. Petrol. Geol.* **73**, 86–116.

TUCKER, M.E. (1985) Shallow-marine carbonate facies and facies models. In: *Sedimentology: Recent Developments and Applied Aspects* (Eds Brenchley, P.J. & Williams, B.P.J.). Spec. Publ. Geol. Soc. Lond. 18, pp. 147–169.

TUCKER, M.E., CALVET, F. & MARSHALL, J.D. (1989) *Triassic (Upper Muschelkalk) reef-mud mounds, Catalan Basin, Spain: early marine diagenesis and dolomitization* (Abstract). 10th IAS Regional Meeting, Budapest, Hungary, pp. 241–242.

TUCKER, M.E. & WRIGHT, V.P. (1990) *Carbonate Sedimentology*. Blackwell Scientific Publications, Oxford, UK.

VAIL, P.R., HARDENBOL, J. & TODD, R.G. (1984) Jurassic unconformities, chronostratigraphy and sea-level changes from seismic stratigraphy and biostratigraphy. In: *Interregional Unconformities and Hydrocarbon Exploration* (Ed. Schlee, J.S.). Mem. Am. Assoc. Petrol. Geol. 36, pp. 129–144.

VAIL, P.R., MITCHUM, Jr., R.M., TODD, R.G., WIDMIER, J.M., THOMPSON III, S., SANGREE, J.B., BUBB, J.N. & HATLELID, W.G. (1977) *Seismic Stratigraphy and Global Changes of Sea Level.* (Ed. Payton, C.E.) Mem. Am. Assoc. Petrol. Geol 26, pp. 49–212.

VAIL, P.R., COLIN, J.P., JAN DU CHENE, R., KUCHLY, J., MEDIAVILLA, F. & TRIFILIEF, V. (1987) La stratigraphie séquentielle et son application aux corrélations chronostratigraphiques dans le Jurassique du bassin de Paris. *Bull. Soc. Géol. France* (8), **III**, 1301–1321.

VEGAS, R. (1975) Wrench (transcurrent) fault system of the south-western Iberian Peninsula, paleogeographic and morphostructural implications. *Geol. Runds.* **64**, 266–278.

VEGAS, R. & BANDA, E. (1982) Tectonic framework and Alpine evolution of the Iberian Peninsula. *Earth Evol. Sci.* **4**, 320–343.

VIA, L. (1987) Artrópodos fósiles Triásicos de Alcover-Montral. II. Limúlidos. *Cuad. Geol. Ibérica* **11**, 281–292.

VIA, L. & VILLALTA, J.F. (1966) Heterolímulus gadeai, nov., gen., nov., sp., repres. *Acta Geol. Hispánica* **1**, 9–11.

VIA, L. & VILLALTA, J.F. (1975) Restos de crustáceos decápodos en el Triásico de Montral-Alcover (Tarragona). *Bol. Inst. Geol. Min. Esp.* **86**, 485–497.

VIA, L. & CALZADA, S. (1987) Artrópodos fósiles Triásicos de Alcover-Montral. I. Insectos. *Cuad. Geol. Ibérica* **11**, 273–280.

VIA, L, VILLALTA, J.F. & ESTEBAN, M. (1977) Paleontología y paleoecología de los yacimientos fosilíferos del Muschelkalk superior entre Alcover y Montral (Montañas de Prades, Provincia de Tarragona). *Cuad. Geol. Ibérica* **4**, 247–256.

VIRGILI, C. (1958) El Triásico de los Catalánides. *Bol. Inst. Geol. Min. Esp.* **69**, 1–856.

VIRGILI, C. (1987) Problemática del Trías y Pérmico superior del Bloque Ibérico. *Cuad. Geol. Ibérica* **11**, 39–52.

VIRGILI, C., PAQUET, H. & MILLOT, G. (1974) Alterations du soubassement de la couverture permo-triassique en Espagne. *Bull. Groupe Franç. Argiles* **XXVI**, 277–285.

VIRGILI, C., SOPEÑA, A., RAMOS, A. & HERNANDO, S. (1977) Problemas de la cronoestratigrafia del Trías en España. *Cuad. Geol. Ibérica* **4**, 57–88.

VIRGILI, C., SOPEÑA, A., ARCHE, A., RAMOS, A. & HERNANDO, S. (1983) Some observations on the Triassic of the Iberian Peninsula. In: *Neue Beitrage zur Biostratigraphie der Tethys-Trias.* Schriftenreihe der Erdwissenschaftlichen Kommissionen 5, 287–294.

WILGUS, C.K. et al. (Eds) (1988) *Sea-level Changes—an Integrated Approach.* Spec. Publ. Soc. Econ. Paleont. Mineral. 42, pp. 407.

WRIGHT, V.P. (1986) Facies sequences on a carbonate ramp: the Carboniferous Limestone of South Wales. *Sedimentology* **33**, 221–241.

ZIEGLER, P.A. (1982) Triassic rifts and facies patterns in Western and central Europe. *Geol. Runds.* **71**, 747–772.

Spec. Publs int. Ass. Sediment. (1990) **9**, 109–144

Stages in the evolution of late Triassic and Jurassic carbonate platforms: the western margin of the Subalpine Basin (Ardèche, France)

S. ELMI

Centre des Sciences de la Terre, UA 11 CNRS, Centre de Paléontologie Stratigraphique et Paléoécologie, 27–43, Bd du 11 Novembre 1918 F-69622 Villeurbanne Cedex, France Programme Géologie Profonde de la France, Thème 11, Subsidence et Diagenesis

ABSTRACT

Comparative sedimentological, palaeontological and structural studies document different kinds of controls on the origin, development and termination of carbonate platforms and adjacent basins. Jurassic sedimentation in the Vivarais–Cévennes area of the French Subalpine Basin was controlled by local tectonics, Tethyan structural evolution, eustacy and climate. The environment changed through time from: (1) sandy alluvial plains; to (2) a shallow proximal platform; (3) distal platform; (4) segmented, small, narrow sub-basins (umbilics); and to (5) a slope of a continental margin. The evolution is illustrated by palinspastic reconstructions, isopach maps and facies analyses of both outcrop and well data.

The vertical arrangement of facies was different in the contiguous blocks where sedimentation was mainly controlled by local tectonics. In contrast, vertical successions of facies were similar when more regional controls were involved. The first type of facies sequence is contemporaneous with a stage of rifting. The second results from more global processes (seafloor spreading, sea-level variations and climatic changes).

At the end of the Jurassic, physiographic, chemical and probably climatic conditions changed as the area went from being segmented to a large and open basin. Afterwards, some particular facies such as iron ore, 'grumeleux' facies and Ammonitico Rosso, disappeared from the region, and from Tethys. Some of these changes could be related to the modification of oceanic water circulation.

INTRODUCTION

The western margin of the Subalpine Basin (southeastern France) has sustained only weak tectonic stresses since Mesozoic sedimentation (Fig. 1). The present distribution of facies and thicknesses is representative of the initial palaeogeography with the exception of some horizontal offsets along strike–slip faults which can be easily evaluated. Along this margin, the basin was the result of intense Triassic and Jurassic faulting related to the differential subsidence of the European basement. The margin bordering the southeastern Massif Central (Ardèche–Vivarais and Cévennes) provides valuable information for studying carbonate platforms and adjacent basins because the palaeostructural setting changed often and rapidly throughout the Mesozoic. The transition between basin and platform has

mostly been studied for a limited time interval. In the Alps and in their forelands, classic studies have dealt with the palinspastic reconstruction of late Jurassic palaeogeography, when oceanic crust was forming in the Internal Zones. These reconstructions allow striking comparison with models from the Recent (Winterer & Bosellini, 1981).

The peri-adriatic platforms and basins have a striking similarity, at least in geometry, to the Bahamas (D'Argenio *et al.*, 1975). However, these models are less appropriate for the earliest stages of platform evolution, especially for areas where siliciclastic sedimentation and carbonate accumulation were associated. Coarse terrigenous deposits are absent in the main part of the southern Tethyan Realm. Exceptions are known in western Algeria and eastern

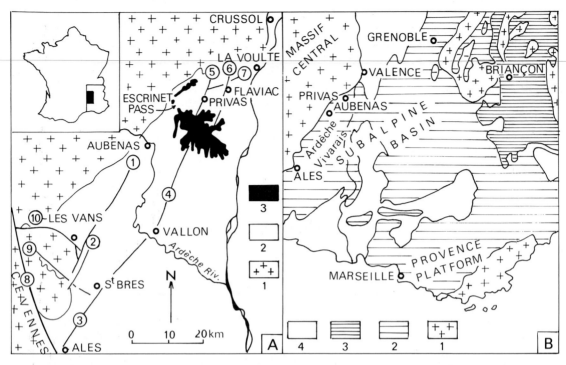

Fig. 1. General location of the study area. (A) The Vivarais Ardèche area along the southeast Massif Central. 1 = pre-Triassic basement, 2 = post-Variscan terrane, 3 = Miocene and Pliocene basalts. Faults (circled numbers): 1 = Païolive F., 2 = Gravières F., 3 = Alès F., 4 = Lagorce F., 5 = Privas F., 6 = Flaviac F., 7 = La Voulte F., 8 = Villefort F., 9 = Bordezac F., 10 = Orcières F. (B) Geological sketch of southeast France. 1 = pre-Triassic basement, 2 = Alpine External Zones, 3 = Alpine Internal Zones, 4 = Recent.

Morocco, where small basement ridges persisted well into the Jurassic (Lucas, 1942; Owodenko, 1946; Elmi, 1978, 1983a).

On the northern rim of Tethys, some large Variscan and pre-Variscan highlands remained prominent for a long time. The western margin of the French Subalpine Basin underwent a strong differential step-like subsidence between the old pre-Triassic basement and the basin. Facies and thickness changes are common, owing to differential subsidence controlled by both local tectonics and global events, such as the birth of the so-called Ligurian Ocean (Bernoulli & Lemoine, 1980; Winterer & Bosellini, 1981) or as general variations of sea-level, well known from previous authors (Haug, 1911; Roman, 1950).

The selected area (Ardèche 'Department' or Vivarais 'Province') has been intensively studied for more than a century and a half. During the last 25 years, detailed stratigraphical, sedimentological and structural research has been completed but no general paper has become available in English. More-over, many data are scattered in unpublished theses. The present paper attempts to synthesize these works as well as the author's own data. Detailed contributions on the Jurassic have been published by Elmi & Mouterde (1965), Elmi (1967), Atrops (1982), Elmi et al. (1984) and Elmi (1985c, 1987). A general presentation can be found in recent contributions by various authors in a volume edited by Enay & Mangold (1980) (see also Debrand-Passard, 1984; Curnelle & Dubois, 1986). The area has been selected as one of the eleven sites of the French Deep Continental Drilling Project ('Géologie profonde: Subsidence and Diagenesis of the Ardèche margin'). It has been pre-selected to be the site of the next well.

METHODOLOGY AND BASIC DATA

Biostratigraphy, lithostratigraphy and sequences

Biostratigraphic correlations have been established on detailed field sections and are based upon ammonite successions. The Ardèche area has furnished many reference sections for the standard Tethyan zonation (Hettangian, Callovian, Kimmeridgian, Tithonian). Microfaunas are of poor stratigraphic value with the exception of the Tithonian calpionellids. Correlations are particularly uncertain for the Sinemurian–Carixian. French names have been kept for the lithostratigraphic nomenclature. The word 'couches' (beds) is used for lithologically heterogeneous formations or members, e.g., alternating marl and limestone sequences. The term 'Series' is here used as a lithostratigraphic designation for related groups of formations depending on their thickness (Fig. 2), and thus the term differs in meaning from North American usage as a major time stratigraphic subdivision. Boundaries between sequences have been placed to highlight significant lithostratigraphic changes: unconformities (stratigraphic gaps due to erosion or to non-deposition or to both processes) or rapid vertical changes in facies. The major variations in the lower sequences are illustrated in Figs 3 and 4.

Sequence S0 coincides with the Rhaetian Croix Blanche Formation which rests uncomformably on the Ucel Variegated Formation (Carnian–Norian). S0 begins with the first appearance of thin laminated sandstones, quartzose dolosparites, black siltites and clays. Some channels are known near the positive areas (Les Vans). They are filled up by coarse sandstones and conglomerates reworked from the Ucel Formation. Oodolosparites and oosparites developed at greatest distance from the shoreline. The environment ranged from intertidal to shallow-subtidal.

The boundary between sequences S0 and S1 is drawn following the disappearance of the sandstones and associated black siltstones. The S1 sequence consists of three facies: basal micrites, median cross-bedded and rippled oosparites and an Upper Mytilid coquina. This evolution documents the change from a protected lagoon to a more open shelf. The S1 sequence has been recognized all along the Ardèche margin with the exception of the Arénier Pass where it has been subsequently eroded (pebbles reworked into Middle and Upper Liassic limestones). In consequence, it is established that the S1 corresponds to the largest extent of the early Jurassic transgression. The next continuous beds are largely younger (Gracilis Zone of the early Callovian).

After the end of S1, erosion and non-deposition prevailed on the ridges. In the basins, the sedimentation was probably continuous but changed abruptly to alternating marls and limestones (lower member of the Calcaires noduleux cendrés Formation or ash-grey nodular limestones, S2). Abundant ammonites (*Psiloceratinae*) were associated with bivalves (*Cardinia, Mactromya*). The facies becomes more calcareous towards the top of the member and several small reef patches have been recognized. The top can be marked by a bored hard-ground or, more commonly, by bioclastic mainly crinoidal, limestones.

The S3 sequence illustrates the evolution from hemipelagic marls and limestones to biocalcarenites and quartzose limestones. The lower part (upper member of the Calcaires noduleux cendrés, S3a) consists of grey marls alternating with wavy-bedded micrites with a mudstone texture. Ammonites are abundant in the lower part but become scarce upward. They document a relatively deep platform to basinal environment. South of Les Vans, the S3a member consists of nodular dolomites without cephalopods.

In the Aubenas sub-basin, the overlying 'Couches du Bosc' Formation (lower part of S3b) consists of micrites with few bioclasts rhythmically alternating with wavy-bedded limestones similar to those of the S3a subsequence. The following 'Couches du Château d'Aubenas' Formation (upper part of S3b) is made up of crinoidal packstones and grainstones with cherts and sandstones. *Gryphaea* coquinas occur rarely near the base. Only a few ammonites have been collected; they are more common in the Privas sub-basin. The S3 sequence ends with the 'Couches de la Garenne' Formation (S3c), siimilar to beds below but the sandstones are coarser and more immature. In the Privas area, the top of the equivalent formation was strongly eroded before the late Domerian. The S3 sequence ranges from the Middle Hettangian to the Lower Pliensbachian (= Carixian).

Along the outcrop, the S4 sequence corresponds to the highly variable and commonly lenticular Vaumalle Formation consisting of crinoidal packstones rich in brachiopods, belemnites and ammonites (late Domerian, early Spinatum Zone). This formation is commonly coarsely detrital. Decimetric pebbles of metamorphic quartzites must have been reworked from the Triassic conglomerates as no Variscan rocks were exhumed at that time. Other pebbles are made of Hettangian limestones and dolomites. The Vaumalle Formation rests on various beds ranging in age from the Triassic (Ucel Formation) to the Carixian (Couches de la Garenne). In subsiding zones, the Domerian is represented by

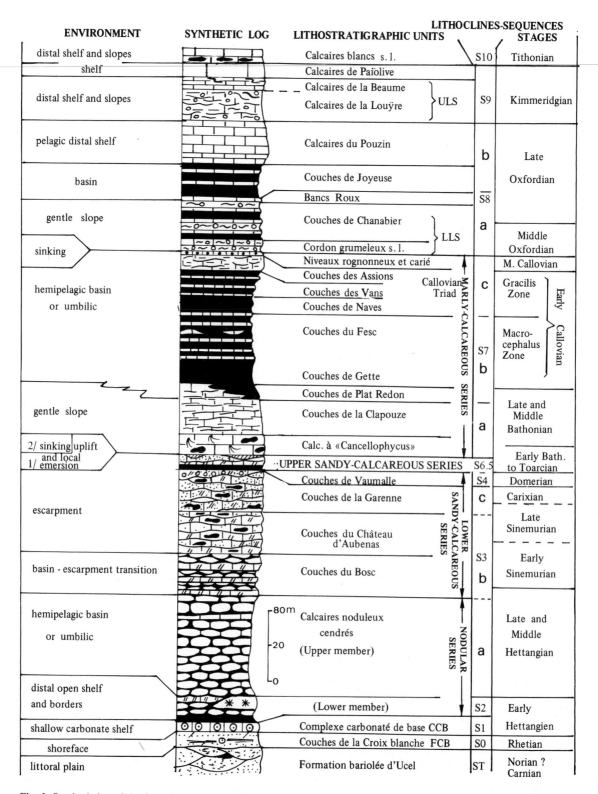

Fig. 2. Synthetic log of the Jurassic Sequence in the Aubenas Sub-basin. Maximum thicknesses have been selected from several measured profiles. Adapted and complemented from Elmi *et al.* (1984).

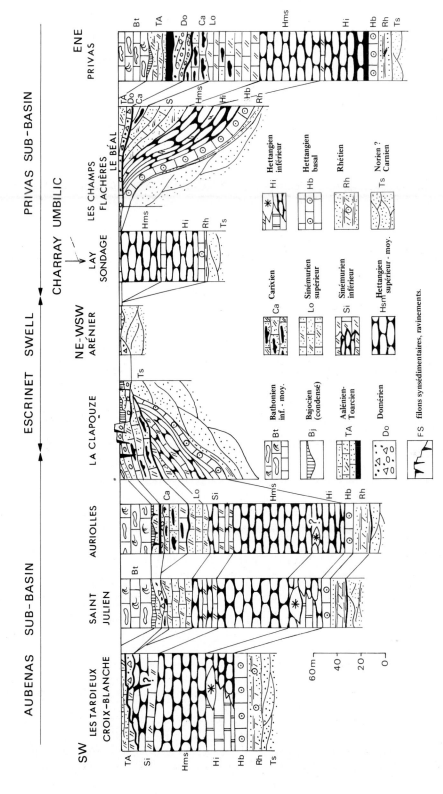

Fig. 3. Profiles across the Escrinet–Arénier Pass. Abbreviations: Bt = early and middle Bathonian, Bj = condensed Bajocian, TA = Toarcian–Aalenian, Do = Domerian, Ca = Carixian, Lo = late Sinemurian, Si = early Sinemurian, Hsm = middle and late Hettangian, Hi = early Hettangian, Hb = lower levels of the Hettangian (CCB); Rh = Rhaetian, TS = Carnian–Norian?, FS = synsedimentary neptunian dykes. Adapted from Elmi *et al.* (1987).

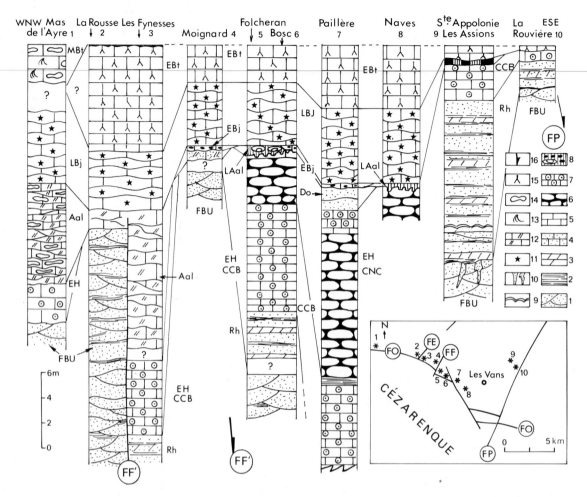

Fig. 4. Profiles along the northern branch of the Orcières Fault and across the Eynesses (FE, FE′), Folcheran (FF, FF′) and Paiolive (FP) Faults. The top of the profiles is drawn below the appearance of the marls: early Callovian east of Bosc, late Bathonian to the west.

Stratigraphical abbreviations: FBU = Carnian–Norian? (Ucel Variegated Formation), Rh = Rhaetian (Croix Blanche Formation), EH = early Hettangian (CCB = Complexe Carbonaté de Base – CNC = Lower Member of the 'Calcaires noduleux cendrés' Formation), Do = Domerian, (L)Aal = (Lower) Aalenian, EBj = early Bajocian, LBj = late Bajocian, EBt = early Bathonian, MBt = middle Bathonian,.

Facies: 1 = channelized sandstones, 2 = clays and siltstones, 3 = dolomites, 4 = quartzitic limestones, 5 = limestones, 6 = nodular limestones, 7 = oolitic limestones, 8 = ferruginous beds and breccias (commonly oolitic or oncolitic), 9 = stromatolites, 10 = giant mud-cracks filled by sandstones, 11 = *Isocrinus* coquinas, 12 = bioclastic limestones, 13 = *Zoophycos*, 14 = cherts, 15 = sponge-spicules, 16 (and circled letters) = synsedimentary faults (see also Fig. 9).

thick marls and micaceous clays, becoming more calcerous upward. The change from the coarse Vaumalle Formation to the basinal marls and clays occurs sharply across some faults (de Brun, 1926; Bourseau & Elmi, 1981).

The S5 sequence is highly variable, thin and commonly missing along the outcrop band. When it is complete, bituminous black paper shales (=

'Schistes-carton') occur at the base. They are overlain by black nodular limestones, passing laterally and vertically to oolitic phosphatic limestones and to oolitic iron ore. The upper part consists of a quartzose bioclastic limestone capped by local breccias and by a widespread hardground. Ammonites are abundant and associated with a rich benthic fauna with the exception of the 'Schistes carton'. The

S5 sequence ranges from the early Toarcian (Serpentinus Zone) to the early Aalenian (Opalinum Zone) or to the beginning of the middle Aalenian (Opalinoides Subzone).

The S6 sequence is made of lenticular beds consisting of oolitic, conglomeratic, commonly ferruginous or phosphatic limestones. The upper beds (early Bathonian) are more regular but their facies remain varied: pelmicrites with 'filaments' (*Bositra* shells) and sponge spicules, oolitic iron ore, oncolitic limestones, synsedimentary breccias. The maximum vertical range of the S6 sequence is from the middle Aalenian to the early Bathonian but its boundaries are heterochronous. On ridges, the top is characterized by phosphatic nodules, hardgrounds or ferruginous layers. On the contrary, the change from S6 to S7 is progressive in the basins. The S5 and S6 sequences are grouped in the Upper Sandy–Calcareous Series owing to the abundance of coarse terrigenous material.

The S7 sequence is made of the 'Marly–Calcareous Series' ranging from the middle (locally early) Bathonian to the middle Callovian (Fig. 5). The end is always marked by a strong unconformity (condensation, erosion surface, non-deposition) following a decrease of the marls and the appearance of glauconite and locally iron ores (La Voulte). The bed-by-bed relation between limestones and marls allow one to define several formations which are summarized in Fig. 5. The lower units are discontinuous until the earliest Callovian (Macrocephalus Zone). The upper part is persistent and forms the 'Callovian Triad' (end of the early Callovian–Gracilis Zone-to the middle Callovian–early Coronatum Zone): thick pyritic marls (Naves Formation), alternating marls and limestones (Les Vans Formation), marly limestones (Les Assions Formation) capped by thin pyritic and glauconitic limestones. The limestones consist of hemipelagic nannomicrites crowded with: *Bositra* 'filaments' associated with sponge spicules,

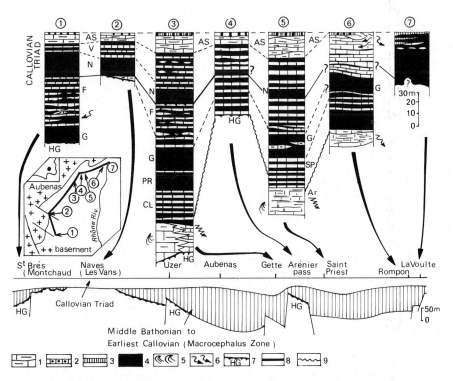

Fig. 5. 'Marly–Calcareous Series' and their changes along the Ardèche margin of the Subalpine Basin. 1 = marly limestones, 2 = glauconitic beds, 3 = ferruginous level (La Voulte iron ore), 4 = marls, 5 = *Zoophycos*, 6 = slumps and slides, 7 = hardground, 8 = base of the Callovian Triad, 9 = base of the upper Bathonian beds.

Abbreviations: Ar = Couches d'Argevillières, AS = C. des Assions, CL = C. de la Clapouze, F = C. du Fesc, G = C. de Gette, N = C. de Naves, PR = C. de Plat-Redon, SP = C. de Saint-Priest, HG = hardground. Adapted from Elmi (1984).

radiolarians and some coccoliths. Ammonites are common but their communities and associations depend strongly on their palaeogeographic origin (Elmi, 1971, 1985a).

There are no continuous Upper Callovian and Lower Oxfordian deposits along the slope. The early Oxfordian is only represented by a centimetric level made of pelagic micrite which is more or less completely eroded before the deposition of the following beds. The S8 sequence is made of nodular marls and limestones ('lower lumpy series' = S8a) changing upward to alternating marls−limestones and to limestones. The missing levels appear sharply along the course of some major faults. The Upper Callovian−Lower Oxfordian deposits are represented by black shales ('Terres Noires') developed in the down-warped areas.

The S9 sequence (Kimmeridgian) shows the evolution from nodular marls and limestones ('upper lumpy series' = ULS), rich in ammonites, to massive limestones poor in cephalopods. The S10 sequence (Tithonian) is made of nodular limestones passing into massive beds.

Figure 6 illustrates the changes presented by the Hettangian to Middle Oxfordian terranes. The difference between a highly variable lower part and a more monotonous upper one is well seen. Facies and thicknesses change rapidly and strongly inside the sequences S0 to S7, documenting the position of the persistent shoals with regard to the subsiding sub-basins. From the early Callovian upward, the changes are less spectacular.

Palaeotectonics

Palinspastic profiles (Fig. 7) have been selected in order to give a coherent summary. They have been traced along the present outcrop. In consequence, some palaeostructures are tangentially cut (Païolive swell and Privas Sub-basin). A method to make clear palaeotectonic influences is to compare the vertical sequence of facies and their horizontal geometry. The standard succession ranges from low-energy marls and carbonates (basin or distal shelf) to high-energy platform limestones; this corresponds to large-scale basin filling or to small-scale shallowing-upward cycles (parasequences). According to Walther's law (Lombard, 1956; see Ferry & Rubino, 1987, for a modern review), this vertical evolution is similar to the horizontal facies change from basin to platform margin. In the Ardèche area, the horizontal development is commonly different from the vertical succession. There are hiatuses or breaks occurring across strike−slip faults with minor horizontal off-sets of a few 100 m. Such interruptions of the logical lateral development of facies can be related to con-temporaneous movement of the faults, which altered the gradual transition from the platform to the basin.

A varied physiography with steep slopes, narrow furrows or swells, and rapid changes from a deep to shallow sea are evident during deposition of the sandy calcareous sequences ranging from early Sinemurian to early Bathonian (greatest extent; Fig. 2). Influx of coarse-grained quartz and feldspar sand

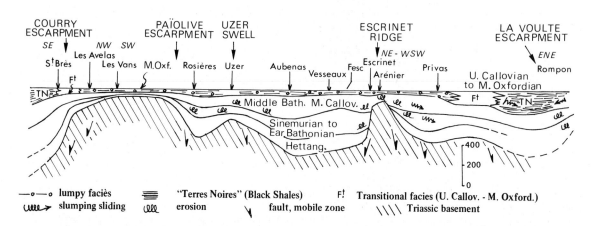

Fig. 6. Diagrammatic cross-section of the Hettangian−Oxfordian terranes along the Ardèche Margin of the Subalpine Basin. The present outcrops are transverse to the Jurassic main trend ('Cévenol' trend) and some of the structures are obliquely cut. Adapted from Elmi (1984).

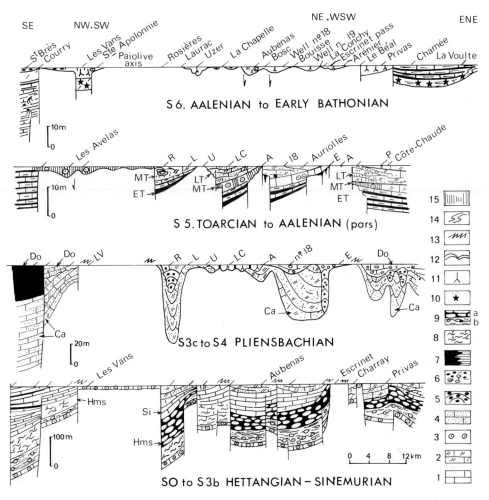

Fig. 7. Palinspastic profiles along the Ardèche Margin of the Subalpine Basin from the early Liassic to the early Bathonian. 1 = fine-grained micrites; 2 = skeletal limestones; 3 = oolitic limestones; 4 = sandy and quartzitic limestones; 5 = coarsely detrital limestones; 6 = coarse, terrigenous clastics and conglomerates; 7 = marls; 8 = nodular limestones without cephalopods; 9 = nodular limestones with cephalopods; a- marly, b- mostly calcareous; 10 = *Isocrinus* breccias and grainstones; 11 = spicules (sponges); 12 = stromatolites; 13 = erosion; 14 = slumps and slides; 15 = ferruginous beds and iron ore.

and quartzite pebbles, reworked from the Triassic sandstones, and of carbonate pebbles sourced from lowest Liassic beds (mainly Hettangian) are widespread along the Ardèche margin. Carbonate matrix as well as bioclasts are genetically independent of the coarse clastic material, especially in regard to transport, as evidenced by calcimetric, granulometric and morphoscopic studies (Talbi, 1984; Elmi *et al.*, 1984; Galien, 1985). Quartz grains or pebbles are immature and have kept their fluvial or littoral characteristics even in the marine environment

where they have been buried. Marine transportation has been fast and short without any modification of the initial fluvial granulometric pattern. For example, the coarsest and more immature quartz material has been found in micritic laminated limestones interbedded in black shales of early Toarcian age. Some of these influxes can be storm-related, but owing to the general environment, they are more commonly related to the existence of steep slopes extending from emergent or very shallow swells, to more-or-less narrow, deep, basins (here called umbilics). There was no flat shelf between the emergent zones and the bottom of the sedimentary basins either on shallow slopes or on basin-borders. Resedimentation took place at the toe of escarpments. Such disturbed successions have been called 'escarpment sequences' (Elmi, 1984) and developed along steep slopes washed by mass-flows. 'Escarpment sequences' are characterized by sliding, slumping of neritic clasts and calcareous or silty turbidites. They can be compared with the steep slope of a tilted block. On more gradual slopes (gentle-ramp side of a block, edge of a platform), hemipelagic marly limestones with *Zoophycos* were deposited.

'Hinge-sequences' developed on top of narrow swells, with multiple internal disconformities, especially during the Domerian–Bathonian. Platforms are usually larger than the swells which are related to the top of tilted blocks (Elmi, 1984). Sedimentary or stratigraphic gaps are associated with high-energy environments: sediments were swept towards the basin or toe of the escarpments. It is striking that, in such high-to-medium energy environments, the rarely preserved lenticular beds consist of micrites with mudstone to packstone textures whereas grainstones are missing. Sediments can only have been trapped during short-lived low-energy times and in small temporarily sheltered areas (of decimetre- to metre-scale). The general depositional setting ranged from peritidal to pelagic, on relatively deep but prominent seamounts. Comparable settings are known during the differentiation stages of several Tethyan basins as, for example, the Italian Central Apennines (Bernoulli, 1971), the Subbetic Ranges of Spain (Seyfried, 1981; Vera, 1984) or the Tlemcenian Domain of western Algeria (Elmi, 1981, 1983a; Alméras & Elmi, 1984). Sedimentation was always highly varied, condensed and/or discontinuous. These conditions are related to the steepness of the slopes at platform borders where they cross the narrow hinge lines. Ferruginous crusts with subordinate, lenticular, ferruginous oolitic beds are common, as well as hardgrounds and firmgrounds.

The vertical change from an escarpment environment to a hinge, or the reverse, can in places be related to general movements of sea-level as during the middle and late Toarcian. However, eustatic sea-level changes have general, or at least regional, effects. If the effects are diverse, and even opposite, from one block to another, one can assume that tectonic controls were also involved.

A comparison between the Crussol outcrops (in the northernmost part of the study area) and the Rosières well (farther south) establishes clearly that differing tectonic behaviour caused the contrasting stratigraphic evolution (Fig. 8). At Crussol a hinge zone existed from the early Liassic to the early Bathonian (gaps can span the Rhaetian–Domerian interval); then middle Bathonian sinking occurred and an escarpment and pelagic basin existed from the Bathonian to the earliest late Oxfordian.

At Rosières, subsidence was strong in the Hettangian and an escarpment formed during the following Sinemurian–Pliensbachian times; a hinge or swell existed from the middle Toarcian to the middle Bathonian (with gaps); pelagic outer shelf to basin occured from the late Bathonian to the middle Callovian; a gap formed with missing Upper Callovian and Lower Oxfordian deposits; and a gentle slope existed during the middle Oxfordian. Conditions became similar to those of Crussol only during the late Oxfordian.

Hinges were commonly separated by distances of a few kilometers and they formed a multidirectional network, delimiting narrow subsiding zones which could deepen strongly by comparison with the surrounding swells (Fig. 7). Such locations are called here 'umbilics' (Elmi, 1981, 1985a; Alméras & Elmi, 1984; Elmi & Ameur, 1986). If the deep zones were elongated, they can be called 'gutters'.

The best palaeotectonic markers are shallow-water deposits or alluvial fans; their change in thickness can only be related to sedimentary progradation and to tectonic controls because bottom-level differences were slight and can be ruled out. Carnian–Norian fluvial sands and Rhaetian–earliest Hettangian shoreline and shallow-marine deposits accumulated along a border recording the local effects of global sea-level changes (late Triassic regression, early Jurassic transgression). These beds thicken rapidly across certain faults. Triassic sandstones (Formation Bariolée d'Ucel) thicken from 5 to 7 m across the les Eynesses fault, a subsidiary of the La Rousse–Planzolles lineament (Fig. 1A, no. 1 and Figs 4 & 9). The sense of such abrupt lateral thickness changes through time, recording modifications of the overall tectonic regime (Colongo et al., 1979; Elmi, 1984).

During accumulation of the Ucel Formation, thickening toward the basin (eastward) happens regularly in each sub-basin. During the Rhaetian, the orientation of the sediment thickening changed and became mainly westward, occurring in the direction of the margin. This evolution is well seen in areas where profiles can be drawn across the margin, such as along the down-dropped northern structural

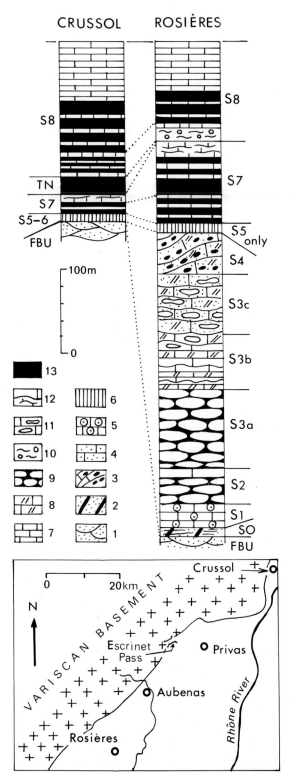

platform segment of the Orcières fault (Fig. 1A, no. 10). On each segment, the beds thicken westward, which is in the opposite direction from the general mean thickening which is eastwards in the direction of the basin centre (see Figs 4 & 9). This is a consequence of tilt-block tectonics, which occurred from the Rhaetian until the late Bathonian.

Palaeobathymetry

Bathymetric data are of prime importance to estimate the amounts of physiographic differences. Many models have been proposed but they are generally suited for limited examples, and generalizations are often unrealistic; one of the best examples of these difficulties is given by the controversy about the depositional depth of the Tethyan 'Ammonitico Rosso' (Farinacci & Elmi, 1981). Similar features such as stromatolitic crusts and nodular bedding may have developed at very different depths (Sturani, 1971). Algal oncolites may have originated in shallow water, but there are similar 'cryptalgal' (bacterial) crusts which occur below the photic zone (Massari, 1979; Dromart & Elmi, 1986). Consequently, relative estimations are often more useful and reliable than rough absolute values.

Estimations used for establishing the palaeostructural and palaeobathymetric reconstructions have been made conservative by using the smallest depth-values in order to avoid exaggerating differences of bottom levels, which could lead to a misinterpretation of the amount of tectonic influence on sedimentation (Figs 7 & 10). Most importance has been given to faunal associations. The deepest deposits are black shales (Terres Noires) of late Callovian–early Oxfordian age which were estimated to have been deposited in water of > 600 m depth, owing to the fossil record (Elmi, 1983b, 1985b; Dromart, 1986). The foraminifera have a marked affinity with the 'abyssal–oceanic' domain: *Epistomina, Globuligerina, Rhizammina?, Glomospirella, Bigerina.*

Fig. 8. Comparison between the stratigraphic columns of Crussol and Rosières. S0 to S10 = sequences as in Fig. 2. 1 = channellized sandstones, 2 = dolomites, sandstones and siltstones, 3 = coarse quartz-pebbles reworked in marine limestone, 4 = quartzitic limestones, 5 = oolitic limestones, 6 = condensed neritic limestones (commonly ferruginous), 7 = fine grained limestones, 8 = bioclastic limestones, 9 = nodular limestones (wavy bedding), 10 = lumpy limestones and marls, 11 = cherty limestones, 12 = marly limestones, 13 = marls.

Abbreviations: TN = Terres Noires (black shales), FBU = Ucel Variegated Formation.

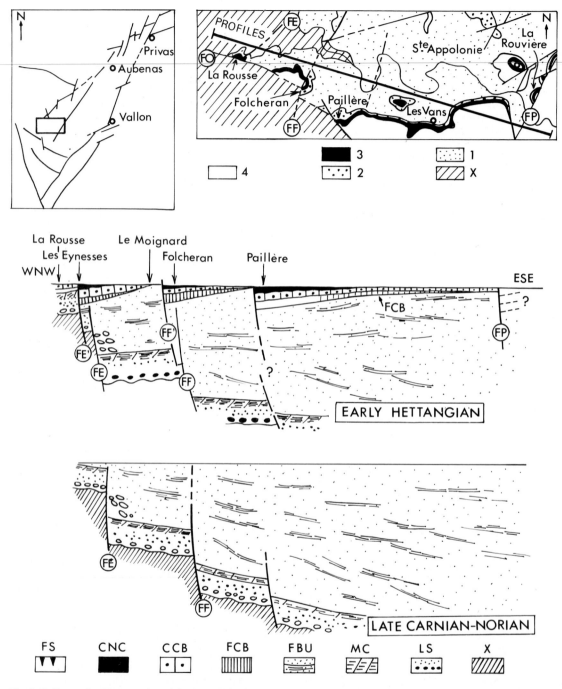

Fig. 9. Palinspastic reconstruction of the latest Triassic and early Jurassic terranes along the Orcières Fault. The change in dynamic style is obvious between the Carnian−Norian (normal faults: FE and FF) and the Rhaetian (antithetic faults: FE′ and FF′). The principal measured sections are located on the map drawn on Fig. 6.

Map: X = metamorphic schists, 1 = pre-Rhaetian Triassic, 2 = Rhaetian and early Hettangian, 3 = reduced Domerian to middle Bathonian beds, 4 = late Bathonian (west) or early Callovian (east) to late Jurassic:

Profiles: X = metamorphic schists, LS = Lower sandstones (Ladinian), MC = middle carbonates (early Carnian), FBU = Ucel Variegated Formation (Carnian−Norian?), FCB = Croix Blanche Formation (Rhaetian), CCB = Complete Carbonaté de Base (éarliest Hettangian), CNC = Calcaires noduleux cendrés Formation (early Hettangian), FS = giant mud-cracks filled by sandstones. Adapted from Colongo *et al.* (1979) and Elmi (1984).

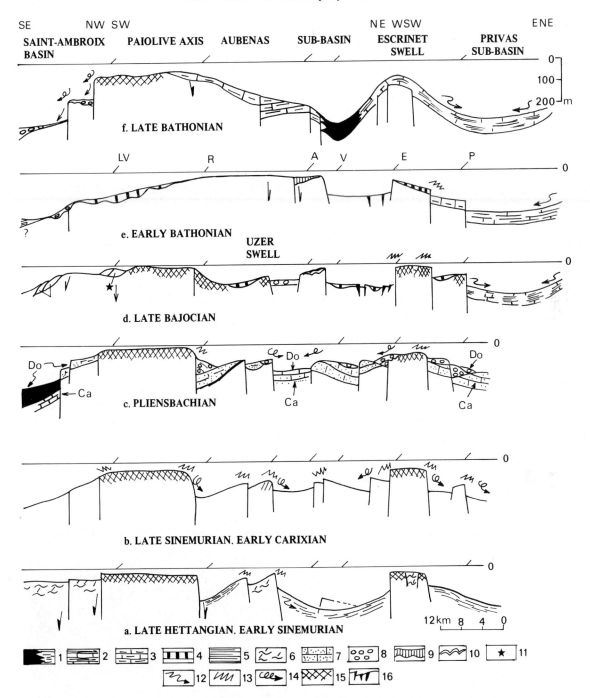

Fig. 10. Palaeobathymetric and palaeostructural evolution of the Ardèche from the Hettangian to the late Bathonian. In Fig. 10a, the tilt block is situated westward (Luol High = lineament TL). Main localities are shown on Fig. 9: A = Aubenas, E = Escrinet Pass, LV = Les Vans, P = Privas, R = Rosières, V = Vesseaux, Ca = Carixian, (early Pliensbachian), Do = Domerian (late Pliensbachian). 1 = alternating marls, 2 = limestones, 3 = marly limestones, 4 = pelagic limestones, 5 = open-sea deposits (Hettangian), 6 = protected deposits (mottled limestones), 7 = sandy limestones, 8 = coarse terrigenous clastics and conglomerate, 9 = oolitic iron ore, 10 = ferruginous stromatolite, 11 = *Isocrinus* coquinas, 12 = turbidites and slumps, 13 = erosion, 14 = sediment winnowing, 15 = non-depositional gap and highly condensed levels, 16 = neptunian dykes.

The crinoids (*Balanocrinus*) seem to have been bound to 'bathyal' environments. Ammonites can be abundant.

Other deep marly deposits, such as the Upper Bathonian 'Couches de Plat Redon' or the Lower Callovian 'Couches de Naves', contain a rich and varied ammonite fauna associated with abundant planktonic bivalves (*Bositra*) but benthic molluscs and foraminifera are scarce. A range within the deeper shelf or the proximal basin, commonly in an umbilic-like physiography, has been accepted for this facies (200–400 m; Elmi, 1985a).

Alternating marls and fine-grained micrites, containing *Bositra*-coquinas, are interpreted as deep outer-shelf deposits. Similar palaeobathymetry and environments can be accepted for the nodular-like marls and limestones of the Hettangian (outer shelf and basin) as documented by the ammonite fauna. However, in the south of the study area (La Cèze Valley), the pseudonodular beds are commonly dolomitized and devoid of cephalopods; they are interpreted to have been deposited in the deeper parts of a limited or protected shelf. Medium- to high-energy deposits (skeletal limestones, sandy limestones, ferruginous or phosphatic oolitic beds) are widespread and give indications of shallow outer-shelf environments. Stratigraphic gaps occurred on highs or at the margin of shoals, in neritic sequences reflecting high-to-moderate energy. In the basinal or outer platform setting, gaps are biostratigraphically documented within pelagic or hemipelagic suites or at the top of pelagic lenticular beds.

The palaeobathymetric estimations used for the construction of the profiles given in Fig. 10 have been tested by Brunet (1985) who has established a model of the amount of distension sustained by the Privas, Aubenas and Saint-Ambroix Sub-basins. Application of Royden's model (1982) allows the calculation of theoretical subsidence curves which are consistent with those obtained from evaluated depths.

Main facies

Jurassic lithologies of the Ardèche margin vary largely as a result of the changing environments especially during Liassic and middle Jurassic times. The use of the classic nomenclature is inappropriate for that time because the physiography is highly differentiated. Seven groups of facies types are distinguished.

Continental facies

Carnian–Norian deposits consist of sandstones, siltstones and clays ('Formation Bariolée d'Ucel'. Variegated clays and siltstones were related either to abandoned meanders of rivers or to episodic, probably seasonal, flooding. Caliche dolocretes and crusts, dolomitic pillars and replacement of silica by dolomite developed during emergence under subaerial weathering (Fig. 11). Dolomicritic horizons developed along the contact between clays and sandstones in the lower part of a phreatic lens of magnesium-rich water (Spy-Anderson, 1980, 1981). The environment is interpreted to have been a bifurcating deltaic braided stream system. Towards the south (La Cèze Valley) and east (Villeneuve-de-Berg borehole), the sequence changes progressively into cyclic sandstones, variegated clays and evaporites of a sabkha environment. The tectonically controlled subsidence was largely balanced by terrigenous input.

Shallow carbonate shelf

Four groups of facies have been recognized in the Rhaetian 'Couches de la Croix Blanche' (Martin, 1985; Avocat, pers. comm.): marginal mud flat, mixed flat (mud and sand), sand flat with flaser-bedded sandstones, carbonate flat (micrites, oomicrites and oosparites).

Shallow inner-shelf environments existed also during the early Hettangian and are represented by

Fig. 11. Pedogenetic dolocretes and pillars (roots?) developed through replacement of clay and quartz by dolomite. Ucel Formation, late Triassic. Terraube near Les Vans.

lagoonal micrites, oomicrosparites and oosparites, and by protected marine facies (heavily bioturbated, mottled limestones), and pelmicrites. At the end of early Hettangian time, small reef-patches developed underlain by mottled protected limestones or by spicule-rich, open-marine limestones. *Thecosmilia* bioherms are surrounded by bedded *Madreporaria* limestones deposited in protected zones. Bioclastic limestones with crinoids, echinoids, bivalves and gastropods developed basinward of the bioherms. Towards the lagoon, the bioclastic limestones change to ostreid coquinas or to *Favreina*-bearing micrites (Fig. 12). These facies are related to the eustatic initial transgression (Fig. 2; S1) and to the following stages of tilting (S2).

Longshore and swell

Condensed beds and lenticular levels are common in the Upper Sandy–Calcareous Series (Fig. 2; Toarcian to Bathonian). They have mainly packstone or wackestone textures with abundant neritic bioclasts and are commonly ferruginous. Locally, they are crowded with ammonite shells. Two kinds of condensed beds have been differentiated:

1 Benthic communities and oxidizing conditions are indicative of moderately shallow bottoms sustaining brief increases of energy (wave base, bottom currents). Waves, storms and currents were active, winnowing most of the sediments. Stromatolites,

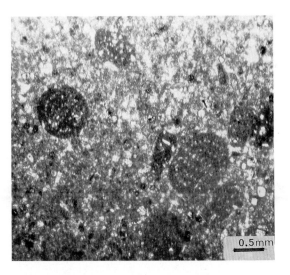

Fig. 12. Biodolomicrosparite. *Favreina* pellets in cross-section. Top of the Lower Member, Calcaires noduleux cendrés Formation (S2), early Hettangian. Soudournas near Aubenas. Millimetre-scale. All thin sections, natural light.

various sized oncolites (from millimetric coated grains to decimetric oncolites or 'snuff-boxes' and micritic and radial oolites are associated (Fig. 13). These rocks are commonly ferruginous or phosphatic. Some bear desiccation cracks. Bioturbation, bioerosion and desiccation have, in some

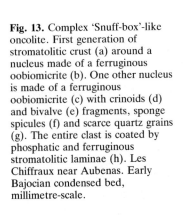

Fig. 13. Complex 'Snuff-box'-like oncolite. First generation of stromatolitic crust (a) around a nucleus made of a ferruginous oobiomicrite (b). One other nucleus is made of a ferruginous oobiomicrite (c) with crinoids (d) and bivalve (e) fragments, sponge spicules (f) and scarce quartz grains (g). The entire clast is coated by phosphatic and ferruginous stromatolitic laminae (h). Les Chiffraux near Aubenas. Early Bajocian condensed bed, millimetre-scale.

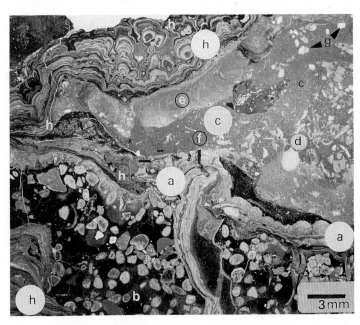

localities, disrupted thin beds facilitating their
erosion by currents and storms. The resulting
unconformities can be very sharp as at Le Folcheran,
near Les Vans, where light-grey micrites of early
Hettangian age are directly overlain by Upper
Aalenian ferruginous breccias (Fig. 14). These
features are related to steep coasts and hinge environments without barriers or a shelf. They appeared when tilt blocks were well differentiated.

2 If the swells were in deeper water, the facies
consist of biomicritic wackestones bearing abundant
crinoids and sponge spicules associated with phosphatic or ferruginous oolites and oncolites. These
facies are indicative of 'hinge sequences' bordered
by steep slopes and a lack of larger perilittoral
shelves.

Proximal external, open shelf

Deposits consist of skeletal limestones oolitic and
crinoidal packstones and grainstones. Iron and phosphate are common and iron ore occurs in the middle
Toarcian of the Privas Sub-basin. In the Les Vans
area, *Isocrinus* grainstones form prograding bodies
with current laminae and ripples (Fig. 15).

Distal shelf or sheltered areas of the shelf

Facies range from sponge-bearing micrites to black
shales ('schistes-carton', Fig. 16). Vicinity to the
shore or to the top of the swell is always noticeable.
Storm or gravity transported quartz grains or pebbles

Fig. 15. Reduced Bajocian–Bathonian sequence (6 m).
The lower part (B) consists of wavy laminated *Isocrinus*-
grainstones (late Bajocian, Subfurcatum Zone). The upper
part (A) is made of irregular beds of wackestones (sponge
spicules, *Bositra*-filaments, oncolites) (early Bathonian,
Zigzag Zone). Between the two: gap of the two topmost
zones of the Bajocian. The *Isocrinus* limestones lie on a
thin lenticular Aalenian ferruginous limestone filling the
irregularities of the Hettangian surface (see also Fig. 14).
The Lower Bathonian beds are directly covered by the
Callovian Triad (Gracilis Zone). Naves near Les Vans.

are abundant, making a transition to the succeeding
environment. In Upper Jurassic strata, well-bedded,
fine hemipelagic micrites can be related to the outer
shelf but in a more uniform environment than during
early and middle Jurassic.

Escarpment or steep slopes

These were developed either between the swell and
the shelf or, without transition, between the swell
and basin. Coarse clastics (quartz, metamorphic and

Fig. 14. Sharp unconformity between Hettangian light-
grey micrite and Upper Aalenian ferruginous breccias,
containing badly sorted Hettangian pebbles. Folcheran.
Overall thickness: 0.50 m.

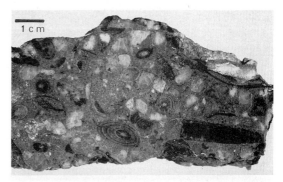

Fig. 17. Synsedimentary breccia. Gravity redeposited quartz grains in a biomicritic limestone with belemnites. Early Domerian. Les Tardieux near La Chapelle-sous-Aubenas.

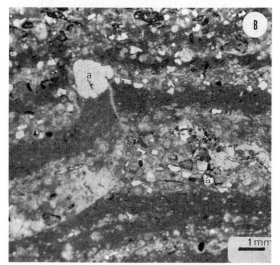

Fig. 16. Laminated limestones (arrow) and black shales, 'schistes−carton' early Toarcian (Serpentinus Zone). Auriolles near Vesseaux. (A) Outcrop. (B) Alternating laminae of silty sparites and micrites with quartz grains (a). Millimetre-scale.

calcareous pebbles) were reworked and buried in sediments deposited in medium to low water energy. In Fig. 17, angular quartz grains, up to 5 mm, are irregularly scattered in a micrite bearing belemnites, ammonite and gastropod phosphate casts. These features indicate significant input of coarse material. This particular example is from one of the rare remnants of Lower Domerian rocks which are preserved in a small synsedimentary graben. Submarine fans and cones were also developed in the Upper Domerian 'Couches de Vaumalle' made of coarse

pebbles embedded in bioclastic packstones and grainstones. They developed at the toe of fault-controlled escarpments. Open fissures and cracks cutting laterally continuous strata of the Marly−Calcareous Series testify to the role of brittle-style tectonics. Millimetric poorly-sorted quartz grains are common in the mudstone to packstone beds ranging from early Sinemurian to late Bathonian age (Fig. 18). More distal environments are indicated by neritic bioclasts resedimented in fine-grained hemipelagic to pelagic limestones (e.g., 'Couches du Bosc', early Sinemurian). Small submarine fans and cones were locally bound to fault-controlled escarpments ('Couches de Vaumalle', Domerian−late Pliensbachian).

Slopes and pelagic facies

Gentle slopes and transitional zones are indicated by slumps, slides and slope-breccias. Facies are fine-grained, somewhat marly limestones. They can be strongly bioturbated and *Zoophycos* (= 'Cancellophycus') is common (from Bajocian to middle Oxfordian). During late Bajocian and early Bathonian times, *Isocrinus* breccias were developed along the slopes bordering the La Voulte escarpment and the Orcières lineament. Depending on the age and on the depth, the fossil content is varied. Sponge spicules and calcitized radiolarians are the most common remains throughout the Jurassic. Protoglobigerinids are abundant in the Oxfordian limestones. So-called 'filaments' (*Bositra* coquinas)

Fig. 18. Biomicritic packstone to wackestone with poorly sorted grains: large multi-crystalline and rounded quartz (a) and small and angular quartz (b). Riou-Petit Formation (Upper Sandy–Calcareous series). Late Toarcian. ARD 14 Well (COGEMA) near Privas (144 m). Millimetre-scale.

are widespread in the 'Marly–Calcareous Series' and coccoliths can be found. 'Lumpy' limestones ('calcaires grumeleux') belonging to the 'Ammonitico Rosso suite' form a particular kind of slope transitional facies (Middle Oxfordian; Bourseau & Elmi, 1981; Dromart & Elmi, 1986; Dromart, 1986).

REGIONAL JURASSIC SETTING

Northern Tethyan margin

At the beginning of the Jurassic, a wide continental block existed over the largest part of the present western Mediterranean region (Chanell *et al.*, 1979; Laubscher & Bernoulli, 1977; Winterer & Bosellini, 1981). Several attempts to reconstruct the opening

of so-called oceanic sub-basins have been presented (see Bernoulli & Lemoine, 1980, for a review; Dercourt *et al.*, 1985). The opening of the Atlantic has been an attractive model but it cannot explain all the kinematics which have been involved in the deformation and uplift of the Mesozoic continental margins. In a wide area stretching from African (or Apulian) to European cratonic zones, Mesozoic evolution began by the opening of many elongate basins. These were related to extensional stresses affecting the same continental basement.

The Subalpine Basin evolved into a complex folded zone (Subalpine Chains) and a narrow westward, slightly deformed fringe (Ardèche–Vivarais and Cévennes area). The pre-Triassic basement crops out in the west (French 'Massif Central') as well as in the east ('Massifs Cristallins Externes' or External Crystalline Massifs). Transition areas to the internal Alps (Piemont Zone) are the 'Subbriançonnais' and 'Briançonnais' Zones. Through early Jurassic times, a tectonic tilting of blocks commonly prevailed (Lemoine, 1983; Elmi, 1980, for the margin) and can be related to a general extension of the European basement leading to a crustal thinning with some Triassic volcanism in the eastern parts of the basin. Moreover, the occurrence of basement massifs both on the external margin (Massif Central) and within the basin (Massifs Cristallins Externes) fits well with the adaptation of Wernicke's model (1981) of detachment faults by Lister *et al.* (1986). The rifted basin appears to have been limited on the east by a 'marginal plateau' situated between two detachment systems. A similar detachment structure was located east of the Briançonnais Zone.

Palaeostructural framework

Thickness data and isopach maps, published by Elmi *et al.* (1984, 1987) are summarized in Figs 19 & 20. They support the palaeostructural interpretation given in Fig. 21. Cartographic, structural and stratigraphical data also have been extensively used. The main structures are oriented SSW–NNE (N10 E to N20 E; longitudinal 'Cevenol' trend), parallel to the structural margin which is obliquely cut by the present strip of outcrop. Transverse trends vary from WNW–ESE to WSW–ENE. The role of these trends changed from time to time during the sedimentary history.

Main divisions

Two main sub-basins were developed along the Ardèche margin. The northern Privas Sub-basin was bordered by an escarpment following the modern La Voulte Fault (Fig. 1A, No. 7; Fig. 21, TV). The junction with southern Aubenas Sub-basin is located along the Escrinet−Arénier Swells (Fig. 21, TA, TE). The influence of these ridges on the sedimentation was considerable throughout the Triassic and Jurassic (gaps, reduced thickness, Fig. 4). In the southern area, the Païolive (Fig. 8, II) and Courry (IV) mobile zones controlled the behaviour of escarpments or slopes which dipped eastward.

Longitudinal trends

These are numbered from I to IX in Fig. 21. The antithetic movement of the La Rousse axis (I) is well documented during the early tilting (Colongo *et al.*, 1979; Elmi, 1984). The Païolive axis (II) commonly marked the true boundary between a narrow western platform and the basin or the deep outer shelf. Rapid thickening prevailed on the eastern, down-dropped compartment and is well documented by outcrop studies and by data from bore holes (Rosières, Saint-André-de-Cruzières) (Fig. 19).

Escrinet and Charray Faults (VI) are recent faults developed on an old mobile zone. During late Triassic and Hettangian times, they had bordered a small weak zone (Charray umbilic; Figs 3 & 21; see also Fig. 7) sustaining a stronger subsidence than the adjacent areas. During that time, the Charray umbilic may have been a small pull-apart basin.

Transverse trends (TO, TB, TU, TA, TV, Fig. 21)

These cut the longitudinal framework producing a mosaic network. The constituent sub-basins are swales undergoing highly differentiated evolution during the early−middle Jurassic rifting stages. The transverse network faded out during the late Jurassic as postrifting conditions succeeded the formation of oceanic crust farther east in the Piemont Zone (so-called Ligurian Ocean). The influence of some of these transverse trends was important during the late Triassic because of the inherited influence of the late structural trends which had not been modified by the rifting. The WSW−ENE Escrinet−La Voulte (TE, TV), the NW−SE Orcières and Bordezac lineaments (TO) and the N−S Uzer Swell (TU) have a major importance in the distribution of sediments.

HISTORY OF THE VIVARAIS MARGIN

Prerift (pre-Rhaetian)

The Triassic sequence is made up of three main lithologic units: Lower Sandstones, Middle Carbonates and Upper Sandstones (Ucel Formation): During late Ladinian and early Carnian times, a general transgression flooded the area. Carbonates and clays were deposited in a shallow sea. Next, regression and restriction of the basins occurred. The Subalpine area became an evaporitic basin bordered by continental alluvial plains and deltas which prograded to the southeast. Calculated subsidence curves reflect extension (Brunet, 1985).

Early differentiation

Synrift stages with strong tectonics (Rhaetian event)

The variations of Rhaetian to earliest Hettangian beds (S0, S1) were largely controlled by local tilt-block tectonics since the calculated subsidence curves show only slight overall extension (Brunet, 1985). The tilting of the Les Eynesses fault is shown in Figs 4 & 9. An arid continental climate prevailed, around the Massif Central, as documented by palynological data (Taugourdeau-Lantz & Lachkar, 1985), and then a change to a humid climate coincided with the extensive humidity of the Hettangian transgression. This could have facilitated the cessation of coarse terrigenous influx related to the general late Triassic denudation.

The inception of the Hettangian shallow marine platform led to the deposition of the 'Complexe Carbonaté de Base' (CCB, S1) containing a widespread oolitic level. The initial flooding was probably related to a general sea-level rise. On the borders, carbonate progradation competed with deepening. The S1 sequence ends with mytilid−coquinas, indicating a stronger marine influence.

General deepening (Hettangian event)

After the accumulation of the mytilid−coquina, the fauna of the lower member of the 'Calcaires noduleux cendrés' indicate a rapid deepening: ammonites, bivalves and sponges became common and no evidence of algal activity has been recorded; benthos was missing except for the burrowing bivalves (*Mactromya*) which caused much bioturbation. These features document a deep outer shelf, re-

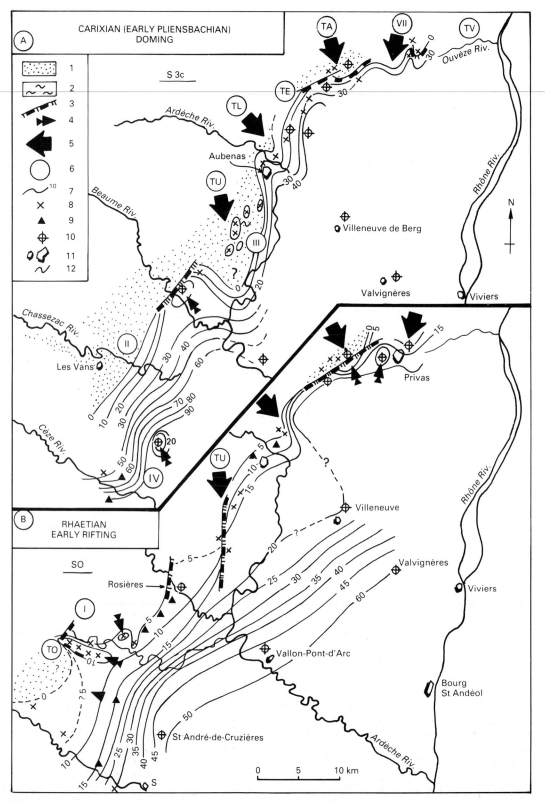

Fig. 19. Simplified isopach maps and palaeostructural interpretation for the Rhaetian (B), the Carixian (A), the late Bathonian (D) and the late Callovian to the beginning of the late Oxfordian (C). 1 = gaps related to non deposition or/and penecontemporaneous erosion, 2 = 'Terres Noires' (black shales): late Callovian and early Oxfordian, 3 = escarpment and

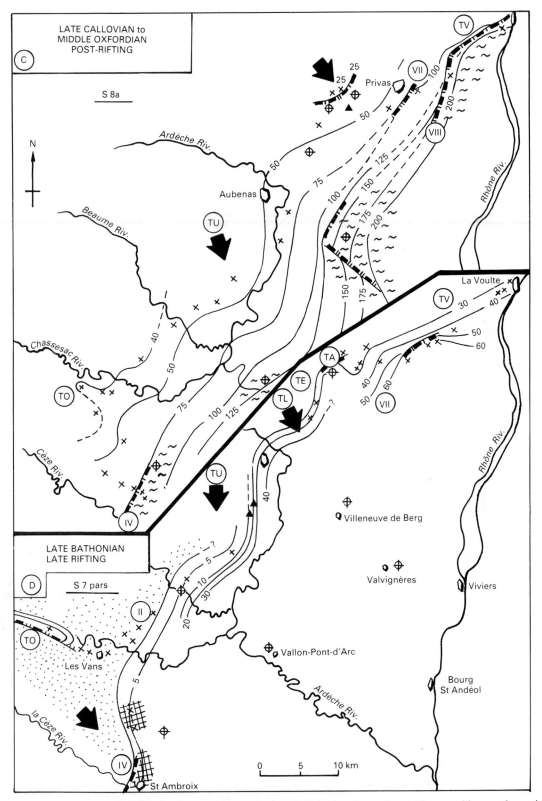

Fig. 19. (*Cont.*) zones of rapid thickening, 4 = umbilic and sense of thickening, 5 = major positive or stable axes, 6 = major palaeostructural lineaments (explanation of abbreviations and code-numbers given in Fig. 21), 7 = isopach line, 8 = measured section (selection), 9 = estimated values, 10 = important boreholes, 11 = localities, 12 = penecontemporaneous erosion. Adapted from Elmi (1967, 1987) and from Elmi *et al.* (1984).

Locality	Tithonian (S 10)	Kimmeridgian (S 9)	Late Oxf. (S 8 b (13))	Middle Oxf. (S 8 a (14))	Early Oxf. / Late Call. (15 et 16)	Middle Call. to (S 7 (17))	Middle Bath. (S 7 (18))	Early Bath. / Toarcian (S 6 - 5)	Domerian (S 4)	Car. Sinem. (S 3 bc (19))	Late and Middle Hettangian (S 3 a (20))	Early Hettangian (S 2 (21))	(S 1)	Rhaetian (S 0)
VALVIGNERES (12)	43	107	88	82	100	114	454	1·259	127 / 175	446	261		18?	50
VILLENEUVE de BERG (12)	49	93	95	85	73?	132	204	247	127	238	150	45	12	20
VALLON PONT D'ARC (12)	32	102	116	67	54	128	250	299	34 / 151	FAULTED				
St BRES St AMBROIX (11)	30 - 35	100	?	94		37	70+	40	15 / 150	100		10	17	30
COURRY (10)	30 - 35	?	?	49	(x)	46	(x)	8	NON EXPOSED					
LES AVELAS (9)	60	?	52	(x)		1·24		3-4	10?	NON EXPOSED				
St ANDRE DE CRUZIERES (8)	?	?	?	?		47·5	(x)	115	25 / 342	59	109		48	
LA ROUSSE LE MOIGNARD (7)	NO OUTCROP			(x)	?	?	?	10-24	(x)	(x)	0-6	0-10		
LES VANS	(30)	100	100+	39	(x)	30+	0	5	(x)	3	10?	0·15	0·10	
ROSIERES	(30)	70	90-100	44	?	57+	29	9	30	150	95	45	30	10
UZER (6)	40	70-80	90-100	40	?	38	55	0-1	0-1	52+	32	48	4-11	12-15
CHAPELLE sous AUBENAS (5)	25	85+	90	57	(x)	28	150	2-3	1-10 / 0·7	100		10	10?	10?
St JULIEN du SERRE (4)	NO OUTCROP					0-5		0-5	45-50	50	35-60	10	10?	10?
PLAT REDON WELL ARD 18 (3)	(25)	120	90	55	(x)	56-70	100	30+	10+	140	121	46	6	15+
AURIOLLES (2)	(40)		50	48	?	66	?	10-15	0-2	NON EXP.		9+	19	19
ANDIGE WELL ARD 19	(20)		75+	48?	(x)	30	190	22	38+ / 0·2	61	25	45	10	19 / 19
ARENIER PASS	10?	?	?	23	?	20+	60	(x)	(x)					
PRIVAS (1)	25?	70	50	74	(x)	70	160	15 / 15	55		100		12-27	
COUX (JAUBERNIE)	30	70?	40?	111	(x)	63	142	15-35				10	8	
FLAVIAC AND CHOU	40+	100	145		(x)	34	40	5-10 / 7-55	26 / 15-35	120+100		15	?	
LA VOULTE	60+	80-100	95 / 194	24?		78 to 110	110	?	NON EXPOSED					
CRUSSOL	135	30+ / 194	23? / 24	4-6		8-9	14-17	(x)	NON EXPOSED					(x)

Legend:

- (22) Sandstones, Conglomerates
- (23) Basinal marls
- (24) Main gaps
- (25) Lenticular beds
- (x) = new escarpment

Fig. 20. Simplified thicknesses-chart of Rhaetian–Jurassic sequences selected in the study- rea (Ardèche and Northern Gard departments, southeast France). Values are given in metres. Data come from both measured sections and boreholes, sometimes supplemented by map evaluations based on personal field work. In several areas, variations are very rapid especially in the pre-Callovian terranes. The range of the thickness variation is indicated in the same column by the two extreme values. Estimations are indicated by question-marks (?). Values in brackets are given from nearby sections. Comments: 1 = South of Privas (Beaudoin) for the Callovian and the late Jurassic, around the town for the lower part; 2 = outcrops near Auriolles and well ARD 12 (COGEMA); 3 = outcrop data for the terranes above the Upper Sandy–Calcareous Series (S5 and S6); 4 = along the Luol transverse lineament (TL); 5 = southern Aubenas Sub-basin, east of the town for the late Jurassic; 6 = outcrops and wells; 7 = located at the toe of the Orcières transverse lineament (TO); 8 = well data at the beginning of the Tithonian; 9 = along the Courry-Saint Brès escarpment (IV); 10 = west of the Courry-Saint Brès escarpment; 11 = east (basinward) of the same escarpment; 12 = new interpretation of wells drilled during the early sixties; 13 = Calcaires du Pouzin and Couches de Joyeuse Formation; 14 = lower lumpy series and Bancs Roux; 15 = lenticular beds along the border; 16 = Terres Noires; 17 = Callovian Triad; 18 = lower part of the Marly–Calcareous Series, beneath the *Gracilis* Zone; 19 = La Garenne, Château d'Aubenas and Bosc Formation; 20 = upper member of the Nodular Series; 21 = Complexe Carbonaté de Base; 22 = sandy or conglomeratic facies of the Vaumalle Fm; 23 = basinal marly facies, Domerian–Aalenian and late Callovian–early Oxfordian only; 24 = major gaps; 25 = lenticular deposits.

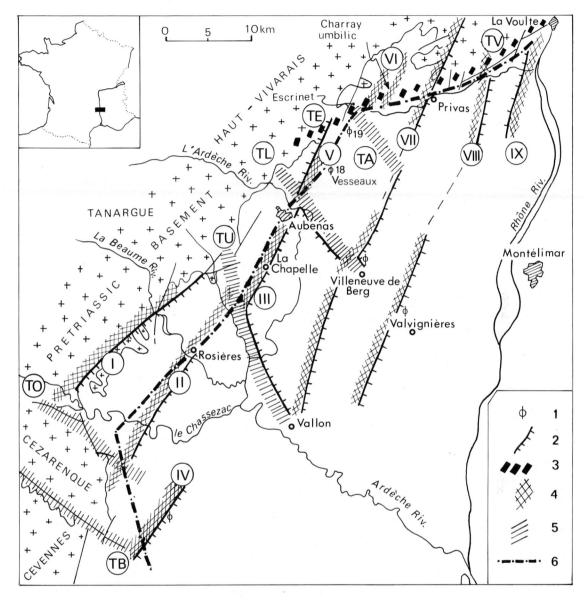

Fig. 21. Palaeostructural map of the Vivarais−Cévennes during the Jurassic (mainly early and middle Jurassic). 'Cévenol' longitudinal lineaments (SSW−NNE): I = La Rousse, II = Païolive, III = Uzer−Aubenas, IV = Couray−Saint Brès, V = Saint Privat−La Conchy, VI = Escrinet−Charray, VII = Privas−Coux, VIII = Flaviac, IX = Le Pouzin. 'Variscan' transverse lineaments (WSW−ENE): TE = Escrinet, TV = La Voulte. 'Velay' transverse lineaments (NW−SE): TB = Bordezac, TO = Orcières, TU = Vinezac-Uzer-Vallon Pont d'Arc, TL = Luol, TA = Escrinet−Arénier.

1 = wells, 2 = escarpments (pre-Gracilis Zone), 3 = Variscan lineaments, 4 = Cévenol lineaments, 5 = Velay lineaments, 6 = traces of the palinspastic profiles.

ceiving hemipelagic sedimentation well below wave base. The depth can be tentatively estimated as 100 m for the deepest zones.

In Ardèche, in areas protected by swells related to tilt-block edges, the general transgression was followed by the differentiation of a protected platform (mottled limestones). Small patch-reefs developed along the outer border but tectonic instability and an influx of clay inhibited extensive progradation. In some areas (Molières, Rosières), sedimentation was nevertheless sufficient to keep pace with subsidence and thick low-energy deposits accumulated on a protected shelf (Figs 22 & 23). Basinward (Villeneuve and Valvignères bore holes), the climax of the extension rate was attained during the Hettangian (Brunet, 1985).

Resumption of rifting

The next sequence (S3a) began after a general change, documented by widespread skeletal limestones (crinoidal limestones), by *Favreina* (Fig. 12)

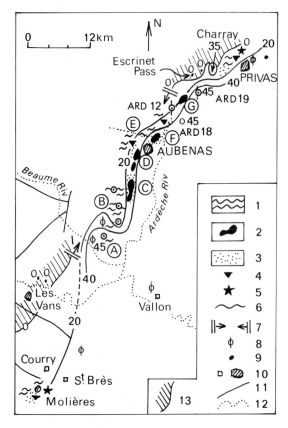

Fig. 22. Isopach map and facies subdivision of the lower member of the Calcaires noduleux cendrés Formation (S2). Early Hettangian (Planorbis Zone). 1 = mottled limestones deposited in protected areas (inner shelf), 2 = small reefs, 3 = bedded limestones with *Madreporaria*, 4 = brachiopods, 5 = crinoids, 6 = isopach curve, 7 = limits of the profile shown on Fig. 23, 8 = wells, 9 = measured profiles or wells, 10 = localities, 11 = present-day faults, 12 = modern rivers, 13 = gaps (erosion, non-deposition).

levels and by *Ostrea* coquinas. This episode of shallowing may have been mainly controlled by changes in sea-level since it is known all along the eastern border of the Massif Central (Mouterde, 1952; Elmi & Vitry, 1987). After this widespread event, the variations were sharp but of less extent, mainly affecting the thickness, local stratigraphic gaps and the quartz supply. The upper member of the 'Calcaires noduleux cendrés' Formation (S3a) was uniform and the change in thicknesses depended mainly on differential subsidence along the structural network inherited from the initial rifting.

Thick successions occurred either in a protected or an open environment. The swells resisted regional subsidence. They were subjected to submarine erosion which is documented by the occurrence of Hettangian limestone pebbles reworked into younger beds (Escrinet–Arénier Swell). The more prominent swells or ridges did not receive any sedimentation (Les Vans area). Sedimentary gaps occur on the more prominent ridges; for example, there is a complete gap from middle Hettangian to early Bajocian in the main part of the Les Vans area at the toe of the Cézarenque horst.

Interaction between rifting and eustacy

Tectonic controls

From the Sinemurian to the early Bathonian, narrow sub-basins (umbilics) were bordered by wide swells (Figs 19 & 21). Coarse clastic influx occurred along steep slopes and prevented the development of a large platform. The tectonic subsidence was only partially compensated by the progradation of skeletal wackestones and packstones. The result was a strong physiographic differentiation. Sediment fill to base level and progradation became predominant only when tectonic movements and sea-level changes were simultaneous as, for example, during late Toarcian–early Aalenian times. Coarse detrital crinoidal limestones covered the related swells. Sedimentation terminated with a general discontinuity (hardgrounds, non-deposition, erosion).

From the Sinemurian on, severe erosion of the swells is documented by several pieces of evidence. Quartz influx resumed, indicating that Triassic beds were being eroded. Erosion was active along the escarpments. At Les Tardieux (near La Chapelle-sous-Aubenas; Figs 7, 17 & 24), the lower Sinemurian Bosc Formation consists of nodular micritic marly limestones alternating with more

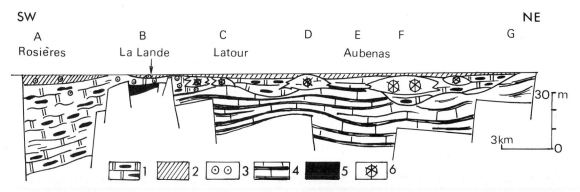

Fig. 23. Profile through the Aubenas Sub-basin during the early Hettangian (lower member of the Calcaires noduleux cendrés Formation; S2). 1 = mottled limestones (inner shelf), 2 = bivalve–coquinas and bioclastic limestones (shallow outer shelf), 3 = oolites, 4 = marls and limestones with ammonites (outer shelf or basin), 5 = marls (protected areas), 6 = reefs. A to G: localities indicated in Fig. 22. Adapted from Elmi *et al.* (1987).

regular and more calcareous beds, characterized by wackestone texture. The Bosc Formation is truncated by a sharp erosion surface and the Upper Toarcian sandy biocalcarenite (Upper Sandy–Calcareous Series) directly overlies the Lower Liassic formation (Fig. 24). Tectonic control is evidenced by the occurrence of a small graben, 100 m wide, filled with Domerian biomicrites bearing abundant quartz grains and limy pebbles, and resting directly on the Bosc Formation.

Tectonic controls are also documented by the occurrence of 'neptunian dykes'. The best example is situated at the southern toe of the Escrinet Swell, along the La Clapouze–La Conchy outcrops (Fig. 25). Domerian beds (Vaumalle Formation) consist of conglomerates made of quartzite pebbles embedded within crinoidal limestones. The pebbles are oriented parallel to the boundary faults (WSW–ENE). Ammonites, (amaltheids, lytoceratids), nautiloids and belemnites, drifted from the open

Fig. 24. Penecontemporaneous erosion. Calcaires du Riou Petit Fm (1) (Upper Sandy–Calcareous Series, late Toarcian) resting on Le Bosc Formation (2) (early Sinemurian). Very irregular contact. Les Tardieux near Aubenas.

Fig. 25. Fissures and neptunian dykes. La Clapouze outcrop near the Escrinet Pass. Domerian breccias (a) are cut by a network of synsedimentary fractures and small faults (b) appearing as diaclases.

sea, are mixed together with benthic gastropods (*Pleurotomaria*) and bivalves (*Aequipecten*). The Vaumalle Formation rests directly upon the eroded Ucel Formation (Triassic) and is overlain by laterally continuous marly limestones (Lower Bathonian La Clapouze Formation), rich in brachiopods and *Bositra*. A hardground marks a discontinuity corresponding to the lack of Toarcian to Bajocian sediments. The surface is affected by several generations of fractures, some of them exposing metre-scale vertical offsets. They are filled with ferruginous micrites. Ammonite occurrences within the dykes establish the fact that these fractures were active from the middle Toarcian (*Hildoceras*) to the early Bajocian (*Docidoceras*). In the northern part of the outcrop, surface irregularites are filled with Upper Bajocian micrites (*Bositra* and spicule-bearing micrites and pelmicrites).

One other striking example of Domerian perturbations occurs at the toe of the La Voulte fault at La Jaubernie near Privas. Carixian, sandy-bioclastic limestones are cut by a narrow and deep channel (of > 50 m) eroded at the end of Carixian and at the beginning of Domerian times. During the late Domerian, the channel was filled by mass gravity-flow deposits, up to 50 m thick. These consist of quartzitic (Triassic) and micritic (Lower Liassic) pebbles within crinoidal packstones and grainstones, containing rare ammonites (*Pleuroceras*) and abundant belemnites. Oblique laminae occur in even the coarsest clastics (up to 0.05 m). The sedimentary body forms a fan; its thickness changes from near 60 m to < 15 m over 1 km. The width is ≈ 300 m. This fan resulted from the progradation of a small submarine deltaic cone, situated at the toe of a steep escarpment bordering a swell over which erosion prevailed. The shelf was relatively deep but seems to have been very narrow. The erosion was related to tectonic movements as it was contemporaneous with a general deepening of the southern France basins (Mouterde, 1952; Gabilly *et al.*, 1985).

Study of the southern Ardèche area (Courry, Saint Brès) suggests that the shelf width did not exceed 5–10 km during the Domerian. A rapid facies change into the basin is well documented across the Courray Escarpment (Fig. 7; Fig. 21, IV). On the western (outer) part, Domerian beds consist of sandy lime-stones; on the eastern side, there is a rapid transition to thick micaceous shales (de Brun, 1926; Bourseau & Elmi, 1981). The faulted margin remained active during the Toarcian–Bathonian interval as indicated by similar abrupt changes.

Well-established relations between erosion and tectonic activities enable one to conclude that the edges of tilt-blocks underwent uplift (shouldering or doming). Basinward, the slope began to be established. Its role became stronger during accumulation of the tremendous thickness of marls at the toe of the southwestern continuation of the Flaviac lineament (Valvignères bore hole; see Figs 20, 21 & 28). The previous framework was disrupted by tectonic activity but sedimentation was also influenced by general sea-level changes.

Sea-level changes and local tectonic controls

A general deepening occurred during the early Toarcian, following an eustatic shallowing at the end of the Domerian on the European Plate and on both rims of western Tethys (for modern reviews, see Vail *et al.*, 1977; Hallam, 1984; Gabilly *et al.*, 1985; Haq *et al.*, 1987). In Ardèche, the sedimentation break lasted into the earliest Toarcian and the sedimentation resumed only during the second Toarcian ammonite Zone (Serpentinus Zone). The resumption in sedimentation was general except on the more positive shoals and ridges; lower Toarcian beds are locally transgressive on the Triassic as in Crussol (Haug, 1911). This local transgression coincided with a general deepening in the basins. The general eustatic model does not fit perfectly with these data, as Haq *et al.* (1987) stated that highstand-controlled transgressions are related to starvation in the basins. However, the model has been worked out for large platforms, very different from the narrow and highly uneven Ardèche border of the Toarcian times. Thus, local patterns exaggerated or diminished the effects of the eustatic fluctuations.

Along the Ardèche margin, local deepening was perceptible only in the sub-basins. Sedimentation mainly consisted of cyclic marl–limestone layers. 'Schistes–carton' (laminated black shales) were also deposited in areas bordering either the shore or the swells. Thus, schistes–carton are associated with sandy limestones and alternations of bioclastic packstones (Fig. 16), indicating that the accumulation occurred in low-energy embayments at the toe of escarpments. When tectonic controls were not involved, the deepening of the sub-basins was coeval with either a sea-level high or a rise, or both processes, on the shoals and along the borders. Similar competitive processes have been well studied by Lucas (1942), illustrating the so-called 'Haug's Law' (Haug, 1900): transgressive events coincide with shallowing of the basins while deepening of the depocentres are coeval with regression along the shorelines.

An important regression occurred during the second part of the middle Toarcian (Variabilis Zone) and during the transition from early to middle Aalenian. Although less frequently reported in the literature than other events, the Variabilis Zone shallowing seems to have been regional (in northwest Europe and on both rims of West Tethys: discontinuity no. 4 in Gabilly *et al.*, 1985; Elmi & Benshili, 1987; see also Haq *et al.*, 1987). The regression is associated with widespread gaps, particularly in Ardèche. It led to a noteworthy differentiation in ammonite speciation which was strong during the second half of the Toarcian (Elmi *et al.*, 1986).

The middle Aalenian eustatic regressive event is well documented and it led to the growth of broad carbonate platforms in northwest Europe (Haug, 1911; Purser, 1975; Contini in Enay & Mangold, 1980; Contini, 1987). In Ardèche, it resulted in a general gap following the progradation of coarse sandy, bioclastic limestones ending the S5 sequence (lower part of the Upper Calcareous Series). Only rare Upper Aalenian–Lower Bajocian lenticular beds have been preserved. Emersion is locally documented and erosion was active along the ridges. The uplift was short lived and the margin subsided again; the slopes were washed by currents which carried the fine sediments to the eastern depocentre where marls and marly limestones accumulated (Elmi, 1967). Data concerning the middle Aalenian event are in agreement with the general eustatic model as the fall of sea-level was followed by non-deposition on the platform and by active marly sedimentation in the depocentre.

The Bajocian deepening (during the S6 sequence) did not result in a resumption of the sedimentation on the margin although the eastward basin depocentre was filled with marly hemipelagic limestones. The change to the western border was probably rapid and the major part of the sediments was winnowed away. Some lenticular beds have been preserved in small protected areas. In the Privas and Aubenas Sub-basins the sediments are made of biopelmicrites with ammonites and sponge spicules. Shallower conditions prevailed in the more stable zones (Les Vans area), leading to accumulation and progradation of crinoidal sands. The winnowing of the remaining highs may be responsible for the supply of a certain amount of iron which precipitated along the shore. All these features can be correlated

to a transgression and to a deepening which occurred on a steep margin. The carbonate platform was reduced to a narrow strip with steep slopes and a relatively high-energy environment which prevented its accretion. The slopes are also characterized by slides and slumps which are, however, better known in slightly younger deposits (late Bajocian and early Bathonian). In contrast, wide carbonate platforms developed at this time in the less mobile platform of north and northeast France (Paris Basin, Jura Mountains, Burgundy). In the basins bordering Tethys, tectonics and eustacy operated together. On more stable continental crust, general tectonic subsidence (Brunet & Le Pichon, 1980) was overcome by the major sea-level changes which controlled the palaeogeographical evolution, and shelf conditions spread over large regions instead of being limited to narrow strips along the shores and the ridges.

Late movements of tilt-blocks

From the middle Bathonian to the beginning of the Callovian (Macrocephalus Zone), rapid changes in thickness occurred. They were complicated by facies variations and by resedimentation processes (gravity-reworked quartz pebbles, slides and slumps; Figs 6 & 10f). During the middle and late Bathonian, a large shelf developed in the Les Vans area as a result of the widening of the Païolive ridge. Sedimentation was missing across large areas of the ridge but some hemipelagic marls and limestones are preserved along the eastern down-dropped segments of some faults. This documents some late tilting movements (Elmi, 1983b, 1984). Transition to the basin normally occurred along gentle slopes (Rosières, Escrinet–Privas) covered by marly limestones (Argevillières Formation, Fig. 5) or by alternating marls and limestones (Saint-Priest Formation, Fig. 5). Some slopes remained steep, such as the Courry Escarpment bordering the Saint-Ambroix Basin (Fig. 10f and Fig. 21, IV). Thin lenticular micrites were deposited on the down-dropped blocks; they contain centimetre-sized quartz pebbles and feldspars associated with ammonites.

In the more subsiding areas marly sedimentation did not keep pace with the subsidence. Small gulleys or umbilics appeared where a nektonic fauna seems to have been trapped, as suggested by the marly Plat Redon Formation of the north Aubenas Sub-basin (Elmi, 1985a). The environment may have been a sinking platform crossed by a network of narrow shelves building a complex bottom morphology which impeded the general circulation of deep waters. Bottom waters became restricted and local anoxic conditions are evidenced by pyritic internal molds of ammonites and nodules. Such protected conditions must have persisted into the earliest Callovian (La Gette Formation), but ammonites as well as foraminifera became very scarce. The scarcity of fauna may have resulted from limited communication with the more open sea and then a subsequent drop in organic productivity.

Post rifting dynamics

The late rifting climax was followed by a global change in the dynamic processes; it can be regarded as a consequence of the spreading occurring at this time in the Piemont Zone (Lemoine, 1985). Along the western margin, the structural mosaic inherited from the early rifting was replaced by a more continuous slope. This evolution is consistent with McKenzie's model (1978). It can be easily compared with the history of the North Sea Basin and of northwest Germany (Pratsch, 1982; Morton, 1987).

During the second part of the early Callovian (Gracilis Zone), 'the Callovian Triad' accumulated all along the Ardèche and Cévennes margin (see Figs 5 & 6). This facies suite (S7c) ends with a hardground which has been recognized for > 200 km along the western borders of the basin. It reflects the end of a shallowing sequence with an upward increase in carbonate, bioclasts, benthic bivalves, glauconite and iron oxides. The final hardground is commonly encrusted but the origin of the coating is not clear. The S7c sequence appears to have recorded a regional eustatic event. Its base overlies a sharp disconformity on the ridges but, in the pre-existing sub-basins, no discontinuity is noticeable. It seems that the basin enlarged towards the west at the beginning of the sequence and that it underwent a subsequent shallowing. The beds thicken strongly eastward.

Revival of the slope: the late Callovian−early Oxfordian event

Major tectonic and sea-level events took place at the transition from middle to late Jurassic times (sequence S8). The margin was probably bordered by a narrow western shelf which is not preserved in Ardèche, because of recent denudation. There are no upper Callovian beds preserved along the outcrop

strip with the exception of the more external zones (La Voulte, Saint-Brès), where the old escarpments (Fig. 21, IV and TV) remained active during post-rifting tectonic activity. In the basinal areas deposition of black shales ('Terres Noires') took place during the early Oxfordian. The 'Terres Noires' overlie the middle Callovian condensed beds. The main area to study these features is around La Voulte and Rompon localities (Sayn & Roman, 1928; Elmi, 1967; Alméras & Elmi, 1984). Black shales with thin evenly-bedded micritic ferruginous levels ('lithoid' iron ore, Athleta Zone) overlie the top of the Lower or Middle Callovian beds, interrupted by sedimentary gaps, small faults with metre-scale offsets and block-sliding. The transition to the 'Terres Noires' Basin was probably controlled by basement faults; evidence is only given by the sliding and the collapse of calcareous beds (La Voulte area) (Elmi, 1980; Bourseau & Elmi, 1981; Dromart & Elmi, 1986; Elmi et al., 1984; Elmi, 1985c). These data seem consistent with the consequences of a general sea-level fall: sediments reduced or missing on the slope, widespread erosion on the seafloor at the top of the Middle Callovian reduced beds and thick marly sedimentation in the basins.

A similar structure prevailed during middle Oxfordian times as sedimentation resumed on the transitional slope. Shelf deposits are not known along the Ardèche border. They may have been similar to those developed on the Jura shelf (French Jura Mountains; see Gaillard, 1983 for a detailed study). On this outer shelf, bedded nodular limestones, rich in sponge spicule-bearing intraclasts accumulated around sponge biotherms ('Couches de Birmensdorf'; Enay, 1966). Transitional slopes were covered by nodular limestones ('grumeleux' facies of the lower lumpy series, LLS, Fig. 2) containing microbial fabrics and mud mounds (Dromart & Elmi, 1986). The slope facies are very similar, except for the marly beds, to the 'Ammonitico Rosso Superiore' of the Trento Plateau in the Italian southern Alps where Sturani (1971) described stromatolites and oncolites. 'Pelagic' oncolites and stromatolites have been extensively studied by Massari (1979, 1981). Similar horizons are known in late Jurassic beds of southeast France and Italy as well as in the Portuguese and Moroccan Pliensbachian. The faunal data are consistent with depositional depths of > 200−300 m (protoglobigerinids, benthic foraminifera, lithistid sponges and hexactinellids, Balanocrinus stalked crinoids, thin-shelled bivalves, radiolarians; Dromart, 1986).

Late Jurassic platform

After the Bifurcatus Zone (beginning of the late Oxfordian), the vertical succession became monotonous (Figs 26 & 27). Oscillations of sea-level can be inferred from slight lithological changes (alternating marls and micrites, cherty limestones, massive limestones); the top of the sequence is shallower than the base as evidenced by the rarity of ammonites and the increase in carbonate. The beds thicken upward. The thickness of the deposits is only slightly reduced on the more resistant ridges (Escrinet–Arénier). Evidence of local tectonic movements is scarce and they only occurred along the slope into the basin. The changes in lithology and sedimentation rate seem to have been controlled by regional subsidence and by sea-level fluctuations. The general setting became progressively that of a large platform.

Fig. 27. Channel and disconformity in the Calcaires de la Beaume Formation. The thick bed is made of nodules (oncolites) and of rock fragments embedded in a mudflow. The transport direction has been roughly from left to right. Kimmeridgian. Crussol near Valence. The bar is 1 m high.

SUMMARY OF THE MAIN EVENTS AND OF THEIR CONTROLS

A diagrammatic and synthetic cross-section illustrates the changes in the nature and style of the

Fig. 26. The Upper Jurassic outcrops of Crussol near Valence, dominating the Rhone Valley. Abbreviations: CP = Calcaires du Pouzin (top of the S8 sequence); CL = Calcaires de la Louyre; CB = Calcaires de la Beaume; CC = Calcaires du Château de Crussol (equivalent to the Calcaires de Paiolive); the S9 sequence is made of CL, CB and CC Formation: at top: Tithonian beds.

factors controlling the physiography of the margin and their influence on sedimentation (Fig. 28). The local variations are diminished on this section as the profiles have been selected both for their palaeogeographic significance and value to a regional model. The localities are not aligned regularly across the present-day margin.

Les Vans illustrates highly reduced early to middle Jurassic deposits; starvation and erosion prevailed from early Hettangian to Bajocian. Local structural features dominated and the sea-level rises were not recorded; the boundary with the adjacent subsiding areas was sharp. The S6 sequence is more largely represented than the S5. Near Rosières, the Jurassic sequence is variable; reduced (or missing) episodes alternate with comparatively thick deposits. The locality is situated at the toe of the Païolive escarpment (Fig. 21, II); the S5 sequence is truncated in the Middle Toarcian and S6 is missing. At Plat Redon, near Vesseaux, relatively thick marginal sequences are located at the toe of the Saint Privat lineament (V). Flaviac and Coux sections have been established from outcrops located across the Flaviac longitudinal (Cevenol) lineament (VIII) and at the toe of the La Voulte transverse structure (TV). Villeneuve-de-Berg is located on the downthrown segments of faults corresponding to the southwestern continuation of the Privas–Coux lineament (VII). The influence of the Cévenol lineaments was different

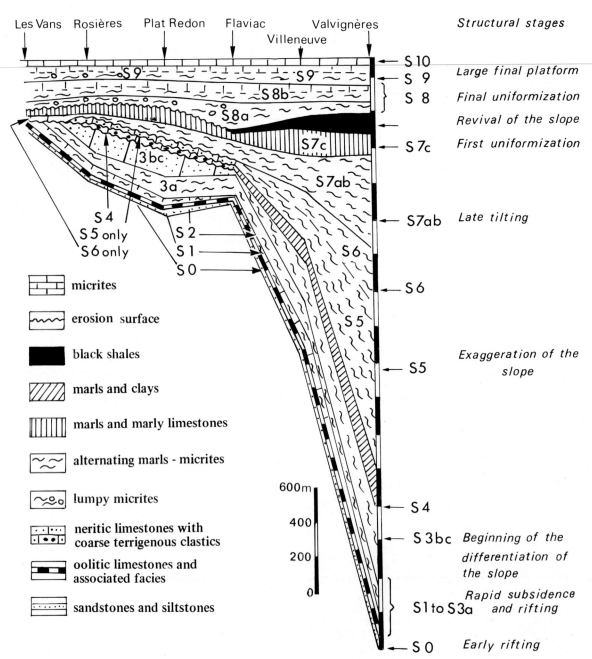

Fig. 28. Simplified cross-section of the Ardèche Margin, without horizontal scale. The thickness of the more reduced sequences has been slightly exaggerated to ease drawing. From measured sections and well data (Rosières, Plat Redon, Villeneuve-de-Berg and Valvignères boreholes).

from NNE to SSW owing to the role of the transverse structures.

A comparison between the cross-section (Fig. 28) and the table of local stratigraphical gaps and discontinuities (Fig. 29) allows an estimation of the respective importance of local tectonics with respect to regional (or general) sea-level variations. The initial transgression (S0, S1, S2) was bound to a general sea-level rise as differences were slight between the basin and the margin. Changes were more noticeable along the margin itself (see Figs 4 & 9).

At the end of the Sinemurian a narrow platform existed along the margin and exposure of the edges of tilt-blocks led to a resumption of the influx of coarse terrigenous material into the small subsiding umbilics. The transition to the basin occurred along major lineaments and it is highlighted by the appearance of micrites and marls (Villeneuve, Vallon, Valvignères boreholes). This differentiation between a marginal western bank and an eastern basin continued to increase from the middle Liassic until the Aalenian (S4 to S6): thick marly deposits in the basin differed strongly from the discontinuous and reduced sedimentation occurring on the platform. The differences lessened during the Callovian and the last important revival of the transitional slope occurred during the late Callovian–early Oxfordian. Afterwards, the physiographic differences diminished and the monotonous sedimentation was only interrupted by occasional slumps along the main Cévenol lineaments.

The amount of precise stratigraphic and sedimentologic data gives an opportunity to test the general models and curves recently summarized by Haq *et al.* (1987). The major supercycles can be recognized. However, some boundaries cannot be fitted exactly to the Haq *et al.* curves:

1 The early Hettangian deepening was rapid and widespread; it resulted mainly from an acceleration of the subsidence rate (Brunet, 1985).

2 There is no evidence of the late Hettangian (Angulata Zone) onlap in Ardèche but biostratigraphical data are poor at this level.

3 The late Domerian–earliest Toarcian shallowing (late Spinatum and Tenuicostatum zones) seems of equal importance to the late Carixian event (pre-Margaritatus Zone).

4 Sedimentary perturbations and associated shallowing were very important during the Variabilis Zone (middle Toarcian), but there is no noticeable corresponding variation of the curves of Haq *et al.* (1987). Since widespread discontinuities occurred at that time in the Tethys and on its borders, the *Variabilis* event seems to be related more to a dynamic phase in the evolution of Tethys.

5 At the end of the early Aalenian (Opalinum Zone) or slightly later (early Murchisonae Zone), the general shallowing attained a climax in harmony with the eustatic curves.

6 Breaks in sedimentation (erosion and non-deposition) were important on the platform during the early Bajocian and the results of a pre-Sauzei event cannot be separated from the general instability; the transitional slope to the basin seems to have been abrupt.

7 The Subfurcatum Zone (late Bajocian) coincided with a deepening which is also well known in much of southern France; it is not recognized as a general event and it must be stressed that in many parts of North Africa a shallowing has been documented (Moroccan Middle-Atlas, for example) contrasting with deepening in other areas (Tlemcen Mountains in Algeria, for instance); these data support the idea that strong tectonics were then dominating the Tethyan Realm.

8 A general instability prevailed during the Bathonian; consequently, the biostratigraphic correlations lack certainty between northwestern Europe and the Tethys; in Ardèche and in other Tethyan areas (for instance, see Elmi, 1972, 1978). It seems that a major event (shallowing, regression) occurred at the beginning of the equivalent of the so-called 'Aspidoides Zone' of northwestern Europe (Tethyan Retrocostatum Zone) rather that at its end.

9 During the early Callovian (equivalent to the Calloviense Zone), the Gracilis Zone does not appear to be a significant time on the curves published by Haq *et al.* (1987). In Ardèche, it coincides with the end of the movement of the tilt-blocks. This led to a widening of the basin; the local evolution is consistent with the general subsidence then occurring in the Tethys.

10 The subsequent local evolution seems to fit well with the eustatic curves; changes are documented at the end of the early Oxfordian, at the beginning and at the end of the Kimmeridgian and during the Tithonian.

A comparison has also been established with the scale of discontinuities recognized by Gabilly *et al.* (1985) in the Jurassic of western France. The fits between the events recognized in Ardèche as well as their scales are good for discontinuities 1, 3, 4, 8, 10 and 13. Discontinuity no. 2 has not been recognized in Ardèche perhaps because the stratigraphic data

STAGES	North Tethyan Zones	Les Vans	Rosières	Aubenas	St Julien	Arénier	Privas	Flaviac-Coux	La Voulte	Crussol	A	B	C
OXF. pars	Transversarium / Cordatum / Mariae										Exaggeration of the slope	15 / 14	
CALLOVIAN	Lamberti / Athleta / Coronatum 1 / Jason / Gracilis 2 / Macrocephalus										Callovian uniform.	13 / 12 / 11	
BATHONIAN	«Discus» / Retrocostatum 3 / Bremeri / Subcontractus 4 / Tenuiplicatus / Zigzag		?		?	?		?			Late tilting	10 / 9	
BAJOCIAN	Parkinsoni / Garantiana / Subfurcatum / Humphriesianum / Sauzei / Laeviuscula / Discites										Deepening	8	
AAL.	Concavum / Murchisonae / Opalinum										Shallow.	7 / 6	
TOARCIAN	Aalensis 5 / Levesquei 5 / Insigne 6 / Thouarsense 5 / Variabilis / Bifrons / Serpentinus / Tenuicostatum						NON EXPOSED					5 / 4	
PLIENSBACHIAN	Spinatum / Margaritatus / Stokesi 5 / Davoei / Ibex / Jamesoni	?			?						Deepening / Shallow.	3 / 2	
SINEMURIAN	Raricostatum / Oxynotum / Obtusum / Turneri / Semicostatum / Bucklandi	?									Quartz inputs		
HETT.	Angulata / Liasicus / Planorbis										Rapid Subsid.	1	
RHAETIAN													

are poor. Discontinuities nos. 5, 6, 7, 9 and 14 are coeval with strong instability on the Ardèche border where they cannot be separated from each other. On the contrary, some of the discontinuities and unconformities well established in Ardèche have not been identified in western France because of a lack of biostratigraphic evidence: end of the early Hettangian, early Sinemurian (Semicostatum Zone) and between the Carixian and the Domerian (pre-Margaritatus). Differences are more noticeable in the early Callovian as the pre-Gracilis event is not documented in western France and the no. 11 (pre-Macrocephalus) and 13 (middle Gracilis) discontinuities are very attenuated in Ardèche.

CONCLUSIONS

During late Triassic to early Callovian times, strong local, tectonic movements led to a segmentation of the physiography along the Ardèche margin. A complicated pattern developed of small sub-basins and wide swells. Authors often call on similar environments to explain perturbations sustained by the early Mesozoic sedimentation. However, there are only a few direct indications. Around the western Tethys, strong tectonic control has been clearly established during the early rifting: southern Alps (Gaetani, 1975), central Apennines (Fazzuolli & Sguazzoni, 1981; Farinacci *et al.*, 1981; Elmi, 1981), western Algeria (Elmi, 1981). These early stages may be comparable with the recent history of the Red Sea (Montenat *et al.*, 1986).

However, strike−slip was probably active in the western Tethys and the structural pattern may have been similar to that of the Recent African Rift as described by Bosworth (1986) from the Kenyan Gregory Rift. The multidirectional palaeostructural pattern of Ardèche during the early Mesozoic seems to be consistent with this model. The modern Tyrrhenian Sea can also furnish a valuable model

from the Recent (Boccaletti & Manetti, 1978). Steep physiography is also known in the Messina Straits where strong dipping slopes and deep plateaus are washed by strong currents active between Charybde and Scilla (Montenat & Barrier, 1985). A similar environment is interpreted to have developed in many parts of the Jurassic Tethys during its early history.

At the end of the Jurassic, changes took place all around the Tethys and led to strong modifications of the sedimentation. One striking result was the disappearance of the physiographic fragmentation which previously obstructed the balance of the ocean water circulation. Oceanic chemical changes can also be deduced from the disappearance of some carbonate facies such as 'grumeleux' (lumpy) and 'Ammonitico Rosso' facies, as rapidly lithified deep-water limestones do not seem to be well documented after the beginning of the Cretaceous. The physiographic evolution must have been interdependent with climatic changes. The noteworthy disappearance of oolitic iron ore was probably influenced by changes in the climate as well as modification of the weathering of the landmasses.

ACKNOWLEDGEMENTS

The English text has been greatly improved thanks to J.L. Wilson (New Braunfels, Texas) and P. Crevello (Marathon Oil Company, Littleton, Colorado). R. Ross (Denver, Colorado) and B.H. Purser (University Paris Sud-Orsay, France) have helpfully reviewed the manuscript. Their editorial remarks have led me to complement the data selected from abundant files and varied sources and to improve the conclusions on the controls. G. Dromart (University Lyon, France) has corrected and discussed the initial manuscript. I wish to thank the Institutions which have given me access to their data, especially the well-cores (Société Nationale

Fig. 29. [*Opposite*] Biochronostratigraphical chart of the major unconformities of the Jurassic sequence along the Vivarais−Cévennes (vertical column A). Correlation with the discontinuities recognized by Gabilly *et al.* (1985) in western France (Centre-Ouest) (vertical column B) and with the major cycles defined by Haq *et al.* (1987) (vertical column C). Oblique lines = gaps, dotted = reduced sequences, circled crosses = lenticular beds; column C: white arrows = megacycle boundaries, black arrows = supercycle boundaries. Biostratigraphical comments: 1 = during the Coronatum Zone, the gap began after the end of the earliest Obductum Subzone; 2 = equivalent to the Calloviense Zone of northwestern Europe; 3 = equivalent to the poorly defined Aspidoides Zone; 4 = the Morrisi 'Zone', considered here as a subzone; the poorly and unclearly defined Progracilis Zone is abandoned; 5 = considered as full zones (continental usage); 6 = equivalent to the Fallaciosum Zone which is considered in France as the lower subdivision of the Insigne Zone; 7 = equivalent to the Falciferum Zone. Main sources of data: Elmi & Mouterde (1965), Elmi (1967), Elmi *et al.* (1987).

Elf Aquitaine, COGEMA, Bureau des Recherches Géologiques et Minières and SMMP Pennaroya).

REFERENCES

ALMÉRAS, Y. & ELMI, S. (1984) Fluctuations des peuplements d'ammonites et de brachiopodes en liaison avec les variations bathymétriques pendant le Jurassique inférieur et moyen en Méditerranée occidentale. In: *Shallow Tethys International Symposium, Padova* 1982: Boll. Soc. Paleontol. Ital. 21, No. 2–3, pp. 1–19.

ATROPS, F. (1982) La sous-famille des Ataxioceratinae (Ammonitina) dans le Kimméridgien inférieur du Sud-Est de la France. Systématique, évolution, chronostratigraphie des genres *Orthosphinctes* et *Ataxioceras*. *Doc. Lab. Géol. Lyon* **83**, 1–463.

BERNOULLI, D. & LEMOINE, M. (1980) Birth and early evolution of the Tethys: the overall situation. In: *Géologie des Chaines Alpines Issues de la Tethys* (Eds Aubouin, J., Debelmas, J. & Latreille, M.) Mém. Bur. Recherch. Géol. Min. 115, pp. 168–179.

BERNOULLI, D. (1971) Redeposited pelagic sediments in the Jurassic of the central mediterranean area. *Ann. Inst. Geol. Hung.* **LIV**(2), 71–90.

BOCCALETTI, M. & MANETTI, P. (1978) The Thyrrhenian Sea and adjoining regions. In: *The Ocean Basins and Margins, vol. 4B, The Western Mediterranean* (Eds Nairn, A.E.M., Kanes, W.H. & Stehli, F.G.) pp. 149–200 Plenum Press, London.

BOSWORTH, W. (1986) Comment and reply on "Detachment faulting and the evolution of passive continental margins", *Geology* **14**, 890–891.

BOURSEAU, J.P. & ELMI, S. (1981) Le passage des faciès de bordure ("calcaires grumeleux") aux faciès de bassin dans l'Oxfordien de la bordure vivaro-cévenole du Massif central français (Ardèche–Gard); *Bull. Soc. Géol. France*, sér. 7, **XXII**(4), 607–611.

BRUN, P., de, (1926) Etude géologique et paléontologique des environs de Saint-Ambroix (Gard). Deuxième partie (Lias inférieur et moyen), *Bull. Soc. d'Etude Sci. Natur. Nimes* **XLII–XLV**, 1–134.

BRUNET, M.F. (1985) Evolution de la subsidence du Trias à l'Oxfordien le long d'une coupe La Voulte-St Ambroix (marge cévenole). *Doc. Bur. Recherch. Géol. Min.* **95–11**, 121–131.

BRUNET, M.F. & LE PICHON, X. (1980) Effet des variations eustatiques sur la subsidence dans le Bassin de Paris. *Bull. Soc. Géol. France*, sér. 7, **XXII**(4), 631–637.

CHANNEL, J.E.T., D'ARGENIO, B. & HORVATH, F. (1979) Adria, the African promontory in Mesozoic Mediterranean paleogeography. *Earth Sci. Rev.* **15**, 213–292.

COLONGO, M., ELMI, S. & SPY-ANDERSON, F.L. (1979) Changements dynamiques dans le comportement tectono-sédimentaire d'un secteur de la marge cévenole au passage Trias–Jurassique (région des Vans, Ardèche). *7ème Réunion Annuelle des Sciences de la Terre*, Lyon, Société Géologique de France, p. 122.

CONTINI, D. (1987) L'influence du milieu sur l'évolution de quelques lignées d'Ammonites à la limite Lias-Dogger.

Cahiers de l'Institut catholique de Lyon, 1, pp. 83–92.

CURNELLE, R. & DUBOIS, P. (1986) Evolution mésozoi'que des grands bassins sédimentaires français; bassins de Paris, d'Aquitaine et du Sud-Est. *Bull. Soc. Géol. France*, sér. 8, **II**(4), 529–546.

D'ARGENIO, B., CASTRO, P., DE EMILIANI, C. & SIMONE, L. (1975) Bahamian and Apenninic limestones of identical lithofacies and age. *Am. Assoc. Petrol. Geol. Bull.* **59**(3), 524–530.

DEBRAND-PASSARD, S. (1984) 11—Conclusions. Grandes lignes et principales étapes de l'evolution géodynamique du Sud-Est de la France. In: *Synthèse géologique du Sud-Est de la France*. Mém. Bur. Recherch. Géol. Min. 125, pp. 581–599.

DERCOURT, J., ZONENSHAIN, L.P., RICOU, L.E., KAZMIN, V.G., LE PICHON, X., KNIPPER, A.L., GRANDJACQUET, C., LEPVRIER, C., BIJU-DUVAL, B., SIBUET, J.C., SAVOSTIN, L.A., WESTPHAL, M. & LAUER, J.P. (1985) Présentation de 9 cartes paléogéographiques au 1/20 000 000e s'étendant de l'Atlantique au Pamir pour la période du Lias à l'Actuel. *Bull. Soc. Géol. France*, sér. 8, **I**(5), 637–652.

DROMART, G. (1986) *Faciès grumeleux, noduleux et cryptalgaires des marges jurassiques de la Téthys nord-occidentale et de l'Atlantique Central: genèse, paléo-environnements et géodynamique associée*. Thèse de Diplôme de Doctorat, Université Lyon I, p. 154.

DROMART, G. & ELMI, S. (1986) Développement de structures cryptalgaires en domaine pélagique au cours de l'ouverture des bassins jurassiques (Atlantique Central, Téthys occidentale). *Comptes Rend. Acad. Sci. Paris*, sér. II, **303**, 311–316.

ELMI, S. (1967) Le Lias supérieur et le Jurassique moyen de l'Ardèche. *Doc. Lab. Géol. Lyon* **19**, 1–845.

ELMI, S. (1971) Les influences mésogéennes dans le Jurassique moyen du Sud-Est de la France, comparaison avec l'Ouest algérien. *Ann. Inst. Geol. Hung.* **LIV**(2), 471–482.

ELMI, S. (1972) L'instabilité des Monts de Tlemcen et de Rhar Roubane (Ouest algérien) pendant le Jurassique, interprétation paléogéographique. *Comptes Rend. Soc. Géol. France* **5**, 220–222.

ELMI, S. (1978) Polarité tectono-sédimentaire pendant l'effritement des marges septentrionales du bâti africain au cours du Mésozoique (Maghreb). *Ann. Soc. Géol. Nord* **XCVII** (1977), 315–323.

ELMI, S. (1980) Jurassique de la bordure ardéchoise du Massif central français. In: *Paléomarge de la Téthys dans les Alpes occidentales du Massif central français aux ophiolites liguro-piémontaises: Géologie alpine* (Eds de Graciansky, P.C. & Lemoine, M.) vol. 56, pp. 126–129.

ELMI, S. (1981) Sédimentation rythmique et organisation séquentielle dans les Ammonitico-Rosso et les faciès associés du Jurassique de la Méditerranée occidentale. Interprétation des grumeaux et des nodules. *Rosso Ammonitico Symposium Proceedings: Tecnoscienza* (Eds Farinacci, A. & Elmi, S.) pp. 251–299.

ELMI, S. (1983a) L'évolution des Monts de Rhar Roubane (Algérie occidentale) au début du Jurassique. In: *Livre jubilaire Gabriel Lucas, Géologie sédimentaire*. Mém. Géol. de l'Université de Dijon 7, pp. 401–412.

ELMI, S. (1983b) La structure du Sud-Est de la France: une

approche à partiré de la bordure vivaro-cévenole du Massif central. *Comptes Rend. Acad. Sci. Paris*, sér. II, **296**, 1615–1620.

ELMI, S. (1984) Tectonique et sédimentation jurassique. In: *Synthèse du Sud-Est de la France*. Mém. Bur. Recherch. Geol. Min., 125, pp. 166–175.

ELMI, S. (1985a) Influence des hauts-fonds sur la composition des peuplements et sur la dispersion des ammonites. In: *Géodynamique des Seuils et des Hauts-fonds*: Bull. Sect. Sci., Commission des Travaux Hist. Sci., Vol. IX, pp. 217–228.

ELMI, S. (1985b) Chronologie et dynamique de l'enfoncement jurassique de la marge ardéchoise le long de la gouttière de Valvignères. *Doc. Bur. Recherch. Geol. Min.* **95**–11, pp. 73–89.

ELMI, S. (1985c) Evolution historique et dynamique de la marge ardéchoise pendant le Mésozoique. *Doc. Bur. Recherch. Géol. Min.* **95**–11, 13–50.

ELMI, S. (1987) In: *Synthèse Géologique Régionale*. Programme Geologie profonde de la France. Troisième phase d'investigation. Theme 11—Subsidence et diagenèse de la bordure ardéchoise du Bassin du Sud-Est. *Doc. Bur. Recherch. Géol. Min.* **123**, 1–143.

ELMI, S. & AMEUR, M. (1986) Quelques environnements des facies noduleux mésogéens *Géol. Rom.* **XXIII** 13–22.

ELMI, S. & BENSHILI, K. (1987) Relations entre la structuration tectonique, la composition des peuplements et l'évolution; exemple du Toarcien du Moyen-Atlas méridional (Maroc). *Boll. Soc. Paleontol. Ital.* 26, No. 1–2, p. 47–62.

ELMI, S., BENSHILI, K. & RULLEAU, L. (1986) Position stratigraphique et systématique des groupes de l'*Ammonites bayani* (*Crassiceras*) et de l'*Ammonites gruneri* (*Gruneria*) dans le Toarcien mésogéen: *Atti i Convegno, Fossili Evoluzione Ambiente, Comitato Centanario Raffaele Piccinini* (Ed. Pallini) pp. 93–109, Pergola, 1984.

ELMI, S., DROMART, G., GALIEN, F. & TALBI, D. (1984) Les contrôles de la structuration précoce de la bordure vivaro-cévenole (Hettangien à Oxfordien), In: *Colloque national*. Programme Géologie profonde de la France. Première phase d'investigation 1983–1984. Thème 11—Subsidence et diagenèse de la bordure ardéchoise du Bassin du Sud-Est. *Doc. Bur. Rech. Géol. Min.* 91–11, pp. 1–22.

ELMI, S. & MOUTERDE, R. (1965) Le Lias inférieur et moyen entre Aubenas et Privas (Ardèche). *Travaux Lab. Géol. Lyon*, **12**, pp. 143–246.

ELMI, S., MOUTERDE, R., RUGET, C., ALMERAS, Y. & NAUD, G. (1987) Le Jurassique inférieur du Bas-Vivarais (Sud-Est de la France). *Cahiers de l'Institut catholique de Lyon* 1, 163–189.

ELMI, S. & VITRY, F. (1987) L'Hettangien inférieur du Beaujolais et du Mont d'Or lyonnais (France, Sud-Est). un système lagune-barrière progradant. *Premier Congrès Français de Sédimentologie*, pp. 167–168.

ENAY, R. (1966) L'Oxfordien dans la moitié sud du Jura Français. Etude stratigraphique. *Nouvelles Archives du Muséum d'Histoire Naturelle de Lyon* **VIII**, 1–624.

ENAY, R. & MANGOLD, C. (1980) Synthèse paléogéographique du Jurassique français. *Doc. Lab. Geol. Lyon* **5**, 210.

FARINACCI, A. & ELMI, S. (Eds) (1981) Preface. *Rosso Ammonitico Symposium Proceedings: Tecnoscienza*, pp. 1–8.

FARINACCI, A., MALANTRUCCO, G., MARIOTTI, N. & NICOSIA, U. (1981). Ammonitico Rosso facies in the framework of the Martani Mountains paleoenvironmental evolution during Jurassic. *Rosso Ammonitico Symposium Proceedings: Tecnoscienza* (Eds Farinacci, A. & Elmi, S.) pp. 311–334.

FAZZUOLI, M. & SGUAZZONI, G. (1981) Presenza di facies tipo "Rosso Ammonitico" e di forme paleocarsiche al tetto dei Marmi in localata' Pianellacio (M. Pisanino—Alpi Apuane): *Boll. Soc. Geol. Ital.* **100**, 555–566.

FERRY, S. & RUBINO, J.L. (1987) Les séquences carbonatées néocomiennes du Sud-Est de la France sont-elles le résultat d'oscillations eustatiques? *Compt. Rend. Acad. Sci. Paris*, sér. II, **304**, 15, 917–922.

GABILLY, J., CARIOU, E. & HANTZPERGUE, P. (1985) Les grandes discontinuités stratigraphiques au Jurassique: témoins d'événements eustatiques, biologiques et sédimentaires. *Bull. Soc. Géol. France*, sér. 8, **I**(3), 391–401.

GAETANI, M. (1975) Jurassic stratigraphy of the Southern Alps. *In Geology of Italy*, (Ed. Squyre, C.) Earth Sciences Society Libyan Arab Republic, pp. 377–402.

GAILLARD, C. (1983) Les biohermes à spongiaires et leur environnement dans l'Oxfordien du Jura méridional. *Doc. Lab. Géol. Lyon* **90**, 429.

GALIEN, F. (1985) *Le Lias inférieur et moyen du bassin de Privas (Ardèche): sédimentologie, interprétations tectono-sédimentaires et paléogéographiques*. Thèse 3ème Cycle, Université Lyon I, p. 203.

HALLAM, A. (1984) Relations between biostratigraphy, magnetostratigraphy and event stratigraphy in the Jurassic and Cretaceous. *Proc. 27th Int. Geol. Congress, Sciences Press* I, pp. 189–212.

HAQ, B.U., HARDENBOL, J. & VAIL, P.R. (1987) Chronology of fluctuating sea levels since Triassic. *Science* **235**, 1156–1166.

HAUG, E. (1900) Les géosynclinaux et les aires continentales. Contribution à l'étude des transgressions et des régressions marines. *Bull. Soc. Géol. France*, sér. 3, **XXVIII**, 617–711.

HAUG, E. (1911) Traité de Géologie II. Les périodes géologiques, fasc. 2. *Jurassique et Crétacé* (Ed. Colin, A.) pp. 929–1396.

LAUBSCHER, H. & BERNOULLI, D. (1977) Mediterranean and Tethys. In: *The Ocean Basins and Margins*, vol. 4A (Eds Nairn, A.E.M., Kanes, W.H. & Stehli, F.G.). Plenum Press, London, p. 1–28.

LEMOINE, M. (1983) Tectonique synsédimentaire mésozoique dans les Alpes occidentales: naissance et évolution d'une marge continentale passive. In: *Livre Jubilaire Gabriel Lucas: Géologie Sédimentaire*. Mém. Géol. l'Université de Dijon 7 pp. 347–361.

LEMOINE, M. (1985) Structuration jurassique des Alpes occidentales et palinspatique de la Téthys ligure. *Bull. Soc. Géol. France*, sér. 8, **I**(1), 126–137.

LISTER, G.S., ETHERIDGE, M.A. & SYMONDS, P.A. (1986) Detachment faulting and the evolution of passive continental margins. *Geology* **14**, 246–250.

LOMBARD, A. (1956) Géologie sédimentaire. Les séries marines (Ed. Masson) p. 722.

LUCAS, G. (1942) Description géologique et pétrographique des Monts de Ghar Rouban et du Sidi el Abed. *Bull. Serv. Géol. d'Algérie*, sér. 2, **16**, 539.

MARTIN, D. (1985) *Modalités de la transgression rheto-hettangienne sur la bordure vivaro-cévenole, dans le sous-bassin d'Aubenas (Ardèche): étude sédimentologique et séquentielle, paléoécologie, paléogéographie.* Thèse 3ème Cycle, Université Lyon I, p. 157.

MASSARI, F. (1979) Oncoliti e stromatoliti pelagiche nel Rosso Ammonitico Veneto. *Mem. Sci. Geol.* **XXXII**, 21.

MASSARI, F. (1981) Cryptalgal fabrics in the Rosso Ammonitico sequences of the Venetian Alps. *Rosso Ammonitico Symposium Proceedings: Tecnoscienza*, 435–469.

MCKENZIE, D. (1978) Some remarks on the development of sedimentary basins. *Earth Plan et Sci. Lett.* **40**, 25–32.

MONTENAT, C. & BARRIER, P. (1985) Dynamique d'un seuil: le détroit de Messine du Pliocène à l'Actuel. *Bull. Sect. Sci., Commission des Travaux Hist. Sci.* **IX**, 11–24.

MONTENAT, C., BUROLLET, P., JARRIGE, J.J., OTT D'ESTEVOU & PURSER, B.H. (1986) La succession des phénomènes tectoniques et sédimentaires néogène,es sur les marges du Rift de Suez et de la Mer Rouge nord-occidentale. *Compt. Rend. Acad. Sci. Paris*, sér. II, 213–218.

MORTON, N. (1987) Jurassic subsidence history in the Hebrides, N.W. Scotland. *Marine Petrol. Geol.* **4**, 226–242.

MOUTERDE, R. (1952) Etudes sur le Lias et le Bajocien des bordures nord et nord est du Massif Central français. *Bull. Serv. Carte Géol. France* **236**, 458.

OWODENKO, B. (1946) Mémoire explicatif de la carte géologique du bassin houiller de Djerada et de la région au Sud d'Oujda (Maroc oriental français). *Ann. Soc. Géol. Belgique* **LXX**, p. 168.

PRATSCH, J.C. (1982) Wedge tectonics along continental margins. In: *Studies in Continental Margin Géology.* Am. Assoc. Petrol. Geol. Memoir 34, 211–220.

PURSER, B.H. (1975) *Sédimentation et diagénèse précoce des séries carbonatées du Jurassique moyen de Bourgogne.* Thèse Doctorat ès Sciences, Université Paris-Sud-Orsay, p. 453.

ROMAN, F. (1950) Le Bas-Vivarais. *Actualités Scientifiques et Industrielles* **1090**, 150.

ROYDEN, L.H. (1982) *The evolution of the Intra-Carpathian basins and their relationship to the Carpathians mountains system.* Thesis, Massassuchets Institute of Technology, p. 256.

SAYN, G. & ROMAN, F. (1928) Monographie stratigraphique et paléontologique du Jurassique moyen de la Voulte-sur-Rhône. *Travaux des Laboratoires de Géologie de Lyon* **XIII**, 165.

SEYFRIED, H. (1981) Genesis of "regressive" and "transgressive" pelagic sequences in the Tethyan Jurassic. *Rosso Ammonitico Symposium Proceedings: Tecnoscienza* (Eds Farinacci A. & Elmi, S.) pp. 547–579.

SPY-ANDERSON, F.L. (1980) *La bordure vivaro-cévenole au Trias dans la région des Vans (Ardèche): histoire tectono-sédimentaire, évolution diagénétique d'encroutements dolomitiques de piémont et de la plaine alluviale.* Thèse 3ème Cycle, Université Lyon I, p. 158.

SPY-ANDERSON, F.L. (1981) Dolocrêtes et nodules dolomitiques, résultats de la dolomitisation directe, en milieu continental, de sédiments terrigènes de la "formation bariolée supérieure" (Keuper) de la région des Vans (Ardèche, Sud-Est de la France). *Bull. Bur. Recherch. Géol. Min.*, sér. 2, **3**, 195–205.

STURANI, C. (1971) Ammonites and stratigraphy of the "*Posidonia alpina*" beds of the Venetian Alps. *Mem. Ist. Geol. Min. Univ.* **XXVIII**, 1–190.

TALBI, D. (1984) *Etude sédimentologique et séquentielle d'une série liasique au Sud de l'Ardèche (bordure vivaro-cévenole).* Thèse 3ème Cycle, Université Lyon I, p. 119.

TAUGOURDEAU-LANTZ, J. & LACHKAR, G. (1985) Stratigraphie par les marqueurs palynologiques sur la bordure ardéchoise du bassin du Sud-Est. Programme Géologie profonde de la France. Deuxième phase d'investigation 1984–1985, GPF 2. Theme 11—Subsidence et diagenèse de la bordure ardéchoise du bassin du Sud-Est. *Doc. Bur. Recherch. Géol. Min.* **95–11**, 149–163.

VAIL, P.R., MITCHUM, R.M. & THOMPSON, S. (1977) Seismic stratigraphy and global changes of sea level, Part 4: Global cycles of relative changes of sea level. In: *Seismic Stratigraphy, Applications to Hydrocarbon Exploration.* Am. Assoc. Petrol. Geol., Mem. 26, p. 83–87.

VERA, J.A. (1984) Aspetos sedimentologicos en la evolucion de los dominios alpinos mediterraneos durante el Mesozoico: Libro homenaje Luis Sanchez de la Torre. *Publ. Geol.* **20**, 25–54.

WERNICKE, B. (1981) Low-angle normal faults in the Basin and Range province: Nappe tectonics in an extending orogen. *Nature* **291**, 645–648.

WINTERER, E.L. & BOSELLINI, A. (1981) Subsidence and sedimentation on Jurassic passive continental margin, Southern Alpe, Italy. *Am. Assoc. Petrol. Geol. Bull.* **65**, 394–421.

Spec. Publs int. Ass. Sediment. (1990) **9**, 145–168

The formation and drowning of isolated carbonate seamounts: tectonic and ecological controls in the northern Apennines

D. M. BICE* and K. ·G. STEWART[†]

Department of Geology and Geophysics, University of California, Berkeley, CA 94720, USA

ABSTRACT

During the early Jurassic, a large carbonate platform—the Calcare Massiccio platform—developed on the Italian continental crust. The subsequent opening of the Liguride Ocean produced a continental margin that thinned and extended along a complex system of intersecting normal faults in the area that is now the northern Apennines. These normal faults fragmented the pre-existing Calcare Massiccio platform, producing a group of small, isolated seamounts separated by interconnected sub-basins up to 1 km deep. The eventual drowning of these seamounts had a profound effect on the later history of the entire basin, and accounts for the stratigraphic differences between the northern Apennines and the adjacent central Apennines, an area that was not broken up to the same extent by normal faulting. Studies of the sedimentologic and stratigraphic record indicate that the drowning resulted from a combination of tectonic and ecologic factors. The drowning appears to have occurred in two stages; an incipient stage, characterized by water depths well within the euphotic zone and abnormally low rates of sediment production and accumulation, and a complete, or terminal, stage in which the seamount tops dropped well below the euphotic zone. The transition between these two stages of drowning was most likely due to the inability of the seamount tops to produce and accumulate enough sediment to maintain a shallow-water position. This inability can be attributed to a variety of tectonic and ecologic factors: the lack of effective reef-builders, the lack of fringing, elevated rims, the small sizes of the seamount tops, the incursion of colder, less saline waters into the basins, crustal subsidence and eustatic sea-level changes. The evolution of these seamounts demonstrates the important controls that plate tectonics and the history of organisms can have on the development of a sedimentary basin.

INTRODUCTION

Carbonate platforms and their adjacent basins are dynamic systems that change through time and space in response to a number of variables. Once initiated, the platforms may expand as their margins grow outward (e.g., the Dolomites; Bosellini, 1984), grow upward while their margins remain stationary (e.g., the Bahamas; Schlager & Ginsburg, 1981), retreat as their margins step backward through time (e.g., parts of the Devonian reefs fringing the Canning Basin; Playford, 1980), or any combination of these basic trends. Among the numerous variables that

influence the evolution of carbonate platforms, James & Mountjoy (1983) have suggested that tectonic setting, sea-level fluctuations, the nature of sedimentation at the margin, variations in reef-building organisms through time and diagenesis are the most important controlling factors. In this paper, it is shown how the tectonic setting and ecological factors have combined to play a key role in the formation and drowning of a group of small, isolated carbonate platforms (seamounts) from the northern Apennines of Italy (Fig. 1).

The tectonic setting is an important factor in producing crustal structures and controlling their patterns. Because crustal structures commonly act as nuclei for the growth of carbonate platforms, the geographical distribution of carbonate platforms may

* Present address: Department of Geology, Carleton College, Northfield, MN 55057, USA.
† Present address: Department of Geology, University of North Carolina, Chapel Hill, NC 27514, USA.

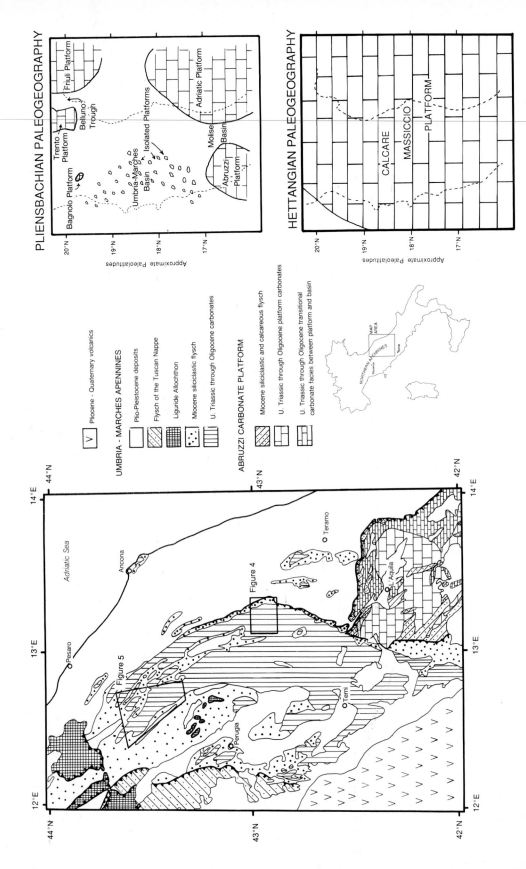

Fig. 1. Index map, generalized geology and palaeogeography of the study area. Hettangian palaeogeography shows the regionally extensive Calcare Massiccio platform. Pliensbachian palaeogeography shows scene after extension and break-up of pre-existing platform, modified from Parotto & Praturlon (1975), Pieri & Mattavelli (1986) and Bosellini et al. (1981). Palaeolatitudes from palaeomagnetic data of Channell et al. (1984) and Westphal et al. (1986).

be a direct result of the tectonic setting. The tectonic setting is also important because it controls the rate and total amount of crustal subsidence, which in turn limits the rate and total amount of carbonate sediment that can accumulate. Tectonically-controlled subsidence rates are also important factors in determining whether a platform will grow predominantly upward or outward (Bice, 1986).

Numerous modern and ancient carbonate platforms have been initiated along crustal structures that provided the starting topography and variations in sediment production rates necessary to create platform and basin systems. Mullins (1983) synthesized geophysical data that argue for structural control on the locations of platforms in the Bahamas, Belize and the Great Barrier Reef of Australia. Hurst & Surlyk (1984) documented the structural control of platform margins in northern Greenland. Castellarin (1972) and Bosellini *et al.* (1981) demonstrated the structural origin of several platform margins in the southern Alps.

The evolution of carbonate platforms is strongly tied to the rate of sediment production, and because most of the sediment is biogenic, the organisms that produce carbonate sediment are also important factors in controlling the development of carbonate platforms. Evolutionary changes in these organisms have played a major role in determining the character of carbonate platform sediments by altering the environmental requirements of these organisms, and the nature and quantity of carbonate they produce (Heckel, 1974; Wilson, 1975; James, 1983). Aside from evolutionary changes, sudden changes in the local or regional environment that affect the productivity of these organisms can also control the development of carbonate platforms (Schlager, 1981).

The Jurassic carbonates of the northern Apennines record the formation of a group of small, fault-bounded structural highs (seamounts) and their subsequent drowning, which represents a critical stage in the evolution of the entire basin. The structural highs have been referred to as seamounts (Centamore *et al.*, 1971; Bernoulli & Jenkyns, 1974), and isolated platforms (Bice & Stewart, 1985). In this paper, the term seamount is used because although it may evoke a submarine volcanic edifice to some, we feel it properly conveys the small sizes of these features. The terms using carbonate platform do not emphasize the small sizes of these features, and they may imply that these features are extremely localized carbonate buildups, much like pinnacle reefs, when in fact the field data discussed below indicate that these features are horsts. These seamounts can also be thought of as remnants of a formerly much more extensive carbonate platform (see Fig. 1).

The purpose of this paper is to assess the roles played by tectonic and ecologic factors in the formation and drowning of these seamounts. These studies of the excellent exposures of Jurassic carbonates in the Umbria-Marches portion of the northern Apennines (Fig. 1) demonstrate that the drowning of these seamounts was a response to subsidence that resulted from extension and thinning of the crust, and ecologic factors that greatly reduced the rate of carbonate sediment production.

Importance of the drowning

In this paper, Schlager's (1981) definition of *drowning* is used, as an event in which the combined effects of crustal subsidence and eustatic sea-level rise outpace the accumulation of carbonate sediment, resulting in the drop of the platform or reef top below the euphotic zone. Many of the organisms that produce carbonate sediment require abundant sunlight that is present only at shallow-water depths. Lowering the sediment surface to greater-water depths can therefore halt the rapid production of carbonate sediment that is needed to keep pace with changes in relative sea-level. The term *incipient drowning* is also used as defined by Kendall & Schlager (1981) to describe the condition where the platform or reef top has been lowered below the zone of maximum carbonate production (1–15 m), but is still within the euphotic zone so that a slight relative sea-level drop or stillstand could enable the platform or reef to return to the zone of maximum sediment production. *Complete* or *terminal drowning* occurs when the platform top drops to depths such that the decreased sedimentation rates cannot bring the platform top back to the shallow-water zone of high sediment production, even with reasonable drops in relative sea-level.

A comparison of the stratigraphy of the Sibillini Mountains (which is representative of the northern Apennines) and the nearby Abruzzi platform (Fig. 2) illustrates the importance of the drowning in the northern Apennines. Before the Jurassic drowning event, these two areas shared a very similar geologic history, but they evolved quite differently after the drowning occurred. In the Sibillini Mountains, the drowning event was followed by the deposition of up

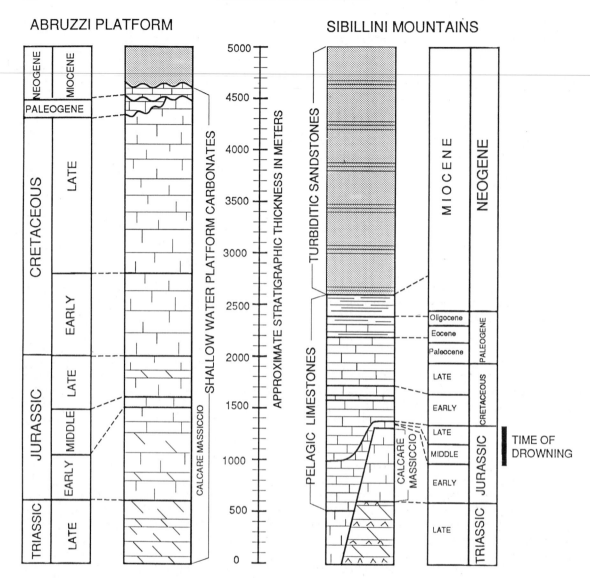

Fig. 2. Stratigraphical columns from the central Apennines (Abruzzi platform, adapted from Parotto & Praturlon, 1975) and the northern Apennines (Sibillini Mountains). Note that the stratigraphy of the two regions is similar until the end of the early Jurassic, but diverges after this time, corresponding to the break-up and drowning of the Calcare Massiccio platform in the northern Apennines.

to 1.5 km of mainly pelagic, deep-water carbonate over a period of about 180 Ma. These carbonates were then buried by as much as 3–5 km of Miocene–Pleistocene siliciclastic flysch. In contrast, after the Sibillini seamounts had drowned, the Abruzzi platform went on to accumulate up to 4 km of shallow-water carbonates during the same 180 Ma. Clearly, the drowning of the seamounts of the Sibillini and the rest of the northern Apennine basin is an important event in the subsequent geological history of this region.

DESCRIPTION OF THE SEAMOUNTS

Tectonic setting

The history of plate motions in the Mediterranean region is complex (Dewey *et al.*, 1973; Alvarez *et al.*, 1974; Alvarez, 1976, Bijou-Duval *et al.*, 1977; Vandenberg, 1979; Dercourt *et al.*, 1986) and the details are still not completely understood, but the general picture for most of the Italian peninsula appears to be fairly well established. The Italian peninsula (often referred to as Apulia or Adria in palaeogeographic reconstructions) was probably connected to southeast Europe in the Triassic (Dercourt *et al.*, 1986). By the middle of the early Jurassic, a small ocean basin, the Liguride Ocean, had begun to form as Italy moved away from southeast Europe. The time of this opening is established by radiometric dates and fossil ages from ophiolites (Bortolotti & Gianelli, 1976) as 160–185 Ma. Similar to the opening of the northern Atlantic Ocean, the Liguride Ocean was preceded by the formation of a system of Triassic rift valleys (Kligfield, 1979; Winterer & Bosellini, 1981; Martini *et al.*, 1986). After the initiation of the Liguride Ocean, the northern Apennines became part of a passive margin that persisted until the Oligocene when it became reactivated as part of a fold and thrust belt.

For the purposes of this paper, two points concerning the tectonic setting should be stressed. The first is that palaeomagnetic evidence (Channell *et al.*, 1984; Westphal *et al.*, 1986) places the study area between 15°N and 30°N latitude and isolated from any significant terrigenous input, making this a favourable setting for the production of carbonate sediments. The second point is that the formation of the Liguride Ocean caused the crust under the study area to thin along a system of extensional faults (D'Argenio & Alvarez, 1980), producing subsidence that persisted from the early Jurassic through the Palaeogene. As will be shown later, this network of extensional faults played a key role in the evolution of the seamounts of the northern Apennines.

Stratigraphic characteristics of seamounts and basins

Detailed stratigraphic and sedimentologic studies of the Jurassic carbonates of the Umbria–Marches portion of the northern Apennines (see Fig. 1) by Colacicchi *et al.* (1970) and Centamore *et al.* (1971) indicate the relatively sudden appearance of pronounced structural highs (seamounts) and lows (basins) in the northern Apennine basin at the end of the Sinemurian. These authors documented the variations in stratigraphic thicknesses and facies in Pliensbachian through early Cretaceous sediments that defined these structures. Centamore *et al.* (1971) introduced the useful terms 'complete sequence', 'condensed sequence' and 'reduced sequence', where the *complete sequence* refers to the thick basinal sediments that appear to be stratigraphically complete, the *condensed sequence* refers to the thinned sequences of sediments deposited on structural highs that appear to be stratigraphically complete, and the *reduced sequence* refers to the greatly thinned sequences of sediments deposited on structural highs that are stratigraphically incomplete. Figure 3 shows the generalized stratigraphic differences between the seamounts and basins. The Pliensbachian through early Cretaceous seamount facies is represented by the Bugarone and Maiolica formations.

The Bugarone Formation consists primarily of light brown carbonate mudstone and wackestone, commonly with a nodular texture, containing skeletal fragments of echinoderms, benthic foraminifera, bivalves and ammonites, with local packstone and grainstone containing peloids, intraclasts, ooids and skeletal fragments (see Centamore *et al.*, 1971 for a more complete description). The first sediments to be deposited on the seamounts are found in the lower portions of the Bugarone Formation, sometimes referred to as the Calcare Stratificati Grigi (Centamore *et al.*, 1971) or the Rosa a Crinoidi (Colacicchi *et al.*, 1970). This part of the sequence has been examined in four locations: Monte Rotondo and Monte Sassotetto from the Sibillini Mountains (Fig. 4), and Monte Cucco and Monte Nerone from the area shown in Fig. 5. The section at Monte Sassotetto has been studied previously by Cecca *et al.* (1981), with particular reference to the Tithonian sediments. The Rosa a Crinoidi here and at nearby Monte Rotondo (Fig. 9a) consists of wackestones and packstones with abundant echinoderm fragments, ammonite fragments, bivalves, ostracods, benthic foraminifera (notably *Vidalina martana*, which is characteristic of the Pliensbachian), peloids and ooids. Echinoderm fragments commonly have syntaxial overgrowths, and many of the skeletal fragments have micritic envelopes. In some areas, the skeletal fragments have been dissolved and the voids filled with a blocky/bladed isopachous rim cement that grades into coarser spar in the void centres. Many of the ooids appear to have been

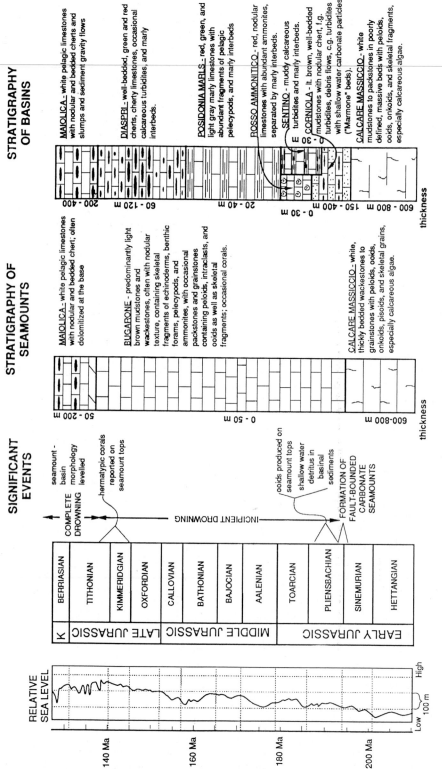

Fig. 3. Composite Jurassic stratigraphy showing lithology and thicknesses on seamounts and basins, estimated eustatic sea-level changes and the timing of significant events in the formation and drowning of the seamounts. Ages and eustatic changes from Haq et al. (1987); stratigraphic ranges of formations adapted from Centamore et al. (1971).

partially micritized, obscuring the concentric lamellae, but others have distinctive lamellae and are locally nucleated on the ostracodes that are typical of the Pliensbachian (Centamore *et al.*, 1971). Ooids also occur in the lowermost Bugarone from Monte Cucco (previously studied by Passeri, 1971), as well as Monte Rotondo, as shown in Fig. 9a. Some of these ooids are nucleated on tests of *Vidalina martana* (Fig. 9b), ensuring that these ooids were formed in the Pliensbachian and were not simply reworked from the underlying Calcare Massiccio.

There are also a number of relevant observations from the upper part of the Bugarone. At Monte Sassotteto (Fig. 4), Cecca *et al.* (1981) reported the presence of corals encrusting the Tithonian Rosso Ammonitico that was deposited on the upper part of the marginal slope of an isolated seamount. These corals are hermatypic, colonial forms that are generally limited to the euphotic zone and do best in water < 20 m deep (Wells, 1957; Buddemeier & Kinzie, 1976). The presence of these corals led Cecca *et al.* (1981) to suggest that the top of this

seamount was at very shallow-water depths as late as the Upper Tithonian. At other locations throughout the northern Apennines, including Monte Acuto and Monte Cucco (Fig. 5), Nicosia & Pallini (1977) found both hermatypic and ahermatypic corals in the Tithonian seamount-top sediments. Mariotti *et al.* (1979) described Kimmeridgian corals that form a small reef structure on the top of an isolated seamount north of Terni (see Fig. 1), but in most cases, these seamount corals did not form typically healthy buildups.

The seamount facies of the Maiolica formation consists of white carbonate mudstones with nodular and bedded cherts. No shallow-water fauna has been observed (Colacicchi *et al.*, 1970), but it is commonly dolomitized at its base.

The Pliensbachian through early Cretaceous basinal facies is represented by the Corniola, Sentino, Rosso Ammonitico, Diaspri and Maiolica formations (Fig. 3). Brief descriptions of the lithofacies of these formation are given in Fig. 3 and the reader is referred to Centamore *et al.* (1971) for a

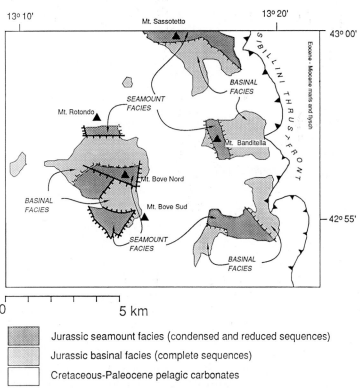

Fig. 4. Simplified geologic map of the Sibillini Mountains, showing distribution of seamounts and basins. Fault pattern suggests a complex system of intersecting normal faults.

Jurassic seamount facies (condensed and reduced sequences)

Jurassic basinal facies (complete sequences)

Cretaceous-Paleocene pelagic carbonates

Approximate trace of Jurassic normal faults drawn as solid lines where observable, dashed where inferred on basis of facies changes.

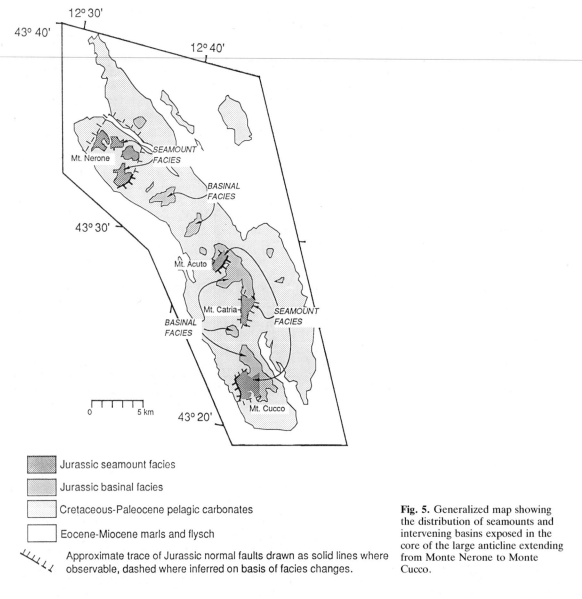

Jurassic seamount facies

Jurassic basinal facies

Cretaceous-Paleocene pelagic carbonates

Eocene-Miocene marls and flysch

Approximate trace of Jurassic normal faults drawn as solid lines where observable, dashed where inferred on basis of facies changes.

Fig. 5. Generalized map showing the distribution of seamounts and intervening basins exposed in the core of the large anticline extending from Monte Nerone to Monte Cucco.

more detailed account. The key features of the basinal lithofacies are: (1) the coarse-grained calcareous turbidites and debris flow deposits of the Corniola formation, called the Marmarone beds which contain detritus derived from the seamount facies of the Bugarone as well as the seamount facies of the Calcare Massiccio (presumably eroded from the marginal escarpment of the seamounts); (2) the lack of such coarse-grained sediment gravity flow deposits in the overlying formations; and (3) the presence of slumps and sediment gravity flows in

basinal Maiolica that demonstrate the persistence of the seamount–basin topography into the early Cretaceous.

Both the seamount facies and the basinal facies are overlain by the Aptian–Albian Fucoid Marls, whose approximately uniform thickness throughout the basin demonstrates that the seamount–basin topography was levelled off by this time (Montanari, 1985). The stratigraphic thickness differences between the seamount and basinal sequences (Fig. 3) indicate that the seamounts ranged from ≈ 1 km to 250 m in height above the adjacent basins.

Throughout the northern Appennines basin, these different sequences representing the seamounts and intervening basins overlie the Hettangian—Sinemurian Calcare Massiccio, which appears to have been deposited in water no deeper than 10—20 m (Colacicchi *et al.*, 1970). In general, the lithofacies of the Calcare Massiccio is a wackestone to grainstone that includes ooids, peloids, oncoids and skeletal fragments of calcareous algae, benthic foraminifera, gastropods, bryozoans and echinoderms. Colacicchi & Pialli (1969), Colacicchi *et al.* (1970) and Centamore *et al.* (1971) recognized that because the thin seamount facies and the thick basinal facies both overlie the shallow-water deposits of the Calcare Massiccio, synsedimentary normal faulting is necessary to explain the sudden formation of structural highs (seamounts) and lows in the basin floor that lasted until the end of the Jurassic.

In addition to these differences in the Pliensbachian through early Cretaceous sediments, there are more subtle differences in the Hettangian—Sinemurian Calcare Massiccio. Colacicchi & Pialli (1969) noted that the Calcare Massiccio found beneath the reduced and condensed Jurassic sequences was deposited in an intertidal setting, whereas the Calcare Massiccio found beneath the complete sequences represents consistently subtidal deposition. Colacicchi & Pialli (1969) interpreted this difference as evidence that the synsedimentary normal faulting that produced these seamounts and basins may actually have started during the Hettangian—Sinemurian. According to Schlager (1981), shallow platform carbonates are capable of producing sediment at rates of $\approx 1-10 \, \text{m} \, 1000 \, \text{yr}^{-1}$, which could outpace reasonable fault-displacement rates and therefore prevent the development of a fault scarp at the surface. Without a substantial fault scarp at the surface, there would be no obvious difference between the sedimentary facies on either side of the fault. Synsedimentary faulting during the deposition of the Calcare Massiccio can only be detected by observing the thicknesses around the basin, but the scarcity of exposures and well data on the base of the Calcare Massiccio prohibit an answer to this question.

Dimensions and map-view pattern of the seamounts

The stratigraphic evidence cited above indicates that the vertical relief of the seamounts was 250—1000 m. The horizontal dimensions and spatial distribution

of the seamounts are more difficult to determine. As mentioned above, the northern Appennines are presently a complex fold and thrust belt that formed in the Neogene. As a result, exposures of the Jurassic seamounts are generally limited to the cores of large anticlines where erosion has removed the overlying sediments. Several of these seamounts and intervening basins are particularly well exposed in the Sibillini Mountains (Fig. 4) and in the anticline that runs from Monte Nerone to Monte Cucco (Fig. 5). Studies of these exposures provide an approximate map-view pattern of the normal faults and the timing of offset on these faults. In a number of localities (Figs 4 & 5), the original marginal escarpments that bounded the seamounts have been located. Several of these seamounts have clearly exposed marginal escarpments, other margins are located on the basis of trends in thickness and facies changes in the post-faulting Jurassic sediments.

The sizes of the seamounts vary throughout the northern Apennines, but few have a surface area > 20 km². Some of the seamount tops are extremely small (of the order of 2 km² at Monte Rotondo) and irregular, perhaps due to dissection by secondary normal faults, whereas other seamounts appear to be relatively broad and flat on top (e.g., Monte Nerone and Monte Cucco). As discussed later, these structurally controlled features of the seamounts (their size and upper surfaces) appear to be important factors in their drowning.

Marginal escarpments of the seamounts

The generalized geological map in Fig. 4 shows two of the seamounts and an intervening basin that has been studied in detail. As described by Bice & Stewart (1985), the marginal escarpments of these seamounts are essentially Jurassic fault scarps that have been modified by submarine erosion from by-passing sediments, producing a series of sub-parallel grooves that can be seen in the photographs of Fig. 6. The seamounts shown in Fig. 6 are composed primarily of the Hettangian—Sinemurian Calcare Massiccio, but as mentioned above, these seamounts cannot be localized carbonate buildups like pinnacle reefs, because the shallow-water Calcare Massiccio also occurs in the grabens separating the seamounts (which are horsts). These escarpments have slopes that range from 25° to 60° and truncate horizontal beds of the Calcare Massiccio. The marginal escarpments are characterized by sub-parallel erosional grooves or gullies that range in depth from $\approx 1-5 \, \text{m}$

Fig. 6. Photographs of preserved portions of the marginal escarpment of seamounts in the Sibillini Mountains. (a) Monte Rotondo, south face: area within the dashed line is the preserved part of the escarpment and is the basis for the diagram in Fig. 8. Height from top of seamount to base of exposed escarpment is 250 m. Young normal fault is approximately perpendicular to marginal escarpment. Horizontally bedded Calcare Massiccio is exposed where the marginal escarpment has been recently eroded. Town of Casali is in foreground.

with a spacing that ranges from 5 to 20 m. These erosional grooves extend from the top of the marginal escarpment to as much as 500 m below the seamount top at Monte Bove; their sub-parallel arrangement is a classic example of a line source of bypassing sediment. The basinal sediments found in the valley between Monte Bove and Monte Rotondo (Fig. 4) consist of coarse-grained calcareous turbidites in the lowermost Corniola Formation, and volumetrically more significant fine-grained calcareous turbidites in the Upper Corniola and Sentino formations. The absence of debris flow deposits or submarine talus deposits derived from the Calcare Massiccio suggests that relatively little erosion of the marginal escarpments has occurred. The relatively minor erosion represented by the grooves is perhaps best explained by the erosive action of bypassing

coarse-grained sediment gravity flows. The bulk of the sediment filling the erosional gullies themselves consists of Pliensbachian–Toarcian aged skeletal fragments. Morphologically similar erosional features have recently been observed on the continental slope off New Jersey (Robb *et al.*, 1983; Farre & Ryan, 1985). Smaller versions of these grooves occur on the southern margin of the isolated seamount located at Monte Acuto (Fig. 5).

Close-up views of these grooved escarpments (Fig. 7) show approximately horizontal, massively bedded Calcare Massiccio truncated by the eroded fault scarp, which is onlapped by Pliensbachian-aged sediments. These Pliensbachian-aged sediments, dated on the basis of ammonites and the benthic foraminifera *Vidalina martana*, occur as crudely bedded, onlapping sediments in the troughs of

Fig. 6. *(Cont.)* (b) Monte Bove, north face, showing preserved marginal escarpment and remnant of the seamount top. Height from seamount top to base of exposed escarpment is 500 m.

grooves, and as thin, scattered veneers on other parts of the escarpment. These exposures allow the dating of the formation of these seamounts as latest Sinemurian to earliest Pliensbachian, clearly demonstrating that these seamounts had abrupt, steep margins (30°–60°), controlled by large normal faults. The stratigraphic relations and topography at this time, as seen in the Sibillini Mountains, are depicted in Fig. 8. One important feature to note is the nature of the margins of the seamount tops; they lack any kind of elevated rim and are simply abrupt slope changes where erosion was taking place. Elevated, fringing rims are important features of many carbonate platforms (Schlager, 1981), and as will be discussed later, their absence may have contributed to the drowning of these seamounts.

The original marginal escarpments and their onlapping sediments are important not only in dating the formation of the seamounts, but also in placing constraints on the displacement rates of the faults that created these seamounts and basins. The rate of displacement or subsidence is important in understanding the events that led to the formation of these seamounts. A minimum displacement rate can be obtained from the sum of the escarpment growth rate plus the basinal sediment accumulation rate, minus the sediment accumulation rate on the seamount top. Using the biostratigraphic results of Cecca *et al.* (1981), the time scale of Haq *et al.* (1987), and measured thicknesses from the map area in Fig. 4, the sediment accumulation rate on the seamount top is approximately $2 \, \text{m Myr}^{-1}$. The basinal sediment accumulation rate is the order of $60 \, \text{m Myr}^{-1}$. The escarpment growth rate is best estimated at Monte Bove (Fig. 4), where erosional grooves on the marginal escarpment extend $\approx 500 \, \text{m}$

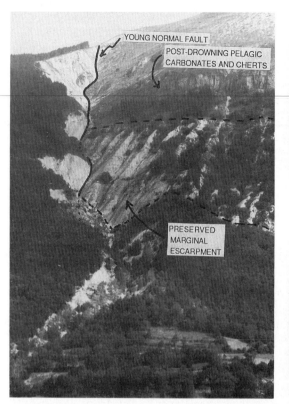

down from the former seamount top, and show no signs of diminishing, so it is estimated they extend at least another 100 m downward. This means that between the very end of the Sinemurian and sometime in the Pliensbachian (no more than 4 or 5 million years using the time scale of Haq *et al.*, 1987), a 600 m escarpment was formed, eroded and onlapped by sediments. This gives a minimum escarpment growth rate of between 120 and 150 m Myr^{-1}. The resulting fault displacement rate is between 178 and 208 m Myr^{-1}, significantly less than the 1000 m Myr^{-1} sediment production rate that is typical of shallow-water carbonate environments (Schlager, 1981). (It should be stressed that these rates are only estimates, and they are strongly dependent on the time spanned by the Pliensbachian.) Two possible conclusions can be drawn from the above rate estimates: (1) the actual fault displacement rates were much greater than the estimated minimum; or (2) environmental factors or subaerial exposure from a sea-level drop greatly reduced or halted the production of carbonate sediment, allowing the basins to form. Of course, these two possibilities are not mutually exclusive. Colacicchi *et al.* (1970), Cecca *et al.* (1981) and Farinacci *et al.* (1981) have inferred subaerial exposure of the seamount tops in the middle Jurassic, but there is no evidence for subaerial exposure of the Calcare Massiccio beneath the basinal areas, which suggests that environmental factors or rapid displacements enabled the basins to form.

DROWNING OF THE SEAMOUNTS

The sedimentary record contained in the seamount top and basinal sediments indicates that the complete drowning of these seamounts was preceded by a prolonged period of incipient drowning that extended through most of the Jurassic for many of the seamounts. The sedimentological and stratigraphical evidence for the stage of incipient drowning is first discussed and then the evidence regarding the timing of complete drowning.

Incipient drowning

From the above discussion of the stratigraphical and sedimentological differences between the seamounts and the basins, it seems quite clear that in the late Sinemurian—early Pliensbachian, the northern Apennines basin was subjected to extension along a network of normal faults. Observations from other continental margins (Steckler & Watts, 1978; Bond & Kominz, 1984) and theoretical studies (McKenzie, 1978) indicate that crustal subsidence is generally rapid immediately after the extension begins, but then tapers off as the crust becomes thermally and isostatically adjusted to the reduction in thickness. One might therefore expect that the northern Appennines would have experienced rapid subsidence immediately after faulting and that this subsidence would be the primary cause of the drowning. Instead, the post-faulting Jurassic sediments deposited on the seamount tops record a prolonged state of incipient drowning in relatively shallow water, which means that there was not significant subsidence immediately after the faulting. In cases where the sediments have been swept off the seamount tops, the detrital sediments found in the adjacent basins contain a similar record of incipient drowning and relatively shallow water on the seamount tops.

The sedimentological and stratigraphical evidence bearing on the drowning of the seamounts can best be understood with reference to a stratigraphic framework for the Jurassic of the northern Apennines, shown in Fig. 3. The first evidence to be considered comes from the post-faulting sediments deposited on the seamount tops, the Bugarone Formation. The second source of evidence is the detrital sediments found in the basinal Corniola Formation, often referred to as the Marmarone beds, and equivalent aged sediments that occur at the bases of the marginal escarpments. The third piece of evidence that will be considered is the presence of hermatypic corals from the upper portion of the Bugarone Formation.

Fig. 7. [*Opposite*] Details of grooved marginal escarpments. Upper left: Closer view of the western portion of the Monte Rotondo escarpment; the grooved surface shows original erosional features of the escarpment. Young normal fault is approximately perpendicular to marginal escarpment. Upper right: View looking up one of the erosional grooves of the marginal escarpment. Dashed line outlines Pliensbachian sediments that rest in the base of a groove eroded into the Calcare Massiccio. The Calcare Massiccio is approximately horizontally bedded here. These sediments contain detritus that was produced on the seamount top. Lower: Closer view of the preserved margin on Monte Bove showing Pliensbachian and younger sediments onlapping the grooved marginal escarpment. The widest part of the groove on the right is about 10−12 m wide.

Sibillini Mountains

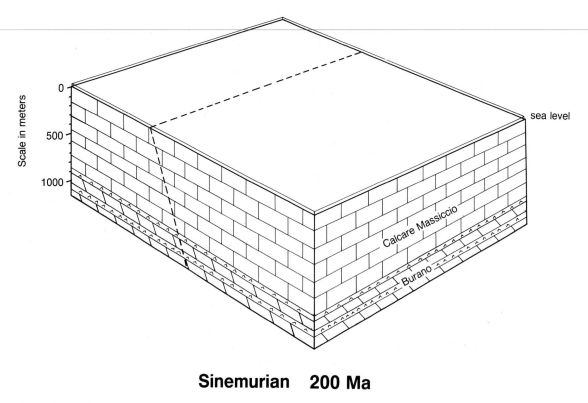

Sinemurian   200 Ma

Fig. 8. Schematic diagrams showing the formation and the approximate dimensions of the isolated seamount at Monte Rotondo. The grooved marginal escarpment can be seen in the photographs in Figs 6a & 7 (upper left).

The first sediments deposited on the seamount tops after the seamounts were created contain ooids, which suggests that the seamount tops at Monte Sassotetto, Monte Nerone, Monte Cucco, Monte Acuto and Monte Rotondo (see Fig. 9a,b) were at shallow-water depths during the Pliensbachian. A few ooids are nucleated on tests of *Vidalina martana* (Fig. 9b), ensuring that these ooids were formed in the Pliensbachian and were not simply reworked from the underlying Calcare Massiccio. Similar ooids have been studied by Jenkyns (1972), who found pelagic nannofossils incorporated into some of the ooids. Jenkyns (1972) considered these ooids to be analogous to micro-oncolites, growing by algal accretion and suggested that the ooids formed somewhere within the euphotic zone. The presence of these ooids in grainstone and packstone textures seems to argue for water that is agitated on a regular

basis, which is more likely in the upper 20−30 m for deep-water waves with a normal range of wavelengths. A similar water depth constraint is imposed by the presence of micritic envelopes, which are probably the result of endolithic algae (Bathurst, 1975). These observations argue that the seamount tops did not drop to great depths and drown immediately after they were formed by the normal faulting. Instead, it seems that the seamount tops remained at a relatively shallow-water depth, probably within the upper part of the euphotic zone.

The diagenetic features of the Lower Bugarone sediments may also be consistent with the hypothesis that the seamount tops were at relatively shallow-water depths during the Pliensbachian. Dissolved skeletal grains, syntaxial overgrowths on echinoderm fragments and the approximately isopachous, bladed to blocky cements, grading to coarser spar

Sibillini Mountains

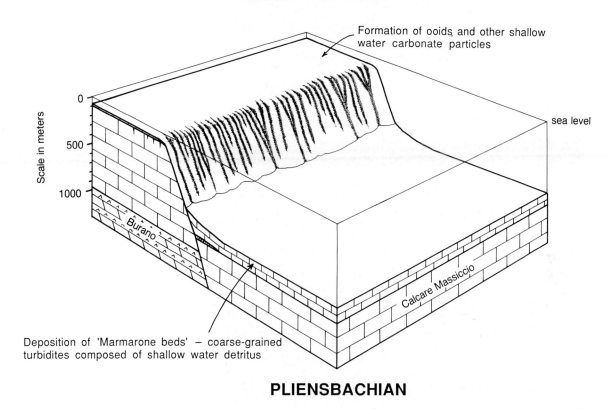

Fig. 8. (*Cont.*)

towards the centres of pores are all characteristic diagenetic features of the fresh-water phreatic zone (Longman, 1980). Since all of the seamounts in the northern Apennines appear to have been isolated from any landmass, the presence of these diagenetic features (and their absence from overlying pelagic sediments of the Cretaceous) seems to require a period of subaerial exposure of the seamount tops. Times of exposure have been postulated by other workers on the basis of independent evidence (see Colacicchi *et al.*, 1970; Farinacci *et al.*, 1981; and Cecca *et al.*, 1981). Given the most recent Jurassic sea-level curve (Haq *et al.*, 1987), which shows mainly low-amplitude fluctuations during this time, it seems quite likely that the seamount tops were probably never too far below sea-level.

The ooids from the seamount tops do not appear to be volumetrically significant in the basinal sedi-

ments. The Marmarone beds seldom exceed 20–30 m in thickness, and the ooids generally make up < 20% of these beds. Part of the reason for the paucity of ooids in the basinal sediments is the small size of most of the seamount tops relative to the sizes of the basins. If the seamount tops were larger and more efficient producers of sediment, one can imagine that the basins would have been completely filled with shallow-water detritus in a manner similar to that described by Bosellini *et al.* (1981). They described a situation in the southern Alps where a somewhat larger fault bounded platform, the Friuli platform (see Fig. 1), produced and exported enough ooids to fill the adjacent Belluno trough with nearly 1 km of reworked ooids. If the small sub-basins of the northern Apennines had been completely filled, there would have been a substantial increase in the total sediment budget, which may have prevented

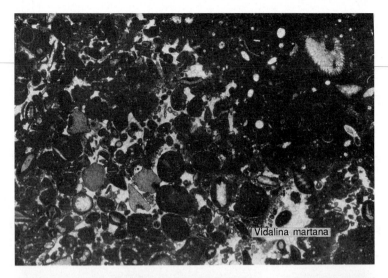

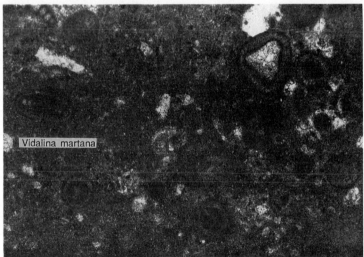

Fig. 9. Thin-section photomicrographs of Pliensbachian-aged sediments from the seamount tops and the basins. (a) Rosa a Crinoidi (lowermost Bugarone) from Monte Rotondo, 20X. (b) Same locality as (a); note ooid with *Vidalina martana* as nucleus, 100X.

the eventual drowning of the entire basin.

The third source of evidence concerning the drowning history of these seamounts comes from the corals of the uppermost Bugarone (Cecca *et al.*, 1981; Mariotti *et al.*, 1979; Nicosia & Pallini, 1977). In most cases, these corals do not occur in typically healthy buildups, which suggests that the ecological conditions were not quite right. Hermatypic corals have been found at depths of 150 m at Bikini Atoll (Wells, 1957), but they are stunted and rare. The corals in the northern Apennines, however, have well-developed, relatively large corallites (Nicosia & Pallini, 1977), which indicate favourable water and light conditions that are found in very shallow water.

The importance of these corals is that they demonstrate that many of the seamounts were still at shallow depths at the end of the Jurassic, which means that the state of incipient drowning lasted from the Pliensbachian to at least the Tithonian.

Although the presence of the hermatypic corals argues that many of the seamounts in the northern Apennines were in a state of incipient drowning until the very end of the Jurassic, not all of the seamounts had identical drowning histories. The Monte Rotondo seamount, for example, appears to have been at shallow depths during the Pliensbachian based on the presence of ooids and freshwater phreatic cements, as discussed above. However, the

Pliensbachian seamount top sediments at Monte Rotondo are succeeded by 50 m of Diaspri cherts and thinly-bedded micrites like those just west of Monte Sassotetto that have been assigned an age of Oxfordian to Kimmeridgian by Cecca *et al.* (1981). These deeper-water sediments are either absent or more condensed on top of other seamounts, such as Monte Sassotetto, and they commonly contain shallow-water fossils that are absent from the Diaspri above Monte Rotondo. The Monte Rotondo seamount seems to have gone from a state of incipient drowning to complete drowning sometime between the Pliensbachian and the Oxfordian. Regardless of the time of complete, or terminal drowning, many of the seamounts remained at shallow depths for a prolonged period of time after they were created by the normal faulting.

Timing of complete drowning

As mentioned above, the seamounts seem to have made the transition from incipient drowning to complete drowning at different times, but a number of seamounts remained in a state of incipient drowning until at least the early Tithonian. The upper Tithonian through lower Cretaceous Maiolica limestones filled the basins to the level of the seamount tops. The very fine-grained texture and abundant planktonic fossils of the Maiolica suggest deposition at substantial water depths in an open-ocean setting, with no evidence of shallow-water fauna (Colacicchi *et al.*, 1970). The evidence from the Maiolica establishes the upper Tithonian as the latest time of complete, irreversible drowning of the seamounts.

CAUSES OF THE DROWNING

Subsidence and sea-level changes

As Schlager (1981) has pointed out, normal rates of carbonate sediment production and accumulation are so high that the drowning of a reef or platform is a paradox. Figure 10 shows how a variety of factors may have combined to cause first the incipient drowning and then the complete drowning. Unusually rapid crustal subsidence or eustatic sea-level rise can conceivably cause drowning, but the preceding discussion discounts the possibility of rapid crustal subsidence. As D'Argenio & Alvarez (1980) pointed out, the nearby Abruzzi platform also experienced crustal thinning and subsidence, but the

platform there was not drowned, so subsidence alone cannot be the cause of the drowning in the northern Apennines. However, it is equally clear that without any subsidence whatsoever, drowning could not occur. The most recent estimate of Jurassic eustatic sea-level changes (Haq *et al.*, 1987) shows no rapid, large-amplitude sea-level changes at the onset of the incipient drowning (see Fig. 3) in the Pliensbachian, so it appears that eustatic sea-level changes were not the cause of the incipient drowning. However, because the rapid eustatic oscillations of the Tithonian (see Fig. 3) correspond to the transition from incipient drowning to complete drowning, it is reasonable to suppose (but difficult to prove) that these sea-level changes played a role in causing the complete, terminal drowning. It is thus necessary to consider what factors may have suppressed sediment production and accumulation rates in order to initiate the incipient drowning of these seamounts.

Configurations of the seamounts

The dimensions and shapes of the seamounts appear to have played a major role in the drowning. The network of extensional faults that broke up the Calcare Massiccio platform created a series of seamounts with relatively small tops. Some of these seamounts appear to have had flat tops (Monte Cucco, for instance) whereas others had irregular tops (Monte Bove) that may have resulted from secondary normal faults dropping down the edges of the seamounts, creating a roughly convex-up surface on the seamount top.

One of the important factors in determining the amount of sediment that accumulates on the seamount tops is the rate of sediment removal (see Fig. 10). The convex-up shape of some of the seamount tops would clearly have enhanced the rate of sediment removal. The small size of the seamount tops also favours the removal of sediment simply by reducing the distance from any point on the seamount top to the edge. The role of platform-top size in sediment removal has been cited by Hine & Steinmetz (1984) as an important factor in the low sediment accumulation rates on Cay Sal Bank, which is a possible modern analogue to these seamounts in the northern Apennines. Cay Sal Bank is relatively small, has no effective fringing reefs, and much of the bank is below 10 m of water, putting it below the zone of maximum carbonate production (Hine & Steinmetz, 1984). Sediment accumulation on the bank top is so low that the bank is not able to build

CAUSES OF DROWNING OF ISOLATED CARBONATE SEAMOUNTS IN THE NORTHERN APENNINES

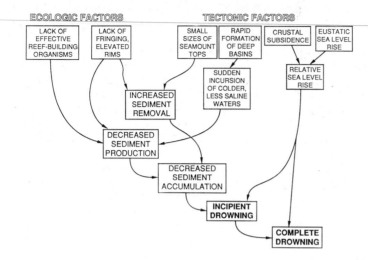

Fig. 10. Schematic diagram illustrating the influence of ecologic and tectonic factors on the drowning of isolated carbonate seamounts in the northern Apennines.

itself back up to shallower water, so Cay Sal Bank is in a state of incipient drowning.

Lack of elevated rims

Kendall & Schlager (1981) and Read (1985) have discussed the importance of elevated rims, commonly organic reefs, in the growth of carbonate platforms. The rims are commonly sites of the most rapid sediment production and they also absorb enough wave energy to control and modify the environment in the platform interior, making it a more efficient site for carbonate production. Without elevated rims, the platform interior is not well protected and can be subjected to environmental changes that decrease productivity. Elevated rims also prevent the easy removal of sediment from the platform interior. Thus, elevated rims enhance sediment accumulation by promoting sediment production and inhibiting sediment removal (see Fig. 10). The margins of the seamounts exposed at Monte Rotondo, Monte Bove, Monte Sassotetto, Monte Acuto and Monte Cucco show no signs of elevated rims. Figure 8 shows the inferred geometry of the seamount margin of Monte Rotondo, an abrupt edge that is the site of erosion rather than accumulation and up-building.

Lack of reef-building organisms

The depositional environment of the Calcare

Massiccio was probably an extensive, shallow platform with migrating and coalescing carbonate sand bodies and no reefal buildups (Colacicchi et al., 1970). In fact, Heckel (1974) and James (1983) noted that there appears to have been a lack of bioherms during early Jurassic time, although such are known in Morocco. The absence of effective reef-building communities in the Calcare Massiccio is significant, because when the seamounts were formed by the normal faulting, there were no reef-builders present to construct rims along the seamount margins. As mentioned above, this lack of rims probably had important consequences for the rate of sediment accumulation on the seamount tops and thus the drowning.

Rapid formation of deep basins

Another factor that probably altered the production of carbonate sediment on the seamount tops is the influence of the deep basins that separated the seamounts. The basins contained oxidized sediments with open marine planktonic fossils, so they were most likely connected to the open Tethys ocean. For this reason, the physical properties of the water in these basins would be similar to those of the open ocean and quite different from those of the water above the Calcare Massiccio platform. The lithofacies and depositional environment of the Calcare Massiccio make it fairly safe to assume that the water above this seamount was significantly warmer and perhaps more saline than normal oceanic water, as observed in the Bahamas (Bathurst, 1975), which is a reasonable actualistic model for the Calcare Massiccio depositional environment. Bathurst (1975) noted that in the Bahamas, warm, saline water is necessary for the effective production of ooids, which are important ingredients in the Calcare Massiccio. The lack of fringing reefs and small sizes of the isolated seamounts tops would allow mixing of the water above the seamounts with the basinal water, leading to cooler, less saline water that could have suppressed the production of carbonate (a condition referred to as 'hypothermia' by Bosellini, 1987). The existence of cooler, less saline water above the seamount tops is consistent with the observations that the seamounts did not produce significant quantities of ooids even though many of them were clearly at shallow enough depths for most of the Jurassic time. The result of this change in the water, brought on by the formation of the deep basins, was a decrease in the growth potential of the seamounts, which made them more susceptible to drowning.

SUMMARY

The opening of the Liguride Ocean in the early Jurassic resulted in crustal thinning through extensional faulting and long-term subsidence of the Italian Peninsula. One portion of this passive margin, the present day northern Apennines, was affected by a complex network of normal faults. These normal faults fragmented the interior of the regionally extensive Calcare Massiccio platform, producing a group of small isolated carbonate seamounts separated by interconnected basins up to 1 km deep. The subsequent drowning of these seamounts had a profound effect on the later history of this basin, and accounts for the important stratigraphic differences between the northern Appennines and the adjacent central Apennines. Studies of the sedimentological and stratigraphical record of this drowning lead to the conclusion that the drowning was the result of a combination of tectonic and ecological factors. Figure 11 shows a series of schematic cross-sections that illustrate a model for the formation and drowning of these seamounts.

Several of these seamounts and intervening basins are particularly well exposed in the Sibillini Mountains. Studies of the original marginal escarpments of these seamounts show that the faulting that created these seamounts and basins occurred during the late Sinemurian–early Pliensbachian (Fig. 11b) and was quite rapid with respect to sediment accumulation rates. Although rapid crustal subsidence would be expected in response to this rapid faulting, there appears to be little subsidence of the seamounts after their formation. The first post-faulting sediments produced on the seamount tops contain ooids, which requires that the seamounts tops were no more than a few meters deep. Despite this shallow depth, the rate of sediment production on the seamount tops (which is the critical factor in determining a platform's survivability) was relatively low, based on the rather small volume of shallow-water detritus found in the basins. The sedimentary record on the seamount tops is incomplete in the middle Jurassic, which Colacicchi et al. (1970), Farinacci et al. (1981) and Cecca et al. (1981) attributed to sub-aerial exposure. Nicosia & Pallini (1977), Mariotti et al. (1979) and Cecca et al. (1981) found small communities of healthy (large) hermatypic corals in the Tithonian seamount top sediments, which places some of the seamount tops in the upper part of the euphotic zone (<20 m) at this time. Some of the seamounts, such as Monte Rotondo, dropped to greater water depths before the Tithonian. The Upper Tithonian Maiolica limestones found above the seamount tops were probably deposited at substantial depths, well below the euphotic zone.

The sedimentary record indicates two stages of drowning. After the faulting and formation of the seamounts (Fig. 11b), they entered a stage of incipient drowning (Fig. 11c) that lasted until the end of the Jurassic, at which time they became completely, terminally drowned (Fig. 11d). The stage of incipient drowning was characterized by water depths within the euphotic zone and low rates of sediment production and accumulation. During this stage, a change in the ecology or environment could have returned the rates of carbonate production and accumulation to high enough levels to prevent drowning, but these rates remained at abnormally low levels, allowing the seamounts to become completely drowned at the end of the Jurassic.

The prolonged stage of incipient drowning indicates that crustal subsidence associated with the normal faulting was not the sole cause of the drowning. The coincidence of apparent rapid sea-level oscillations with the transition from incipient drowning to complete drowning suggests a cause-and-effect relationship. However, there do not appear to be any dramatic sea-level changes that correspond to the beginning of the stage of incipient drowning, so

CONCEPTUAL MODEL FOR THE FORMATION AND DROWNING OF CARBONATE SEAMOUNTS IN THE NORTHERN APENNINES

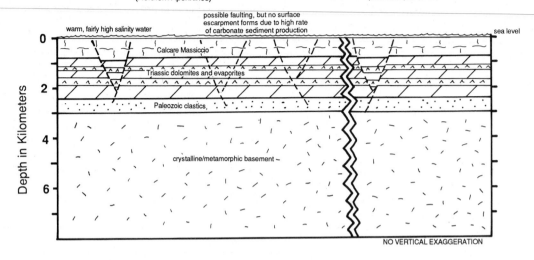

Sinemurian

(b) FORMATION OF SEAMOUNTS

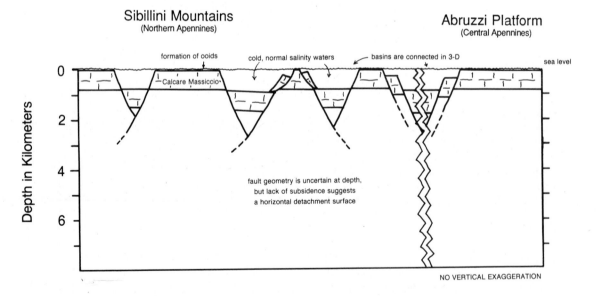

Pliensbachian

Fig. 11. Schematic chronological evolution of the northern Apennines basin in the Jurassic. The precise palaeogeographic connection between the Sibillini Mountains and the Abruzzi platform is uncertain, but detritus and megabreccias in the Sibillini Mountains are believed to have been derived from the Abruzzi platform (Castellarin *et al.*, 1978). (a) Pre-faulting stratigraphy; Calcare Massiccio platform covers both the northern and central Apennines. (b) Pliensbachian scenario; extension and relatively rapid normal faulting affects only the northern Apennines and forms small seamounts separated by deep interconnected basins. Seamounts begin stage of incipient drowning; limited production of ooids on seamount tops.

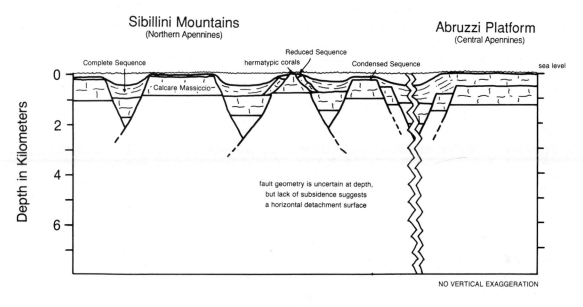

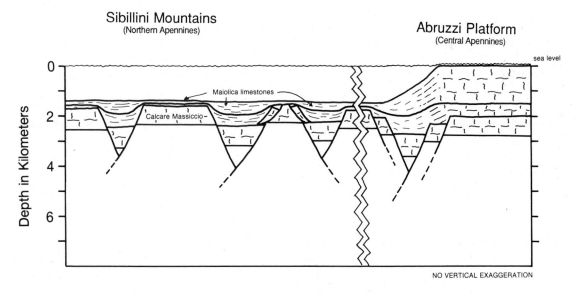

Fig. 11. (*Cont.*) (c) Stage of incipient drowning continues into the Tithonian. Basins fill with pelagic sediments; seamount tops are still rather shallow as shown by hermatypic corals but carbonate production is severely reduced. (d) Complete drowning of the seamounts occurs and the basin is levelled off by the Aptian. Entire northern Apennines Basin subsides to water depths of over a kilometre; Abruzzi platform remains at shallow depths.

the important causal factors appear to be those that disabled the system of carbonate production and accumulation. The small sizes of these seamounts facilitated the removal of sediment, reducing the rate of sediment accumulation. The lack of fringing organic reefs, which normally form elevated rims around the margins of seamounts, probably had two important effects: it contributed to the ease of sediment removal since elevated rims commonly act as efficient sediment traps, and it exposed the interiors of the seamount tops to the water and waves of the open ocean, preventing the development of the warm, high salinity water that is typical of the interiors of healthy carbonate platforms. The deep, rapidly formed basins that surrounded the seamounts appear to have been connected to the open ocean. The basinal water was probably much colder and less saline than the pre-existing water above the healthy Calcare Massiccio seamount, and the mixing of this basinal water with the water of the seamount tops may have been an important factor in lowering the rates of carbonate production, thus contributing to the drowning.

The small sizes of these seamounts, the sudden incursion of cold, less saline water and the crustal subsidence that eventually drowned these seamounts are all closely related to the system of extensional faults, so they are structurally controlled. These structures are in turn related to the plate tectonic history of the Italian Peninsula. To some extent, the lack of elevated rims around the seamounts is also related to the normal faults in that they isolated fragments of a former platform interior where there were no effective reef-building communities in residence. Perhaps to some extent, the absence of effective reef-builders is related to a world-wide scarcity of such organisms at the time of formation of the seamounts (Heckel, 1974; James, 1983).

In summary, the drowning of the seamounts of the northern Apennines represents a critical stage in the history of this basin. The drowning of these seamounts is mainly the result of a dramatic decrease in the rate of sediment accumulation on the seamount tops, which is the net result of increased sediment removal and decreased sediment production. The underlying causes of these changes in the sediment budget and the eventual drowning are: (1) the system of extensional faults related to the opening of the Liguride Ocean; and (2) the general scarcity of reef-building organisms during the Lower Jurassic, when these seamounts were created. The evolution of these seamounts from the northern Apennines demonstrates the powerful controls that plate tectonics and the history of organisms can have over the development of a sedimentary basin.

ACKNOWLEDGEMENTS

Many of the ideas in this paper are the products of close collaboration with Walter Alvarez, Sandro Montanari, Lung Chan and Mark Anders. The authors are especially indebted to Walter Alvarez for his enthusiasm and encouragement. This paper benefitted greatly from thorough reviews by G. Pialli, J.F. Read, M. Santantonio and J.L. Wilson. We thank the Club Alpino Italiano, San Severino Chapter and the Comune di Piobbico for field support; Carolee Bice for field assistance. The authors acknowledge support from NSF grants EAR80−22846 and EAR83-18660, grants from Chevron Overseas Petroleum, Inc., and Geol. Soc. America Research Grants awarded to Kevin Stewart and David Bice.

REFERENCES

ALVAREZ, W. (1976) A former continuation of the Alps. *Geol. Soc. Am. Bull.* **87**, 891−896.

ALVAREZ, W., COCOZZA, T. & WEZEL, F.C. (1974) Fragmentation of the Alpine orogenic belt by microplate dispersal. *Nature* **248**, 309−314.

BATHURST, R.G.C. (1975) Carbonate sediments and their diagenesis. *Developments in Sedimentology* 12, 2nd edn. Elsevier, Amsterdam, p. 658.

BERNOULLI, D. & JENKYNS, H.C. (1974) Alpine, Mediterranean, and central Atlantic Mesozoic facies in relation to the early evolution of the Tethys. In: *Modern and Ancient Geosynclinal Sedimentation* (Eds Dott, R.H., Jr. & Shaver, R.H.) Spec. Publ. Soc. Econ. Paleont. Mineral. no. 19, pp. 129−160.

BICE, D.M. (1986) Computer simulation of prograding and retrograding carbonate platform margins. *Geol. Soc. Am. Abs. Prog.*, vol. 18, no. 6, pp. 540−541.

BICE, D.M. & STEWART, K.G. (1985) Ancient erosional grooves on exhumed bypass margins of carbonate platforms: Examples from the Apennines. *Geology* **13**, 565−568.

BIJOU-DUVAL, B., DERCOURT, J. & LE PICHON, X. (1977) From the Tethys ocean to the Mediterranean seas: A plate tectonic model of the evolution of the western Alpine system. In: *Histoire Structurale Des Bassins Mediterraneens* (Eds Bijou-Duval, B. & Montadert, L.) Editions Technip, Paris, pp. 143−164.

BOND, G.C. & KOMINZ, M.A. (1984) Construction of tectonic subsidence curves for the early Paleozoic miogeocline, southern Canadian Rocky Mountains: Implications for subsidence mechanisms, age of break-up, and crustal thinning. *Geol. Soc. Am. Bull.* **95**, 155−173.

BORTOLOTTI, V. & GIANELLI, G. (1976) Le rocce gabroiche dell'Appennino settentrionale: 1. Dati recenti su rapporti primari posizione stratigrafica ed evoluzione tettonica. *Ofioliti* **2**, 99–105.

BOSELLINI, A. (1984) Progradational geometries of carbonate platforms: Examples from the Triassic of the Dolomites, northern Italy. *Sedimentology* **31**, 1–24.

BOSELLINI, A. (1987) Dynamics of Tethyan carbonate platforms. *Am. Assoc. Petrol. Geol. Bull.* **71**, 532.

BOSELLINI, A., MASETTI, D. & SARTI, M. (1981) A Jurassic "Tongue of the Ocean" infilled with oolitic sands: The Belluno trough, Venetian Alps, Italy. *Marine Geol.* **44**, 59–95.

BUDDEMEIER, R.W. & KINZIE, R.A., III (1976) Coral growth. In: *Oceanography and Marine Biology, Annual Review* (Ed. Barnes, H.) Vol. 14, pp. 183–225.

CASTELLARIN, A. (1972) Evoluzione paleotettonica sinsedimentaria del limite tra "piattaforma venetia" e "bacino lombardo", a nord di riva del garda. *Giornale di Geologia* **38**, 11–212.

CASTELLARIN, A., COLACICCHI, R. & PRATURLON, A. (1978) Fasi distensive, trascorrenze e sovrascorrimenti lungo la "Linea Ancona-Anzio" da Lias medio al Pliocene. *Geologica Romana* **17**, 161–189.

CECCA, F., CRESTA, S., GIOVAGNOLI, M. C., MANNI, R., MARIOTTI, N., NICOSIA, U. & SANTANTONIO, M. (1981) Tithonian "ammonitico rosso" near Bolognola (Marche-Central Apennines): A shallow water nodular limestone. In: *Rosso Ammonitico Symposium Proceedings. Tecnoscienza* (Eds Farinacci, A. & Elmi, S.) pp. 91–112.

CENTAMORE, E., CHIOCCHINI, M., DEIANA, G., MICARELLI, A. & PIERUCCINI, U. (1971) Contributo alla conoscenza del Giurassicao dell'Appennino umbromarchigiano. *Studi Geologici Camerti* **1**, 7–89.

CHANNELL, J.E.T., LOWRIE, W., PIALLI, G. & VENTURI, F. (1984) Jurassic magnetic stratigraphy from Umbrian (Italy) land section. *Earth Planet. Sci. Lett.* **68**, 305–325.

COLACICCHI, R. & PIALLI, G. (1969) Relationship between some peculiar features of Jurassic sedimentation and paleogeography in the Umbro-Marchigiano basin (Central Italy). *Ann. Inst. Geol. Publ. Hung.* **54**, 195–207.

COLACICCHI, R., PASSERI, L. & PIALLI, G. (1970) Nuovi dati sul Giurese Umbro-Marchigiano ed ipotesi per un suo inquadramento regionale. *Soc. Geol. Ital. Mem.* **9**, 839–874.

D'ARGENIO, B. & ALVAREZ, W. (1980) Stratigraphic evidence for crustal thickness changes on the southern Tethyan margin during the Alpine cycle. *Geol. Soc. Am. Bull.* **91**, 681–689.

DERCOURT, J., ZONENSHAIN, L.P., RICOU, L.E., KAZMIN, V.G., LE PICHON, X., KNIPPER, A.L., GRANDJAQUET, C., SBORTSHIKOV, D.H., BOULIN, J., SIBUET, J.C., SAVOSTIN, L.A., SOROKHTIN, O., WESTPHAL, M., BAZHENOV, M.L., LAUER, J.P. & BIJOU-DUVAL, B. (1986) Geological evolution of the Tethys belt from the Atlantic to the Pamirs since the Lias. *Tectonophys.* **123**, 241–315.

DEWEY, J.F., PITMAN, W.C., III, RYAN, W.B.F. & BONNIN, J. (1973) Plate tectonics and the evolution of the Alpine system. *Geol. Soc. Am. Bull.* **84**, 3137–3180.

FARINACCI, A., MARIOTTI, N., NICOSIA, U., PALLINI, G. & SCHIAVINOTTO, F. (1981) Jurassic "pelagic sediments" in Central Apennines and their paleoenvironmental significance. In: *Rosso Ammonitico Symposium Proceedings: Tecnoscienza* (Eds Farinacci, A. & Elmi, S.).

FARRE, J.A. & RYAN, W.B.F. (1985) 3-D view of erosional scars on U.S. mid-Atlantic continental margin. *Am. Assoc. Petrol. Geol. Bull.* **69**, 923–932.

HAQ, B.U., HARDENBOL, J. & VAIL, P.R. (1987) Chronology of fluctuating sea levels since the Triassic. *Science* **235**, 1156–1167.

HECKEL, P.H. (1974) Carbonate buildups in the geologic record: A review. In: *Reefs in Time and Space* (Ed. Laporte, L.F.) Soc. Econ. Paleont. Mineral Spec. Publ. 18, pp. 90–154.

HINE, A.C. & STEINMETZ, J.C. (1984) Cay Sal bank, Bahamas—a partially drowned carbonate platform. *Marine Geol.* **59**, 135–164.

HURST, J.M. & SURLYK, F. (1984) Tectonic control of Silurian carbonate-shelf margin morphology and facies: North Greenland. *Am. Assoc. Petrol. Geol. Bull.* **68**, 1–17.

JAMES, N.P. (1983) Reef environment. In: *Carbonate Depositional Environments* (Eds Schoole, P., Bebout, D. & Moore, C.) Am. Assoc. Petrol. Geol. Mem. 33, pp. 345–440.

JAMES, N.P. & MOUNTJOY, E.W. (1983) Shelf-slope break in fossil carbonate platforms: an overview. In: *The Shelf-break: Critical Interface on Continental Margins* (Eds Stanley, D.J. Moore, G.T.) Spec. Publ. Soc. Econ. Paleont. Mineral. 33, pp. 189–206.

JENKYNS, H.C. (1972) Pelagic oolites from the Tethyan Jurassic: *J. Geol.* **80**, 21–33.

KENDALL, C.G. ST. G. & SCHLAGER, W. (1981) Carbonates and relative changes in sea level. *Marine Geol* **44**, 181–212.

KLIGFIELD, R. (1979) The northern Apennines as a collisional orogen. *Am. J. Sci.* **279**, 676–691.

LONGMAN, M.W. (1980) Carbonate diagenetic textures from nearshore diagenetic environments. *Am. Assoc. Petrol. Geol. Bull.* **64**, 461–487.

MARIOTTI, N., NICOSIA, U., PALLINI, G. & SCHIAVINOTTO, F. (1979) Kimmeridgiano recifale presso Case Canepine (Monti Martani, Umbria): Ipotesi paleogeografiche. *Geol. Rom.* **18**, 295–315.

MARTINI, I.P., RAU, A. & TONGIORGI, M. (1986) Syntectonic sedimentation in a middle Triassic rift, northern Apennines, Italy. *Sed. Geol.* **47**, 191–219.

MCKENZIE, D. (1978) Some remarks on the development of sedimentary basins. *Earth Planet. Sci. Lett.* **40**, 25–32.

MONTANARI, A. (1985) Cenomaniam amoxic foreslope inferred from turbiditic cherts in the pelagic basin of the northern Apennines. *Geol. Soc. Am. Abs. Prog.* **17**, 667.

MULLINS, H.T. (1983) Structural controls of contemporary carbonate continental margins: Bahamas, Belize, Australia, In: *Platform Margin and Deep Water Carbonates* (Eds Cook, H.E., Hine, A.C. & Mullins, H.T.) Soc. Econ. Paleont. Mineral. Short Course Notes 12, pp. 2–1–2–57.

NICOSIA, U. & PALLINI, G. (1977) Hermatypic corals in the Tethyan pelagic facies of central Apennines: Evidence of upper Jurassic sea level changes. *Geol. Rom.* **16**, 243–261.

PAROTTO, M. & PRATURLON, A. (1975) Geological summary of the Central Apennines. In: *Estrato dalla collana: Structural model of Italy, Rome*, CNR, pp. 1–57.

PASSERI, L. (1971) Stratigrafia e sedimentologia dei calcari giurassici del M. Cucco (Appennino Umbro). *Geol. Rom.* **10**, 93–130.

PIERI, M. & MATTAVELLI, L. (1986) Geologic framework of Italian petroleum resources. *Am. Assoc. Petrol. Geol. Bull.* **70**, 103–130.

PLAYFORD, P.E. (1980) Devonian "Great Barrier Reef" of Canning basin, western Australia. *Am. Assoc. Petrol. Geol. Bull.* **64**, 814–840.

READ, J.F. (1985) Carbonate platform facies models. *Am. Assoc. Petrol. Geol. Bull.* **69**, 1–21.

ROBB, J.M., KIRBY, J.R., HAMPSON, J.C., JR., GIBSON, P.R. & HECKER, B. (1983) Furrowed Eocene chalk on the lower continental slope offshore New Jersey. *Geology* **11**, 182–186.

SCHLAGER, W. (1981) The paradox of drowned reefs and carbonate platforms. *Geol. Soc. Am. Bull.* **92**, 197–211.

SCHLAGER, W. & GINSBURG, R.N. (1981) Bahamas carbon-ate platforms—the deep and the past. *Marine Geol.* **44**, 1–24.

STECKLER, M.S. & WATTS, A.B. (1978) Subsidence of Atlantic-type continental margins off New York. *Earth Planet. Sci. Lett.* **41**, 1–13.

VANDENBERG, J. (1979) Reconstructions of the western Mediterranean area for the Mesozoic and Tertiary time-span. *Geol. Mijnb.* **58**, 153–160.

WELLS, J.W. (1957) Coral reefs. In: *Treatise on Marine Ecology and Paleoecology* (Ed. Hedgpeth, J.W.) Geol. Soc. Am. Mem. 67, pp. 1087–1104.

WESTPHAL, M., BAZHENOV, M., LAUER, J.P., PECHERSKY, D.M. & SIBUET, J.C. (1986) Paleomagnetic implications on the evolution of the Tethys belt from the Atlantic ocean to the Pamirs since the Triassic. *Tectonophys.* **123**, 37–82.

WILSON, J.L. (1975) *Carbonate Facies in Geologic History.* Springer Verlag, New York, p. 471.

WINTERER, E.L. & BOSELLINI, A. (1981) Subsidence and sedimentation on Jurassic passive continental margin, southern Alps, Italy. *Am. Assoc. Petrol. Geol. Bull.* **65**, 394–421.

Spec. Publs int. Ass. Sediment. (1990) **9**, 169–202

Controls on Upper Jurassic carbonate buildup development in the Lusitanian Basin, Portugal

P. M. ELLIS*, R. C. L. WILSON* *and* R. R. LEINFELDER[†]

**Department of Earth Sciences, The Open University, Walton Hall, Milton Keynes MK7 6AA, UK; and
[†]Institute für Geowissenschaften, Universität Mainz, Saastrasse 21, D-6500 Mainz, West Germany*

ABSTRACT

A variety of carbonate buildups developed in the Lusitanian Basin during the late Jurassic. During an Oxfordian–Kimmeridgian rift phase evidence can be seen for both fault and diapiric control of buildup development. Fault-controlled buildups occur on the east side of the basin. They exhibit shelf profiles, are relatively thin (200–500 m), show well-developed lateral facies zonation and are dominated by lime mudstones and wackestones, with only minor amounts of packstones and grainstones. Salt-controlled buildups on the northwest margin of the basin are relatively thick (500–1500 m), show only gradual lateral facies variation with no distinct shelf break facies, and are dominated by grainstones and packstones.

During the latest Oxfordian–early Kimmeridgian, a sudden relative sea-level rise drowned or partially drowned the earlier buildups, and this was quickly followed by a major influx of siliciclastic sediments. In the centre of the basin, a thin grainstone-dominated carbonate sequence of middle Kimmeridgian age developed on top of a prograding siliciclastic slope system. In the siliciclastic-starved southern part of the basin, a prograding low-energy ramp sequence of Kimmeridgian–Berriasian age was deposited.

Carbonate facies associations described from Portugal also occur in Mesozoic carbonate bank sequences of eastern America. Data from recent US wells, and comparisons with Portugal, suggest that the eastern American Atlantic 'reef trend' is largely composed of grainstone–packstone dominated shelf-break sediments with only relatively minor amounts of biogenic reefal framework.

INTRODUCTION

Upper Jurassic carbonate buildups occur in marginal basins on both sides of the southern North Atlantic. Seismic sections show that a buried carbonate bank complex or 'reef trend' lies under the eastern US continental shelf from the Blake Plateau, northwards for a distance of over 1200 km to Georges Bank (Schlee & Grow, 1982). A similar carbonate development is known from the Nova Scotian Shelf (Eliuk, 1978). Upper Jurassic buildups also occur along the eastern Atlantic margin, including Morocco (Jansa & Wiedmann, 1982), the southern Western Approaches (Masson & Roberts, 1981) and in Portugal.

The Lusitanian Basin of Portugal provides a unique opportunity to study Atlantic margin buildups both at outcrop and in the subsurface. This paper describes the composition, structure and developmental controls of four carbonate sequences that developed in contrasting tectonic and palaeogeographic settings. Recent well data from the eastern US continental shelf are also described and commented upon in the light of information gathered from Portugal.

Figure 1 shows a palaeogeographic reconstruction of the Atlantic during the late Oxfordian. The Lusitanian Basin was not in an ocean margin location at this time, but was situated to the east of the newly opened southern North Atlantic, which was between 1300 and 1500 km wide (Jansa, 1986). The basin was located to the northeast of a postulated short segment of an ocean ridge and associated major transform that displaced spreading southeastwards into Tethys. Rifting that initiated the separation of Iberia

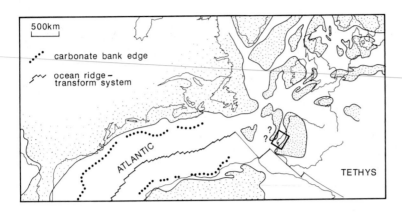

Fig. 1. Palaeogeographic sketch map for the North Atlantic region during the late Oxfordian. Box on western margin of Iberia shows the location of the Lusitanian Basin. Adapted from plate 9C of Vogt & Tucholke (1986).

from North America began in early Oxfordian times (Wilson, 1979, 1988; Wilson *et al.*, 1989), suggesting significant tectonic controls on carbonate buildup development during this period.

GEOLOGICAL SETTING AND TYPES OF BUILDUP

Stratigraphy

The Mesozoic of the Lusitanian Basin comprises four unconformity-bounded megasequences, the development of which was related to events in the evolution of the North Atlantic (Wilson, 1988; Wilson *et al.*, 1989). The first two of these megasequences, and part of the third are shown in Fig. 2 as a simplified, and as yet, informal lithostratigraphic summary chart for the southern part of the Lusitanian Basin. The three megasequences are described below.

Triassic–Callovian

This sequence is typical of the early rift–sag successions encountered in most North Atlantic margin basins. Triassic red fluvial siliciclastics (Silves formation) are capped by Hettangian evaporites (Dagorda formation) which subsequently influenced the manner in which reactivation of Hercynian basement faults affected the cover of younger sediments. Where the evaporites were thick, halokinetic structures formed above basement faults, but where they were thin, the faults propagated through younger formations. The Triassic and Hettangian sediments

accumulated in grabens and half grabens, though the later Lower and Middle Jurassic sediments (Coimbra, Brenha and Candeiros formations) blanketed the basin and exhibit simple facies geometries with relatively minor indications of contemporaneous faulting.

Middle Oxfordian–Berriasian

The base of the Upper Jurassic is marked by a basin-wide hiatus, spanning the whole of the Lower Oxfordian and the early part of the Middle Oxfordian (Mouterde *et al.*, 1971). In places, this boundary is unconformable and accompanied by karst surfaces developed on the underlying Middle Jurassic limestones (Ruget-Perrot, 1961; Wright & Wilson, 1987). Throughout the basin, the earliest Oxfordian sediments consist predominantly of lacustrine ostracod–charophyte lime mudstones (Wright & Wilson, 1985; Wright, 1985) comprising the Cabaços formation. This formation contains highly bituminous horizons and is considered to be the major source for the many hydrocarbon shows in the southern part of the Lusitanian Basin.

Fully marine carbonate deposition returned to most of the southern and central parts of the Lusitanian Basin in the late Oxfordian, associated with a relative rise in sea-level. This was the main carbonate buildup phase of the basin's history, when the limestones of the Montejunto formation and its equivalents were deposited. The initiation of buildup development in the Lusitanian Basin is somewhat later than those in other southern North Atlantic-marginal basins, such as the western High Atlas Basin of Morocco and in the Nova Scotian Shelf Basin.

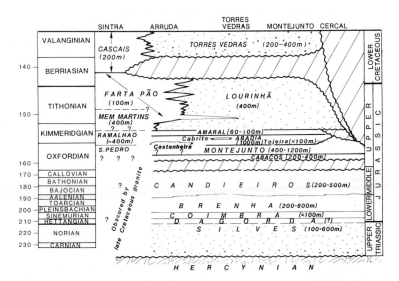

Fig. 2. Summary chart of an informal lithostratigraphic nomenclature currently under discussion for the Triassic−Lower Cretaceous of the Lusitanian Basin (as the scheme is informal, 'formation' and 'member' are not capitalized in the text). Note that the vertical time scale from Kent and Gradstein, (1985) is doubled from the base of the Oxfordian upwards in order to accommodate the Upper Jurassic units described in the paper. Formations are shown in capitals, members in lower case letters. Hiatuses are shown by diagonal ruling.

An abrupt rise in relative sea-level, accompanied by uplift of marginal basement highs during the early Kimmeridgian, ended carbonate deposition in many parts of the Lusitanian Basin and was followed by a sudden influx of siliciclastics, represented by the Abadia formation. In the central part of the basin, this event is marked by a sequence of shales and limestone breccias of the Tojeira member at the base of the clastic Abadia formation. However, on some structural highs, carbonate sedimentation continued into the Kimmeridgian.

The Abadia formation is ≈ 1000 m thick and consists predominantly of shales and siltstones. On seismic sections, the upper 550 m show southward dipping clinoforms (see Fig. 14; and Wilson, 1988, fig. 7) indicating a prograding slope deposit. Coarse sandstones of the Cabrito member accumulated at the base of this prograding slope (Ellwood, 1987). The slope deposits of the Abadia formation are capped by the Amaral Limestone formation, which contains up to 80 m of high-energy carbonates similar to the Upper Jurassic Smackover Formation of the Gulf Coast (Crevello & Harris, 1982) and is overlain by the fluvial and marginal marine clastics of the Lourinhã formation. Around salt structures, and on the eastern margin of the basin (Fig. 2), the Lourinhã formation rests unconformably on earlier Upper Jurassic strata.

The equivalents of the upper part of the Montejunto formation, and the Abadia formation, can be traced southwards towards Lisbon (where they were penetrated by the Monsanto well), and

southwestwards towards Sintra (Fig. 3). In the latter region they are brought to the surface by a small diapir-like granitic intrusion (the Sintra granite) dated at 76−80 Ma (i.e., very late Cretaceous). The San Pedro formation is a sequence of thermally-metamorphozed marbles exposed around the granite, dated as Oxfordian (Mouterde *et al.*, 1971; Ramalho, 1971) and probably a lateral equivalent of the Montejunto formation. The overlying Ramalhão and Mem Martins formations are suggested to be slope deposits, with carbonate debris flows forming an important component of the Mem Martins formation.

Rapid subsidence and synsedimentary fault and halokinetic movements characteristic of megasequence 2 suggest a rifting episode that may have heralded a pre-Aptian phase of ocean opening to the southwest of the Basin (Wilson *et al.*, 1989; Mauffret *et al.*, 1989)

Berriasian−Lower Aptian

The generalized facies pattern of siliciclastics being replaced southwards by carbonates seen at the end of megasequence 2 continues into megasequence 3 (Fig. 2). The Farta Pão formation, which contains brackish faunal elements of Tithonian age at its top (Ramalho, 1971) is overlain, probably without a break, by a series of carbonate units described by Rey (1972) which are grouped here into the Cascais formation. Elsewhere in the basin where megasequences 2 and 3 are developed in dominantly

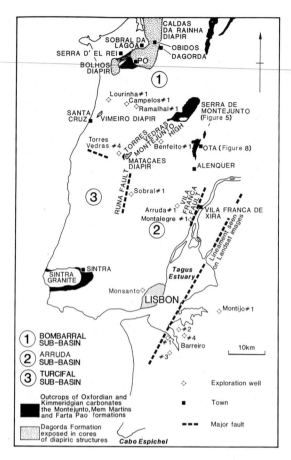

Fig. 3. Map of the southern part of the Lusitanian showing locations, exploration wells and structures mentioned in the paper.

Structural setting

Seismic sections show the Brenha and Candeiros formations to be relatively constant in thickness over the southern half of the Lusitanian Basin, indicating little or no differential subsidence during the early and middle Jurassic. In contrast, thickness variations in the Cabacos, Montejunto and Abadia formations suggest major changes in Basin configuration in this area during the late Jurassic. Reactivation of NNE–SSW trending strike–slip faults in the Hercynian basement, along with the formation of diapiric structures resulted in the formation of three sub-basins (Fig. 3).

The Bombarral sub-basin developed as a withdrawal structure as salt migrated into the Vimeiro–Caldas da Rainha diapir line along its western margin. To the southeast the sub-basin is bounded by the Torres Vedras–Montejunto anticline (Fig. 2) which may be a fault-associated salt structure that lies on the southern limit of thick Dagorda evaporites (Wilson, 1988; Wilson *et al.*, 1989.)

The Arruda sub-basin is bounded on the east by the Vila Franca Fault. It is filled with > 2 km of arkoses of the Castanheira Member of the Abadia formation, and capped by the southward prograding slope silts of the top part of the formation. The northern margin of the sub-basin is formed by the Torres Vedras–Montejunto high, but its southern margin cannot be delineated, as it is obscured by the Cretaceous and younger rocks of the Tagus estuary.

The Turcifal sub-basin is bounded to the east by the Runa Fault zone, and to the north by an unnamed E–W trending fault. The nature of its Oxfordian–Kimmeridgian fill has yet to be determined by drilling.

In the southern-most part of the Lusitanian Basin, the tectonic setting of the Upper Jurassic sediments is more difficult to determine. To the southeast, the carbonate buildup at Barreiro occurs beneath a Neogene basin containing sediments up to 1 km thick. The Barreiro wells are situated immediately to the southeast of a major NNE–SSW lineament observed on Landsat images, the trend of which is coincident with the Lisbon canyon offshore. The lineament probably reflects a major basement fault. To the northwest of this lineament, the Monsanto borehole penetrated ≈ 1000 m of Abadia formation twice the thickness seen at Barreiro. The tectonic setting of the Upper Jurassic sediments brought to the surface around the late Cretaceous Sintra granite

siliciclastic facies, Rey (1972) suggested that Berriasian strata are missing, and so the siliciclastics of the Torres Vedras formation rest unconformably on older strata. In the Cercal area, Rey (1972) showed not only the Torres Vedras formation but also a later Aptian siliciclastic unit cutting across older strata, thus indicating continued tectonic activity along this part of the eastern margin of the Bombarral Sub-basin (Fig. 2).

Megasequence 3 is equivalent to the synrift sequences that preceded ocean opening drilled off the northwestern margin of Iberia (Sibuet & Ryan, 1979; Wilson *et al.*, 1989), but is much thinner and contains no deep-water sediments.

can only be inferred from sedimentological data discussed in a later section.

Buildup types and locations

The use of the term 'buildup' in this paper follows that of Wilson (1975) for 'locally formed... carbonate sediment which possesses topographic relief'. The term carries no inference about origins, geometry (i.e., ramp, shelf, etc.) or internal composition. We follow Tucker's (1985) definition of the terms 'platform', 'shelf' and 'ramp'. None of the carbonate depositional systems in Portugal are extensive enough to be termed platforms.

Figure 4 summarizes the age and principal facies characteristics of the Upper Jurassic carbonate sequences occurring within megasequence 2. Four types are recognized, which developed in distinct geographic regions within the basin.

Fault-controlled carbonate buildups

These are relatively thin (up to 500 m) shelves, contain lime-mud dominated facies with strong lateral variations, and contain distinct shelf-break facies. They are Upper Oxfordian to Upper Kimmeridgian in age (Montejunto formation and Ota limestones) developed only on the east side of the Lusitanian Basin (Fig. 4, columns 5–8).

Salt-controlled carbonate buildups

These are relatively thick (up to 1500 m) shelves, dominated by grainstones and packstones with little lateral facies variation and do not display a distinct shelf-break facies. They are largely of Upper Oxfordian–Lower Kimmeridgian (Montejunto formation) age and occur on the northwestern side of the basin (Fig. 4, columns 2 and 3).

Postrift passive basin fill sequence

This occurs on the southwest side of the Basin in strata exposed around the Sintra granite (Fig. 3; Fig. 4, column 2) and represents a prograding carbonate ramp system. Of Upper Kimmeridgian–Tithonian age (Mem Martins and Farta Pâo formations), it is dominated by carbonate mud facies with debris flow deposits common in the lower half of the sequence.

Limestone cap on prograding siliciclastics

This very thin (< 100 m) unit (the Amaral formation) occurs over much of the central and southern part of the basin. It caps the prograding slope deposits of the Abadia formation and is dominated by high-energy ooid grainstones (Fig. 4, columns 2–4, 6 & 9).

Nine carbonate facies associations are recognized within these sequences and are summarized in Table 1. The environmental interpretations of the facies associations are based on lithological and palaeontological data and the stratigraphical and spatial setting of individual sequences that are described in later sections. They are probably applicable to time-equivalent carbonates in other Atlantic settings.

FAULT-CONTROLLED BUILDUPS

These buildups are exposed on the northeast side of the basin at the Serra de Montejunto and at Ota, and were encountered to the south in the boreholes Montalegre #1 and Barriero #1–4.

The Serra de Montejunto

Setting

The buildup is bounded to the east by a WNW–ESE trending normal fault, which is parallel to an important basement structural trend affecting the Mesozoic development of the basin (Wilson *et al.*, 1989), from which the limestones of the Montejunto formation dip gently to the northwest (Fig. 5). The outcrop is terminated by a fault (the Pragança Fault) which passes northwards into a monoclinal flexure. Fluvial sandstones of the Lourinhā and Torres Vedras formations overstep the limestones on the northern side of the buildup, whereas the southern part of it is deformed by a large asymmetric anticline exposing Middle Jurassic carbonates in its core.

Age

Lime mudstones exposed to the southwest of the Montejunto anticline (Fig. 5) have yielded ammonites indicating the Bifurcatus to Planula Zones of the Upper Oxfordian (Mouterde *et al.*, 1971). The underlying Cabaços formation is considered to be middle Oxfordian in age (Ribeiro *et al.*, 1979), and the overlying Tojeira member has yielded ammonites

Table 1. Characteristics, interpretation and distribution of Upper Jurassic carbonate facies association of the Lusitanian Basin

Facies associations	Description	Principal biota	Environmental interpretation
1 Ammonitic lime mudstone	Dark clay-rich mudstones, associated with shales, small-scale slumps and carbonate turbidites	Ammonites, belemites, *Zoophycos*	Periplatform and basinal oozes
2 Limestone breccias	Oligomictic, matrix and clast supported breccias, generally with a matrix of calcareous shales	Rare or absent	Debris flows with resedimented carbonates
3 Thrombolitic bindstones	0·5−1·5 m thick, framework of thrombolitic or micritic crusts with stromatactis-like cavities and unbound wackestones	Hexactinellid and lithistid sponges, *Thartharella*, encrusting foraminifera	Deep-water/low light intensity biohermal mounds
4 *Tubiphytes* wackestones	Dark, fine-grained bioclastic wackestones, commonly with oncolites	Thin-shelled bivalves, corals. Tubiphytes	Lower-energy deposition, largely below wave-base and in shallower but sheltered locations
5 Reefal limestones	Wide variety of types including framestones, bindstones and bafflestones, often associated with reefal debris	Diverse: corals, stromatoporids *Solenopora*, chaetetids, nerineids bivalves, lithophagid bivalves, encrusting algae, encrusting foraminifera	High−moderate energy, fully marine sediments
6 Nerineid wackestones and packstones	Varied, highly fossiliferous limestones showing considerable variation on both a cm and dm scale	Diverse: nerineids, corals, bivalves, forams and dasycladaceans	Shallow marine, normal salinity lagoonal sediments; lithological variation caused principally by winnowing
7 Intraclastic and bioclastic packstones and grainstones	Many clasts with thin oolitic coats, oncolite-rich horizons common	Diverse: *Solenopora*, corals, stromatoporoids, diceratids, nerineids, foraminifera and thick-walled dasycladaceans	Moderate−high energy, fully-marine sediments
8 Lime mudstones and wackestones	Dense micritic limestones containing disseminated bioclasts and small micritic intraclasts	Restricted, low diversity: bivalves, nerineids (including *Ptygmatis*), dasycladaceans, miliolid and cyclinind foraminifera	Subtidal lagoonal deposits with some levels representing restricted conditions Shelf interiors
9 Fenestral limestones	Mudstones, wackestones and packstones, commonly oncolitic, frequently showing 'Lofer-like cycles', occasionally associated with soil horizons and omission surfaces	*In situ* Megalodontids (?*Pachyrismella*); nerineids *Ptygmatis*)	Shallow-subtidal, intertidal and supratidal sediments

M: Montejunto.
O: Ota.
B: Barriero.

Table 1. (*Cont.*)

Montejunto formations (east side of Basin)	Montejunto formations (northwest side of Basin)	Mem Martins & Farta Pão formations (southwest side of basin)	Amaral formations (central basinal area)
			Absent
	Deeper parts and basinward of the buildups and capping drowned buildups	Lower parts of the Mem Martins formation (up to 350 m on Fig. 18)	Absent
			Absent
M: thick sequences basinward of the shelf edge. O: lower reef-front and sheltered areas within the reef B: lower part of buildup	Bulk of the lower part of the buildup	Uppermost part of the Mem Martins formation (380−420 m on Fig. 18)	Absent
M: small coraliferous framestone knolls in front of the shelf break; small coral−chaetetid patch reefs within the shelf interior. O: fairly continuous framestone and bafflestone reefs at the shelf break, with coral bafflestones within the shelf interior B: towards top of buildup above association 4 and below the deep-water cap (associations 1−3)	Small coral−stromatoporid−chaetetid framestones most common in the intraclast grainstones (association 7)	Small bindstones & bafflestones of varying composition throughout the upper part of the Mem Martins formation (420−440 m on Fig. 18)	Coral bindstones and bafflestones just basinward of shelf break, and in lagoonal settings
Shelf interiors	Thin sequences associated with association 7, becoming more common towards the salt structures	Uppermost part of the Mem Martins formation (420−440 m on Fig. 18)	Inter-biostrome areas in lagoons
M: thin, discontinuous sequences at the shelf break O: behind and interfingering with reef (association 5), and thin cap to buildup B: with association 5	Thin and widespread sequence forming the upper half of the buildup	Absent	Fringing biostromal structure: cross-bedded ooid grainstones overly biostromes: deposited at shelf break
Shelf interiors	Absent	Farta Pão formation	Absent
Shelf interiors	Absent	Absent	Absent

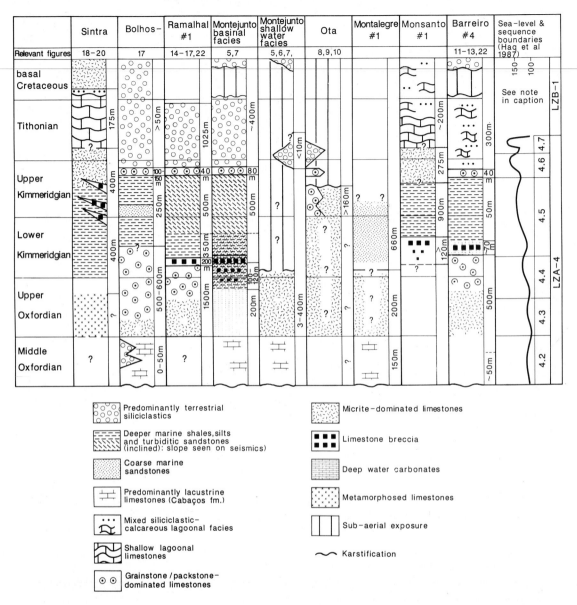

Fig. 4. Summary chart showing the stratigraphic distribution of the Upper Jurassic carbonate formations of the Lusitanian Basin. The correlation between the sea-level curve and sequence boundaries of Haq *et al.* (1987) with the Portuguese sequences cannot be made precisely due to lack of good biostratigraphic control. It is particularly difficult to identify the Oxfordian–Kimmeridgian boundary in shallow-water carbonate sequences. No attempt is made to show LZB-1 cycles, as precise dating of the Tithonian successions is not possible.

(Atrops & Marques, 1986) and foraminifera (Stam, 1985) indicating a topmost Upper Oxfordian to Lower Kimmeridgian age.

The dating of the shallow-water shelf carbonates of the Montejunto formation exposed to the north of the Montejunto anticline is less certain. They are conventionally placed entirely within the Oxfordian (Zbyszewski *et al.*, 1966) and are believed to be underlain by the Middle Oxfordian Cabaços formation. However, characteristic assemblages of

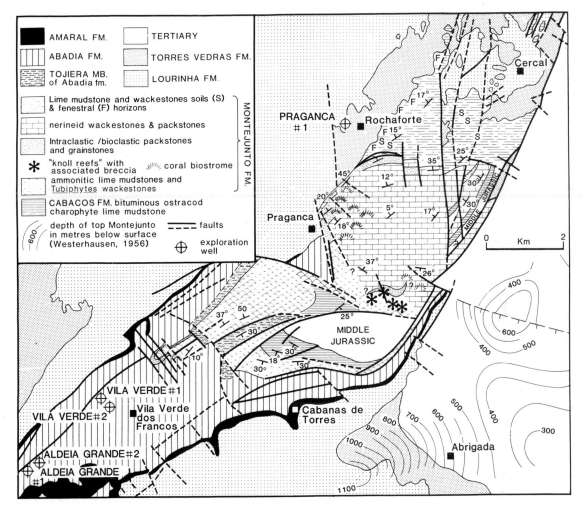

Fig. 5. Geological map of the area around the Serra de Montejunto showing distribution of carbonate facies associations in the Montejunto formation. Post-Montejunto outcrops are taken from Zbyszewski *et al.* (1966)

large diceratid rudists (*Diceras* and *Epidiceras*) found within them strongly suggest, but do not prove, an age no older than Lower Kimmeridgian for some of the limestones (P.W. Skelton, pers. comm.). The apparent absence of diagnostic microfossils such as *Campbelliella striata* and *Clypeina jurassica* suggest that Upper Kimmeridgian limestones are probably not present. Therefore, it is probable that although much of the shallower-water shelf carbonates are coeval with the ammonitic lime mudstones situated to the southwest, some may be of Lower, or even middle Kimmeridgian age (Guéry, 1984, suggested an age range up to middle Kimmeridgian).

Description and interpretation

Figure 6 is a schematic SW−NE cross-section across the Montejunto buildup summarizing the lateral facies changes mapped in the area, and incorporating our interpretation of its formation.

Periplatform and basinal facies are represented by ≈ 200 m of ammonitic lime mudstones exposed to the west and southwest of the Serra de Montejunto (Ruget-Perrot, 1961). Turbiditic packstones and wackestones, small-scale slump horizons and trace fossils (including *Zoophycos*) occur within the mudstones. To the southwest of Montejunto, the ammonitic lime mudstones are overlain by allochthonous oolitic packstones and grainstones.

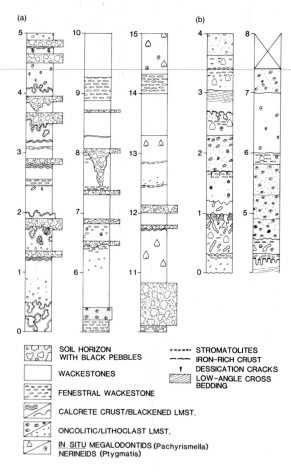

(a)   (b)

SOIL HORIZON
WITH BLACK PEBBLES

WACKESTONES

FENESTRAL WACKESTONE

CALCRETE CRUST/BLACKENED LMST.

ONCOLITIC/LITHOCLAST LMST.

IN SITU MEGALODONTIDS (Pachyrismella)
NERINEIDS (Ptygmatis)

STROMATOLITES
IRON–RICH CRUST
DESSICATION CRACKS
LOW–ANGLE CROSS
BEDDING

Fig. 6. Sample logs of sequences within the lime mudstone and wackestone facies (association 8) of the Montejunto buildup near Rocharforte (for location, see Fig. 5). (a) Soil-rich sequence; (b) peritidal 'loferitic' sequence. Scale on side of logs is in metres.

The Tojeira member overlies the ammonitic lime mudstones, but it is never found above the shallow-water shelf facies of the Montejunto formation to the northeast. Limestone breccias and shales are the principal lithologies of this member. A well-developed thrombolite boundstone is also exposed within its lower part, which incorporated angular pink feldspars and other basement fragments during its growth, indicating that basement was exposed nearby during this time. Basement clasts, along with contemporaneously-karstified limestones also occur within the breccias, and become more common in the upper parts of the Tojeira.

To the north of the Montejunto anticline, towards the inferred shelf-edge, the ammonitic lime mudstones pass into a thick sequence of wackestones and *Tubiphytes*-wackestones which were probably deposited on a low-angle slope. A number of small, well-developed coral framestone knolls occurs within the wackestones and are associated with redeposited beds of reef breccia. The dominance of micrite within this facies, and its basinward location, suggest deposition below normal wave-base, with the reef breccias deposited as storm deposits.

Thin, discontinuous sequences of oolitic and bioclastic grainstones are taken to mark the shelf-break (Fig. 5). Grainstones also occur farther to the northeast, where, such as at Rochaforte, they are particularly fossiliferous, containing heads of corals, stromatoporoids and *Solenopora*.

The shelf interior is largely composed of the nerineid wackestone facies (association 6). These limestones contain a wide range of fossils and show considerable lithological variation. Nerineids are concentrated in some horizons, as are heads of the chaetitid *Ptychochaetetes*, which in places combine with corals to form small patch-reef framestones and bindstones. Close to the shelf-edge grainstone shoals, a coral biostrome composed of abundant massive coral heads and phaceloid coral bushes occurs.

The nerineid wackestones pass abruptly northwards into lime mudstones and wackestones (association 8). Though there are some levels within the mudstones exhibiting a normal marine biota the majority of limestones contain few megafossils (largely nerineids when present) though some contain foraminifera and dasycladaceans indicating a very shallow, restricted environment.

Fenestral wackestones occur in the upper part of the mudstones around the northwestern margin of Montejunto (Fig. 5) and are formed of crudely-developed lofer-like cycles (Fig. 6). Supratidal conditions are marked by omission surfaces and rare stromatolite and soil horizons. The fenestral limestones are associated with a thick sequence of black pebble soil horizons exposed at Rochaforte (Fig. 6), which suggests that the area was frequently emergent.

The Montejunto buildup probably formed as an aggradational carbonate shelf (*sensu* Read, 1982). Both the shelf break and the boundary between facies associations 6 and 8 trend NW–SE, which is parallel to an important secondary element in the tectonic evolution of the Lusitanian Basin. The shelf break is marked by relatively minor thinning of the

Montejunto formation, in contrast to the dramatic thinning of the underlying Cabaços formation (Fig. 7). This, together with the fact that the Montejunto formation (in both its shallow- and deep-water developments) does not contain any limestone breccias suggestive of an abrupt shelf-edge, suggest that the slope break was located over a gentle tectonic flexure (Fig. 7).

However, it seems that this flexure did develop into a fault scarp at the end of the late Oxfordian and in early Kimmeridgian times, shedding limestone breccias basinward to form the Tojeira member, while shallow-water carbonate sedimentation probably continued over the shelf. The steep dips and monoclinal flexures between the facies blocks shown in Fig. 5 are probably related to Tertiary inversion tectonics.

Away from the Serra de Montejunto, the Praganca #1 well penetrated 500 m into the buildup, but did not reach the underlying Cabaços formation. Other exploratory wells to the southwest show that the basinal ammonitic lime mudstones are only ≈ 200 m thick; this southwestward thinning trend is also shown by the Cabaços formation.

Ota

Setting

The Ota buildup is situated ≈ 5 km S−SE of the Montejunto buildup (see Fig. 3), though the structural relationship between the two is obscured by the Lourinhã and Torres Vedras formations. The exposed part of the Ota buildup is some 6 km long, 2 km wide and up to 160 m thick (Fig. 8), with a tectonic dip a few degrees to the east. It is bounded on its western side by a N−S trending fault and is presumed to extend eastwards under a cover of Tertiary sediments. In places the Lourinhã formation rests unconformably on the karstified surface of the buildup.

As Ota is cut by deep quarries and natural valleys, lateral facies relationships are much better displayed than those at Montejunto. The presence of middle Oxfordian Cabaços formation pebbles within the Ota limestones indicates the uplift of a nearby region after the middle Oxfordian, but before the late Kimmeridgian. It is possible that the fault-bounded block of Cabaços formation situated to the northeast

MONTEJUNTO (U.OXFORDIAN)

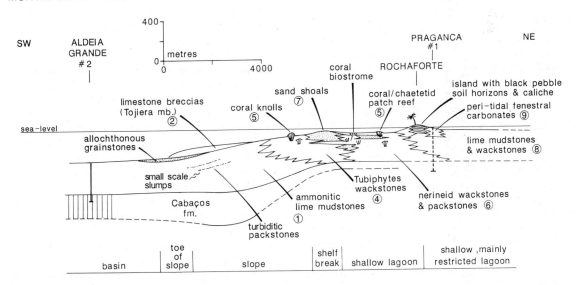

Fig. 7. Interpretative cross-section across the Montejunto buildup showing distribution of the facies associations described in Table 1.

of the Ota buildup (Fig. 8) was such an uplifted region.

Age

The association of the dasyclads *Clypeina jurassica*, and *Campbellina striata*, together with the foraminifera *Labyrinthina mirabalis* and *Alveosepta jaccardi* is typical of the Upper Kimmeridgian of Portugal (Ramalho, 1971). As the Amaral formation partly drapes the Ota limestones, and also occurs as an internal fill to palaeokarst within it, it is suggested that the buildup was deposited contemporaneously with the top part of the Abadia formation. The fault-bounded block of Cabaços formation to the northeast of Monte Redondo (Fig. 8) indicates that marine sedimentation did not occur prior to the late Oxfordian. A 250 m deep uncored water well drilled in the Ota limestone apparently did not reach the base of the unit (Manuppella & Balacó Moreira, 1984), so it is possible that the unexposed base of the buildup may be Lower Kimmeridgian or even Upper Oxfordian in age.

Description and interpretation

Facies zonation indicative of an aggradational shelf carbonate buildup is clearly shown in Fig. 10. A narrow high-energy reefal barrier zone occurs on its western margin, behind which are situated back-reef sands, tidal limestones and lagoonal, low-energy lime mudstones and wackestones. The reefal belt exhibits a high proportion of coral framestones and algal-stabilized debris, together with algal bindstones and intraclastic–bioclastic grainstones (Fig. 9a). The interfingering of these sediment types suggests the development of a reefal spur-and-groove system (Fig. 9b).

Leeward, back-reef sands (association 7) are differentiated into bimodally-sorted sand flat intraclastic grainstones, partly exhibiting beachrock cements and poorly-sorted lagoonal grapestone facies. Tidal flat limestones are characterized by shallowing-upwards sequences, mostly composed of bioturbated lime mudstones and wackestones with an upward increasing number of irregular birdseyes, which grade into thin sheets of laminoid fenestral limestones (association 9). The presence of oncolitic channel lag deposits indicates the development of a discontinuous tidal flat zone crossed by tidal channels.

The lagoonal sediments are chiefly composed of

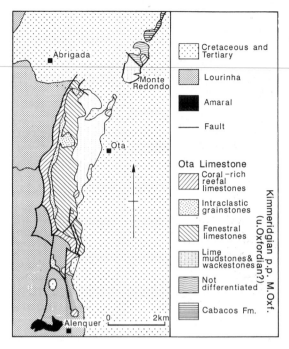

Fig. 8. Geological map of the area around Ota. From Leinfelder (1987).

restricted lime mud–wackestone facies (association 8) with local enrichment of skeletal or inorganic particles due to winnowing. Local mud-crack horizons and early diagenetic freshwater influence may indicate the development of shallow, stabilized linear mud banks as in the modern Florida Bay (*cf.* Enos & Perkins, 1979).

A prominent black pebble horizon related to intralate Kimmeridgian subaerial exposure (Leinfelder, 1987; see Fig. 9) allows accurate facies correlation. It is clear that the facies pattern did not shift laterally (at least throughout the exposed part of the unit), despite local intraformational subaerial exposure and possible eustatic and tectonic oscillations. This may be explained by the existence of a bypass escarpment margin, preventing progradational reef growth despite its high productivity (Fig. 10). The position of this palaeoescarpment coincided with the present-day fault on the western limit of the reefal zone (Fig. 8). On seismic sections, this fault can be seen to have been active during the Kimmeridgian (Leinfelder & Wilson, 1989).

The Ota Limestone is probably located on a very narrow uplifted basement high and apparently represents the major part of a fairly small, isolated

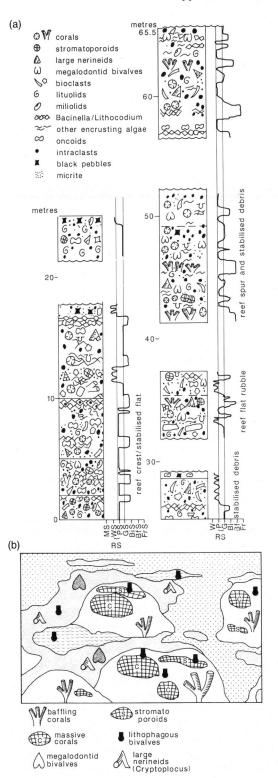

Bahamian-like carbonate bank, which, during the Kimmeridgian, had no direct connection with similar, though older buildups to the north (Montejunto) and south (Leinfelder & Wilson, 1989). During the later Upper Kimmeridgian, the Ota block stopped subsiding, and intraclast and ooid grainstones spread over the structure and filled karst features. From the early Tithonian to at least the early Cretaceous, the block was subjected to subaerial exposure (Leinfelder, 1985).

The Barreiro buildup

Setting

First recognized as a stratigraphic anomaly on seismic sections, the Barreiro buildup occurs just to the south of the Tagus estuary (Fig. 3). In it the normal reflection character of the Upper Jurassic, comprising moderate amplitude, moderately continuous flat-lying reflectors is replaced by a lensoid anomalous zone of discontinuous to chaotic reflectors, with relatively high-amplitude dipping reflectors with a short lateral extent at its margins (Fig. 11). Seismic reflections beneath the anomaly are obscured, so that a clear picture of its structural setting cannot be gained. However, as the Barreiro anomaly is one of several seismic anomalies aligned along a line just to the south of and parallel to the major NNW−SSE trending lineament identified on Landsat images, it seems reasonable to suggest a crestal location on a titled-block dipping gently to the southeast.

Four petroleum exploration wells have been drilled at Barreiro (Figs 11c and 12). Barreiro #4 penetrated the main part of the seismic anomaly, and #2 and #1 were situated to the north and west of it respectively, but #3 was drilled close to another small anomaly to the southwest.

The location of the cores taken from the three wells that penetrated the Barreiro buildup is shown

Fig. 9. Facies variation in the reefal zone in the Ota area. (a) Lithological log. Key: WS = wackestone; PS = packstone; GS = grainstone; RS = rudstone; BiS = bindstone; BfS = bafflestone; FrS = framestone. The relative abundances of lithologies within this sequence by volume are: packstones (including algal stabilized sediment) = 46%; bindstone + bafflestone + framestone = 25%; grainstone = 22%; wackestone = 7%. (b) Summary diagram illustrating the interfingering of facies types. A scale bar is not shown, as the variations shown may occur across widths ranging from 1 to 10 m.

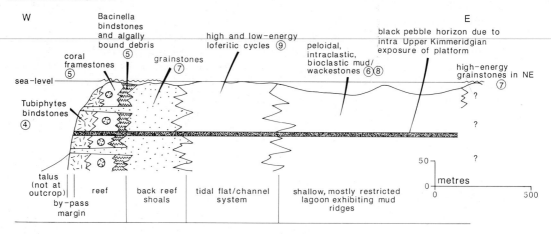

Fig. 10. Interpretative cross-section across the Ota buildup showing distribution of the facies associations described in Table 1.

in Fig. 12. Clearly, the data base is much less comprehensive than that for outcrops, but cores are available from both the central part of the seismic anomaly (#4) and its flanks (#1 and 2).

Age

On the basis of thin-section studies of algae and foraminifera, Ramalho (1971) concluded that the carbonate sequence beneath the Abadia formation in Barreiro #1-3 is Upper Oxfordian in age. However, the much thicker sequence encountered in Barreiro #4 may extend into the Kimmeridgian (E. Matos, pers. comm.). Ramalho's (1971) identification of the Amaral formation overlying the buildup places its upper age limit in the early part of the Upper Kimmeridgian.

Barreiro #4

The buildup in this well consists of three distinct units (Fig. 13). The lower unit consists largely of low-energy oncolitic wackestones (association 4) and rests directly on Middle Jurassic limestones, with the Cabaços formation apparently missing. The middle unit is composed dominantly of coral–stromatoporoid framestones and is overlain by an upper unit of deeper-water limestones and limestone breccias.

Bioclasts are common and varied in the lower wackestone facies, including fragments of stylinid and microsolenid corals, nerineid gastropods, brachiopods, inozoans, oysters and large cyclinid foraminifera. Small *in situ* heads of chaetetids and corals, some of which are oyster-encrusted, are also present.

The second unit, between 2358 and 2363 m, consists of a framework containing *in situ* heads of corals, microsolenid corals, stromatoporoids and ?*Solenopora* interbedded with light-coloured bioclastic nerineid wackestones, containing brachiopods, nerineids and foraminifera and small *in situ* microsolenids. The microsolenid corals, which form the main frame builder, are platey and are encrusted with a variety of organisms, chiefly thrombolitic crusts, with lesser numbers of inozoans, bryozoans and skeletal algae. Pervasive dolomitization of the lower parts of this core has destroyed its structure, though has greatly increased its porosity. The wackestones and packstones of the middle unit are less clay-rich than the wackestones of the lower unit, and may have been deposited in higher-energy, shallower-water conditions.

The transition between the shallow-water sediments of the buildup and the deeper-water carbonates of the upper unit that cap it is represented in core 2. *Tubiphytes* − wackestones and packstones occur at the base of the core, which, like the *Tubiphytes* − wackestones occurring at Montejunto, may have been deposited on slopes just basinward of the main buildup. A number of large angular limestone lithoclasts (largely consisting of association

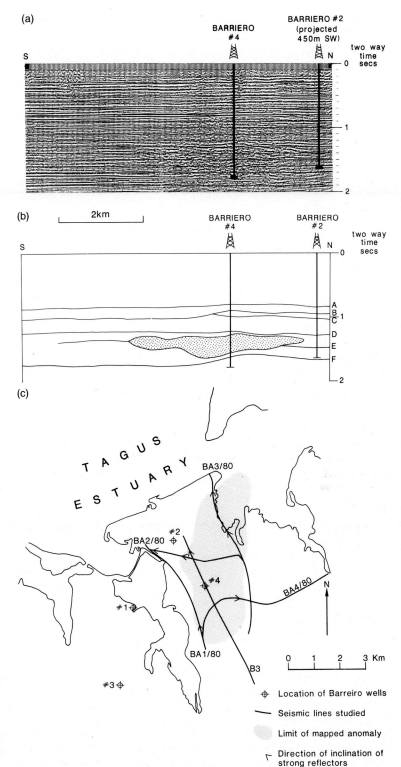

Fig. 11. The seismic anomaly produced by the Barreiro carbonate buildup. (a) Unintepreted migrated 24-fold vibroseis seismic line B3. (b) Interpretation of seismic line shown in (a) with anomaly shown as stippled area. Stratigraphic identification of reflectors is as follows: A = base Miocene; B = base Tertiary; C = top Portlandian; D = top Abadia formation; E = base Abadia formation; F = top Middle Jurassic. (c) Map of seismic anomaly showing positions of inclined reflectors at its margins and the location of the wells discussed in the text.

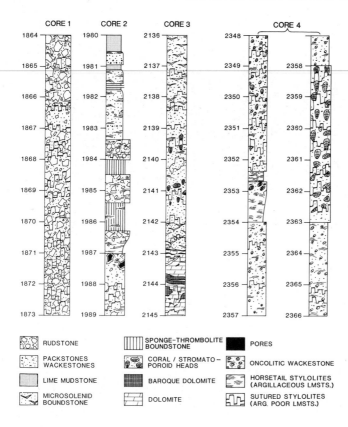

CORE 1   CORE 2   CORE 3   CORE 4

RUDSTONE

PACKSTONES
WACKESTONES

LIME MUDSTONE

MICROSOLENID
BOUNDSTONE

SPONGE-THROMBOLITE
BOUNDSTONE

CORAL / STROMATO-
POROID HEADS

BAROQUE DOLOMITE

DOLOMITE

PORES

ONCOLITIC WACKESTONE

HORSETAIL STYLOLITES
(ARGILLACEOUS LMSTS.)

SUTURED STYLOLITES
(ARG. POOR LMSTS.)

Fig. 12. Summary logs of cores taken from Barreiro #4. The positions of the cores within the buildup are shown in Fig. 13.

4) float within it, some of which were leached before their incorporation into the wackestones. The *Tubiphytes* — wackestones are overlain by a series of argillaceous limestone breccias and packstones, representing the deposits of debris and turbidity flows into deeper water, along with *in situ* thrombolitic bindstones.

Shales and allochthonous limestones form a 150 m thick cap above the buildup. These occur within core 1, which is composed entirely of limestone breccia formed of shallow-marine limestone clasts (largely fine-grained association 4, and some 6) set in a deeper-water mud matrix.

Barreiro #2

This well encountered half the thickness of Montejunto formation than that penetrated by Barreiro #4. All three units described from Barreiro #4 also occur within this well, suggesting some progradation over basinal facies. The initial argillaceous mound stage is represented by light, fine-grained bioclastic wackestones and lime mudstones

containing numerous siliceous sponge spicules, associated with argillaceous breccia beds and massive diagenetic anhydrite. Only ≈ 50 m of the shallow-water carbonates occur in Barreiro #2 (core 7), and are formed of fossiliferous packstones and wackestones with *in situ* heads of corals. The packstones have moderate vuggy porosity with numerous oxidized oil shows. As in Barreiro #4, these are overlain by deeper-water shales, limestone breccia beds and thrombolitic bindstones.

Barreiro #1

Situated in a more basinal location than Barreiro #2, this well contains carbonates similar to the Cabaços formation at Montejunto. These consist of oil-impregnated thin-bedded ostracod—charophyte mudstones with numerous thin packstones and wackestones (cores 8—12). The overlying Montejunto limestones are more basinal than those in Barreiro #2 and #4. Dark, argillaceous wackestones (core 6) are followed by light, fossiliferous wackestones with good vuggy porosity (cores 3—4),

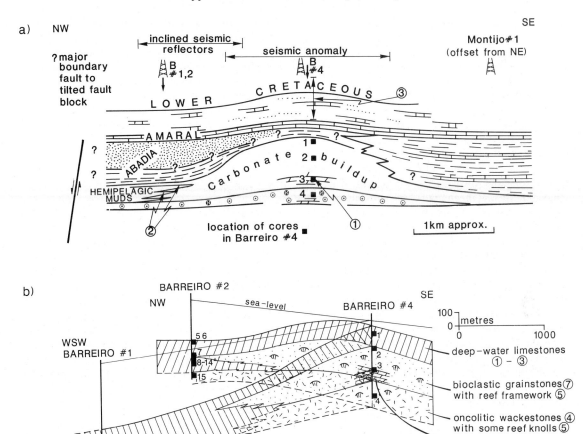

Fig. 13. Interpretative cross-sections of the Barreiro buildup. (a) Sketch section showing the relationship of the buildup to other stratigraphic units, and the location of cores in Barreiro #4. The circled numbers refer to features relevant to petroleum exploration as follows: (1) dolomitized porous zone within buildup; (2) dead oil shows in porosity within limestone breccias; (3) tar sands within the Upper Jurassic and Lower Cretaceous that accumulated in a trap formed by compactional drape over the buildup. (b) Interpretative cross-sections across the buildup showing position of cores within Barreiro #1, #2 and #4 and the distribution of facies associations described in Table 1.

which are overlain by argillaceous wackestones with limestone lithoclasts (core 2).

Barreiro #3

Barreiro #3 is located to the southwest of the main buildup, and reveals a very different sequence to those encountered in the other wells. Here the limestones are much thinner (just over 160 m thick) and composed largely of limestone breccias. In the lower part, the clasts are set in a matrix of light and interbedded bioclastic wackestones, possibly deposited under relatively shallow-water conditions. The breccia beds become more argillaceous up the sequence, and are interbedded with argillaceous and pyritic ammonitic lime mudstones, suggesting an increase in depth of deposition. As no major faults are revealed on seismic sections, a fault scarp talus origin for the Barreiro #3 sequence seems unlikely. However, the well is located very close to another seismic anomaly (not shown in Fig. 11, but see Fig. 21) and so the sequence could be a talus slope derived from a nearby buildup.

Discussion

Core data from Barreiro #4 indicates that the seismic anomaly shown in Fig. 11 is composed of a shallowing-up sequence consisting principally of facies associations 4, 5 and 7, which is capped by a deep-water sequence of associations 1, 2 and 3. In the more basinal areas to the west and northwest of the anomaly, Barreiro #1 and 2 also show a shallowing trend overlain by a deep-water cap, but the deeper-water facies associations 1, 2 and 3 predominate throughout the sequence (Fig. 12). The inclined reflectors at the margin of the anomaly suggest slight progradation to the west and northwest — a conclusion consistent with the shallowing trends seen in the wells. The deep-water limestones and breccias that cap the Barreiro buildup suggest that shallow-water carbonate sedimentation was terminated by drowning and the partial break-up of the buildup. Breccias of equivalent age encountered in Monsanto #1, 15 km across the Tagus estuary in Lisbon, may have been derived from the Barreiro buildup.

Speculation concerning the nature of the carbonate facies to the east of the Barreiro buildup can only be based on data available from Montijo #1 and outcrops at Cabo Espichel (for locations, see Fig. 3). At both locations, mudstones and wackestones of facies association 8 and 9 occur interdigitated with siliciclastic fluvial sands.

Compactional drape of Tithonian and Lower Cretaceous siliciclastic sands over the Barreiro buildup formed a trap in which an oil column > 200 m high accumulated, but was later oxidized to produce tar-impregnated sands (Fig. 12).

SALT-CONTROLLED BUILDUPS

Setting and age

A large buildup developed on the southeast side of the Caldas da Rainha—Vimeiro salt structure, which extends over an area of ≈ 200 km^2. It has been brought to the surface around the salt structure, and was penetrated by the Lourinhã, Campelos and Ramalhal wells (see Fig. 3 for locations). The wells show that the buildup thickens eastwards into the Bombarral Sub-basin, which was subsiding due to salt withdrawal. Ramalhal #1, which did not reach the Cabaços Formation, showed the buildup reached thicknesses of at least 1500 m, but at outcrop it is only a few hundred metres thick.

Dating of the Ramalhal—Serra d'el Rei buildup is uncertain. The presence of the Abadia formation over the buildup in the subsurface means that it is certainly Oxfordian in age, though no detailed faunal and floral information is available to state its upper age limit with precision, which may extend into the Kimmeridgian. At outcrop, Abadia-equivalent clastics are extremely thin and carbonate sedimentation more probably continued into the Kimmeridgian.

Seismic features

Figure 14 shows two seismic lines through the buildup near Ramalhal #1. Both sections show that the top is characterized by high amplitude mounded reflectors, whereas the lower part does not show such strong reflectors. In Figs 14a & b, mounded reflectors also show a progradational pattern. The positive relief on the apparent slope into the basin is of the order of 200 m, and beyond this on both Figs 14a & c the thinner zone of parallel high amplitude reflectors may be caused by deep-water ammonitic lime mudstones of the Montejunto formation and the overlying Tojeira member. This interpretation has not been substantiated by drilling.

Core data

Ramalhal #1

The 1500 m sequence penetrated by Ramalhal #1 represents the thickest development of the Montejunto formation in the Basin. As at Barreiro, the Ramalhal buildup shows a three-fold division consisting of a generally shallowing-upwards sequence from relatively fine-grained limestones (facies association 4, core 1), passing into more fossiliferous packstones and grainstones (association 5 and 7), which in turn are capped by deeper-water limestones (Fig. 15).

Biohermal limestones occur within the lower part of the buildup and these consist of heads of microsolenid corals, stromatoporoids and some *?Solenopora* in a fine-grained wackestone matrix. Numerous types of encrusters are present, including thrombolitic crusts, skeletal algae, stromatoporoids and bryozoa. The biohermal facies shows little vuggy porosity, but is cut by numerous small fractures. The upper part of the shallow-water buildup sequence (represented by core 7 starting at 2057 m, to core 18 down to 2460 m) consists of higher-energy fossi-

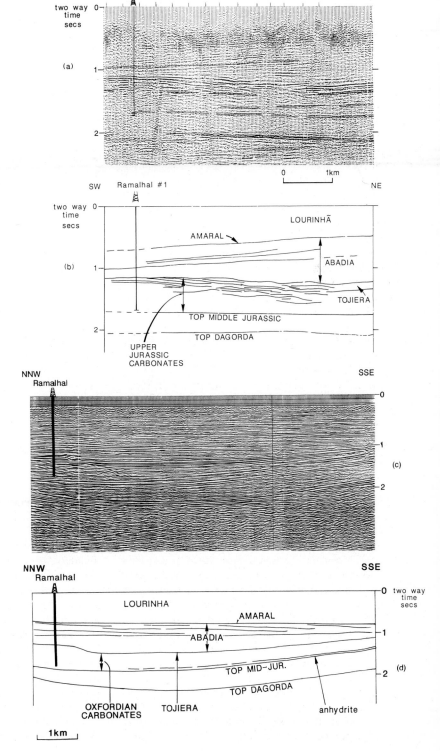

Fig. 14. Seismic sections across the Ramalhal part of the salt structure related buildup on the northwest side of the basin. (a) Unmigrated 24-fold dynamite line. Note that some interpretative lines were drawn on this section prior to its receipt by the authors. (b) Interpretation of line shown in (a) showing mounded clinoforms at the top of the Upper Jurassic carbonate section. (c) Migrated 48-fold vibroseis line. (d) Interpretation of line shown in (c).

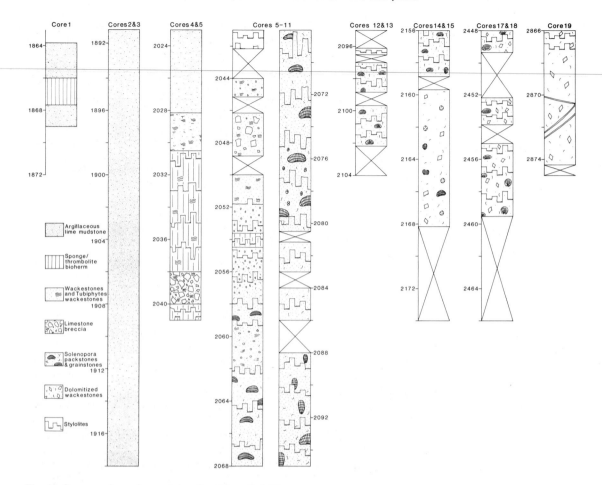

Fig. 15. Summary logs of cores taken from Ramalhal #1.

liferous *Solenopora* grainstones, packstones and wackestones, ≈ 400 m thick, containing heads of corals, stromatoporoids and *Solenopora*, and some *in situ* specimens of *Trichites*. A wide variety of clasts is contained within the packstones, many with thin oolitic coats, and the sediment shows good vuggy porosity, some of which is partially filled with baroque dolomite. The deeper-water limestones which cap the buildup consist of pyritic ammonitic lime mudstone, intercalated with matrix-rich limestone breccias and some minor thrombolitic bioherms (Fig. 15).

Campelos #1

The Cabaços formation is 400 m thick, and consists of lacustrine ostracod−charophyte mudstones inter-

calated with feldspathic sandstones. The Montejunto formation is significantly thinner, at ≈ 600 m, than in Ramalhal #1. Though no extensive coring was undertaken in this well, the three-fold division of the buildup can be recognized. The lower 400 m of the overlying Montejunto formation consists predominantly of lower-energy wackestones, above which occur 120−150 m of porous *Solenopora* grainstones. Deeper-water shales and limestones cap the buildup and can be recognized as a distinctive zone on electric logs by an increase in caliper and spontaneous potential and surprisingly, considering its clay content, a decrease of gamma-ray (Fig. 16).

Lourinha #1

A cored interval in the Cabaços formation consists

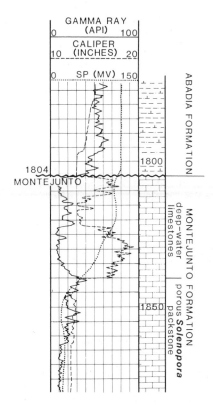

Fig. 16. Log signature of the top of the salt structure related buildup obtained from Campelos #1 showing the deeper-water limestone cap.

of feldspathic sandstones. These also contain angular evaporite clasts, possibly derived from the rising Bolhos diapir to the northwest. The three-fold division of the buildup cannot be recognized in this well, where the overlying marine carbonates consist of a thin sequence of fossiliferous packstones containing *Solenopora*, as well as corals and stromatoporoids, which are overlain by fine-grained, darker wackestones containing gastropods, serpulids, large cyclinid foraminifera, dasycladaceans and *Cayeuxia*, as well as *in situ* stick-like corals and stromatoporoids. The Abadia formation is not present in this well, so it seems possible that carbonate deposition continued near the crest of the salt high during the early Kimmeridgian, whilst siliciclastic sedimentation occurred farther into the Bombarral Sub-basin.

Outcrop data

The northwestward continuation of the Ramalhal

buildup is exposed at the surface at several locations along the Vimeiro–Caldas diapiric structure and may extend in age into the Upper Kimmeridgian (Guéry, 1984). At these outcrops, Upper Jurassic carbonates show considerable variations in thickness, due to contemporaneous diapiric movement. At Vimeiro (Fig. 2), ≈ 300 m of limestone are exposed, consisting chiefly of oncolitic wackestones and oolitic packstones, with some minor amounts of *Solenopora* packstones and framestones. To the north, at Sobral de Lagoa and Dagorda, no more than 150 m occur associated with a thick Upper Jurassic karst fill in the underlying Middle Jurassic limestones (Ellis *et al.*, 1987). This may be the site of a diapir which reached the surface during the Jurassic (Fig. 3).

Between the Bolhos diapir and Pó, to the southeast of the town of Serra d'el Rei (Fig. 3), the Montejunto formation is 500–600 m thick, with a thin development of the Abadia formation sandwiched between the Oxfordian carbonates and 150 m of limestones of Upper Kimmeridgian age that are probably lateral equivalents of the Amaral formation. In the Serra d'el Rei area, the Cabaços appears to be only locally developed and, as in Lourinha #1, consists of intercalated ostracod–charophyte lime mudstones and arkosic sandstones. The lower part of the Montejunto formation, as at Ramalhal, is micrite-dominated incorporating a richly fossiliferous and varied sequence of rocks similar to the nerineid wackestone facies (association 6) at Montejunto. These are overlain by thick sequences of bioclastic, oolitic and oncolitic packstones and grainstones with abundant corals, stromatoporoids and *Solenopora*. The corals and stromatoporoids locally form framestones up to 4 m thick within the grainstones. The limestones at Serra d'el Rei show no evidence of drowning, and it is probable that shallow-water carbonate sedimentation continued into the Kimmeridgian.

Summary

Figure 17 shows an interpretative cross-section linking the subsurface and outcrop sequences of the Ramalhal–Serra d'el Rei buildup described above. The coarsening-up trend from wackestones to grainstones is common to nearly all the sequences studied, but the deep-water cap appears to be confined to basinal locations (Campelos #1, Ramalhal #1). In contrast to the Montejunto and Ota buildups, a well-defined reef zone or shelf-break facies did not

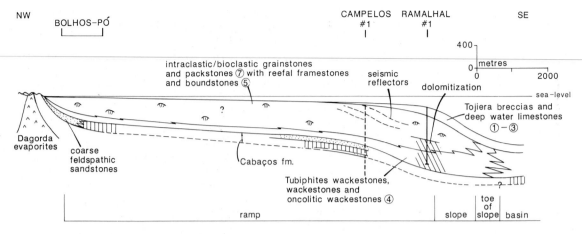

Fig. 17. Interpretative cross-section across the Oxfordian–Kimmeridgian salt structure related buildup on the northwest side of the basin, showing the distribution of the facies associations described in Table 1.

develop, and this may have prevented the development of a distinct lateral facies zonation. The buildup appears to have developed initially on a ramp produced as the Vimeiro–Caldas da Rainha structure rose to produce a broad salt pillow. The development of mounded, slightly progradational seismic reflectors at the top of the buildup, and the increase in slope angle in distal locations suggest that it was beginning to develop a shelf profile before it was drowned towards the end of the Oxfordian.

POSTRIFT PASSIVE BASIN FILL SEQUENCE

Setting and age

Dramatically increased subsidence rates associated with a rifting event in the Kimmeridgian (Wilson, 1979, 1988; Wilson *et al.*, 1989) halted carbonate sedimentation throughout much of the basin, though shallow-water carbonate sedimentation continued on palaeohighs into the Middle Kimmeridgian. This event was accompanied by the drowning and partial break-up of the older Montejunto formation buildups (of which the major examples have already been described) and was followed by the influx of Abadia clastics.

The southern part of the basin, however, remained starved of coarse clastics throughout the remainder of the Jurassic. The Monsanto borehole (Fig. 3) shows the basin filled with shales which are capped by a shallowing-upwards limestone sequence. To the west, near Sintra, a carbonate–shale slope deposit prograded into the newly deepened basin. These Upper Jurassic carbonates are now exposed around the diapir-like late Cretaceous Sintra granite.

The sequence exposed on the coast at Praia Abano is > 1000 m thick and ranges in age from Upper Oxfordian to Berriasian. It is divided into four lithostratigraphic units (see Fig. 2), the top three of which are shown on the summary log of Fig. 18. The lowermost Upper Oxfordian San Pedro formation consists of thermally-metamorphozed limestones in which characteristic lithofacies types cannot be discerned but its implied age and purity imply an equivalence to shallow-water Montejunto limestones.

On the basis of studies of foraminifera and algae, Ramalho (1971) assigned the following ages to the formations above the San Pedro:

Ramalhão formation:
  Lower to Middle Kimmeridgian
Mem Martins formation:
  Middle Kimmeridgian–Tithonian
Fartã Pão formation:
  Tithonian–lowermost Berriasian

The sequence was studied in detail by Ellis (1984).

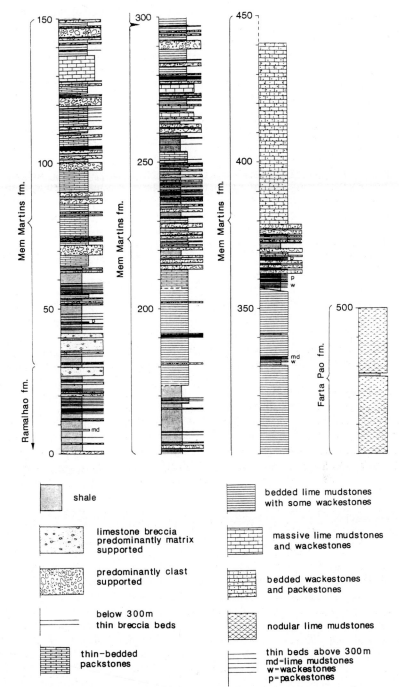

Fig. 18. Summary log of Kimmeridgian–Tithonian Sequence exposed north of Praia do Abano, on the coast southwest of Sintra.

The Ramalhão formation

This formation is dominated by shales (thermally metamorphozed to slates), but resedimented carbonates occur at the base as thin breccia beds and turbiditic wackestones. The formation is ≈ 400 m thick.

The Mem Martins formation

The Mem Martins formation is formed of a mixed carbonate–shale sequence, which passes upwards into pure limestones (Fig. 18). The base of this formation is marked by the first appearance of thick limestone breccias containing large allochthonous blocks of limestone (Ramalho, pers. comm.). Ranging from 0.2 m to over 4 m in thickness, the limestone breccias are composed of angular limestone lithoclasts largely composed of marine wackestones (facies association 4), as well as some transported corals and stromatoporoids, arranged chaotically within a matrix of calcareous shale. They were deposited as sheet or lobate debris flows (Ellis, 1984). None of the clasts show conclusive evidence of contemporaneous karstification. Thin-bedded packstones and grainstones occur in association with the breccias and were deposited as thin-bedded debris or turbidity flows. The background sediments consist of shales passing up into ammonitic lime mudstones (175–207 m, 280–355 m in Fig. 18), similar to Read's (1980, 1982) deep-water ribbon carbonates. Ellis (1984) concluded that the lack of sedimentary boudinage structures and other slope-related structures in the breccia-dominated part of the Mem Martins formation indicate that it was deposited on a slope inclined at less than one degree. The lack of breccias in the inland exposures to the east suggests that their disappearance was caused by a change in inclination at the toe of slope (Fig. 19a & 20). Tool marks at the base of one debris flow, and flute marks at the bases of carbonate turbidites suggest that the source area lay to the northwest.

The presence of abundant clasts of previously lithified limestone member is similar to that found in the Tojeira member at Montejunto and the deep-water caps of both the Barreiro and Ramalhal build-ups. The age of the lithified limestones is not known but is probably virtually contemporaneous with the background sediments. As lithified reefal debris is not a significant component of the breccias, a source area consisting of a reef bypass margin must be ruled out, and a fault scarp origin favoured as depicted in Fig. 19a. Transported corals and stromatoporoids and other bioclasts within the breccias, however, indicate that small patch-reefs were growing on the lithified limestones in the source area.

The top part of the Mem Martins formation (above 297 m in Fig. 18) does not contain Tojeira-like limestone breccias. Ammonitic lime mudstones continue to 355 m, and they contain thrombolitic bindstones and algal microsolenid bindstones to the east of the coastal sequences. An erosional surface above the ammonitic lime mudstones on the coast is overlain by shales and allochthonous packstones and wackestones and coraliferous rubble beds which have a limited lateral extent. These are overlain by fossiliferous *Tubiphytes* wackestones (association 4) with phaceloid coral bushes which pass up into nerineid wackestones and packstones (association 6), containing coral–chaetitid framestone patch-reefs and coral biostromes.

The top part of the Mem Martins formation is interpreted as a prograding carbonate ramp, across which the faunal diversity of bioherms increased from thrombolitic bindstones, through algal–microsolenid bindstones to coral–chaetitid framestones up the ramp into shallower waters. Small slump scars and debris flows also occurred (Fig. 19b).

The Farta Pão formation

Nodular lime mudstones containing a restricted biota of cyclinid foraminifera *Anchispirocyclina lusitanica*, miliolid foraminifera and dasyclads (*Salpingoporella annulata, Campbelliella striata*) (facies association 8) are the main constituent of this formation. The top of the formation contains condensed horizons, minor sandstones and oyster-encrusted firmgrounds (Ellis, 1984).

The restricted lagoonal environment represented by the Farta Pão formations marks an abrupt change in conditions from the fully marine Mem Martins formation and heralds the platform conditions of the Berriasian.

Summary

The Ramalhão and lower part of the Mem Martins formations are lateral equivalents of the Abadia formation, which prograded southwards during a relative sea-level stillstand. In the Sintra area, the deep-water basin was progressively and passively filled by shales and debris flows originating from a postulated fault scarp to the northwest (Figs 19 & 20). Once the basin had almost filled, a carbonate ramp system developed represented by the upper part of the Mem Martins formation (Fig. 20). A similar pattern of sedimentation occurs 25 km to the east beneath Lisbon, where the Monsanto well revealed a shallowing-up carbonate sequence with lithologies similar to the upper part of the Mem

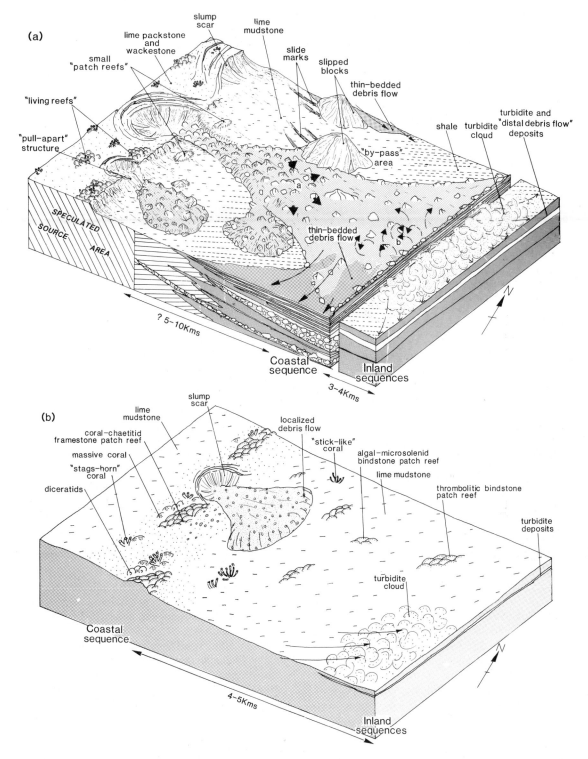

Fig. 19. Depositional models for the Kimmeridgian–Tithonian Sequences in the Sintra area (from Ellis, 1984). (a) Model for the breccia-dominated lower part of the Mem Martins formation. (b) Model for the upper part of the Mem Martins formation.

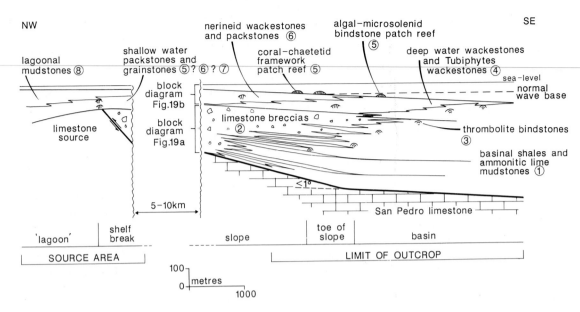

Fig. 20. Interpretative cross-section of the Kimmeridgian–Berriasian buildup on the southwest side of the basin, showing the distribution of facies associations described in Table 1.

Martins and Farta Pão formation overlying 900 m of Abadia marls.

LIMESTONE CAP ON PROGRADING SILICICLASTICS

Setting and age

The Amaral formation forms a thin (≈60 m) limestone cap to the southward prograding slope marls and silts of the Abadia formation. It extends over an area of > 1000 km² from the southern part of the Bombarral Sub-basin into the Arruda Sub-basin (Wilson, 1988, fig. 7) and is seen on seismic sections in the Turcifal Sub-basin.

The top part of the Abadia formation contains ammonites characteristic of the *tenuilobatum* and *pseudomutabalis* Zones which, together with the occurrence in the Amaral formation of the foram *Kurnubia palastiniensis* and the dasyclad *Clypeina jurassica* clearly indicates an Upper Kimmeridgian age for the latter formation (Dölher in Leinfelder, 1986).

Description and interpretation

The Amaral formation represents a carbonate sheet

sand and exhibits fewer facies associations than the other buildups described in this paper (Table 1). High-energy grainstones and bioherms (associations 5 and 7) dominate, but lower energy fossiliferous wackestones (association 6) occur in lagoonal areas.

In the centre of the Arruda Sub-basin, the basal 30 m of the Amaral formation consists of coral bindstones forming individual structures up to 10 m thick. These contain a diverse biota of baffling and massive corals and stromatoporoids, with molluscs, echinoids and encrusting algae and forams. This reefal facies grades vertically and laterally into coral-rich bioclastic oncolitic packstones and grainstones, with minor nerineid packstones and bioturbated bioclastic wackestones. Significant amounts of ooids also occur in these sediments.

Around the Torres Vedras–Montejunto high, and in the Campelos and Ramalhal wells, the Amaral consists predominantly of ooid grainstones up to 30 m thick. This facies also occurs above the lower biostromal unit in the centre of the Arruda sub-basin. At outcrop, the grainstones are cross-bedded, with sets up to 10 m thick. The amount of detrital quartz-forming ooid nuclei increases upwards, and siliciclastic sandstones and oyster patch-reefs up to 2 m thick are intercalated with the grainstones towards the top of the sequence.

In the Torres Vedras−Montejunto area, the ooid grainstones of the Amaral formation were interpreted by Ellwood (1987) as an ooid bar system that formed at the shelf break of the southern prograding Abadia formation slope system. To the south, in the Arruda region, the coral biostromes appear to have developed slightly downslope from the shelf break ooid sands.

POSSIBLE CONTROLS ON CARBONATE DEVELOPMENT IN PORTUGAL

Recapitulation

Four carbonate sequence types occur in the Upper Jurassic of the Lusitanian Basin; the geographical distribution of three of them is shown in Fig. 21, and their stratigraphical relationships in Fig. 4.

1 On the east side of the basin, faults exerted a significant influence on carbonate facies distribution, resulting in the development of carbonate buildups with shelf geometries during the Oxfordian − Kimmeridgian. These developed distinct lateral facies zonations, and in the case of the Montejunto and Ota buildups were aggradational. Carbonate muds form a significant proportion of the structures as they were protected by a distinct shelf-break facies. The buildups are relatively thin over fault blocks (200−500 m, Figs 7, 10 & 13) indicating relatively low subsidence rates.

2 On the northwest side of the basin, a thick (up to 1500 m) Oxfordian−Kimmeridgian coarsening-up grainstone-dominated buildup formed on the flank of a rising salt pillow structure. No distinct shelf-break facies or significant lateral facies zonation developed. The top of the buildup shows progradational features and the development of a shelf-like profile on seismic sections (Fig. 17).

3 Following the Kimmeridgian rifting event a thick (≈1000 m) Kimmeridgian−Tithonian Sequence of shales, debris flows and low- to moderate-energy wackestones and packstones developed in the southwest of the basin. This sequence, which lacks high-energy grainstones, developed as a prograding ramp deposit and is interpreted as a passive basin fill

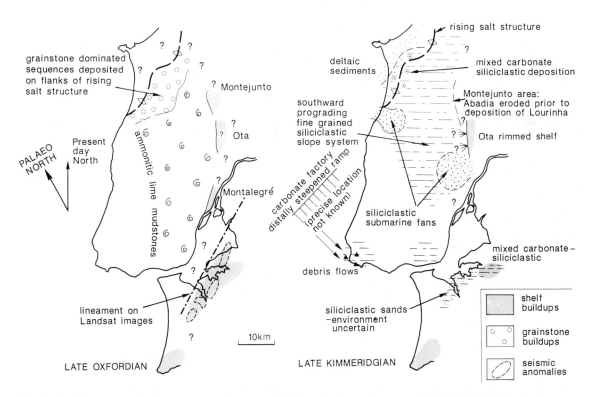

Fig. 21. Sketch maps showing the occurence of carbonate buildups in the southern part of the Lusitanian Basin during the late Oxfordian and late Kimmeridgian.

in a siliciclastic starved part of the basin (Fig. 20).

4 A thin ($\approx$60 m) Upper Kimmeridgian coral bio-strome and ooid grainstone developed as a sheet-like cap to the southward prograding slope marls and silts of the Abadia Formation.

The possible controls of the formation and demise of these buildups is now reviewed under the headings of palaeogeography, tectonic setting and sea-level changes.

Palaeogeography

As indicated at the beginning of the paper, during the Late Jurassic the Lusitanian Basin opened southwestwards on to the newly opened southern North Atlantic (see Fig. 1). Therefore the eastern and northern parts of the Basin were likely to have experienced the highest wave energies, as waves approaching from the southwest would have had the opportunity to develop over the longest fetch, indeed along the whole 4−5000 km length of the North Atlantic as it then existed. Thus it seems probable that the southwest margin of the present-day basin, which during the Upper Jurassic faced southeast (see Fig. 20) was the most sheltered, thus explaining the development of the lowest energy carbonate sequence in the Sintra area.

Tectonic setting

The differences between the nature of buildups on the eastern and northwestern margins of the basin during the Oxfordian−Kimmeridgian are clearly related to their tectonic settings. The grainstone-dominated buildups to the northwest developed on the flanks of a rising salt structure, whereas those on the east were bounded on their basinward sides by flexures or faults. For this reason, the eastern build-ups, such as Montejunto and Ota, developed good lateral facies zonation patterns, which were unable to migrate laterally because tectonically-controlled shelf break prevented basinward progradation. The subsidence histories of the two settings account for the difference in thickness of the two buildup types (Fig. 22). It is probable that both regions experienced comparable tectonic subsidence during the Upper Jurassic, but in the Ramalhal area this was augmented by salt withdrawal. This effectively added Triassic−Hettangian tectonic subsidence to that occurring during the late Jurassic. Clearly, carbonate sedimentation at Ramalhal was able to keep up with a subsidence rate of some 75 mm yr^{-1} (at least 1500 m

of compacted carbonate accumulated during the late Oxfordian timespan of 2 Myr).

Sea-level changes

The eustatic sea-level curve of Haq *et al.* (1987) is plotted on the summary diagram of buildup development presented in Fig. 4. Lack of precision concerning the ages of parts of the Upper Jurassic sections in Portugal means that caution must be exercised in correlating the Haq curve and its associated sequence boundaries with depositional events in the Lusitanian Basin. An additional problem is that during the latest Oxfordian and early Kimmeridgian, parts of the Lusitanian Basin experienced extremely rapid rates of tectonic subsidence which could have obscured effects produced by eustatic sea-level changes (Wilson *et al.*, 1989).

The most significant change of relative sea-level during the late Jurassic in Portugal was the 700 m rise followed by a relative stillstand that permitted the Abadia slope system to prograde southwards and the Mem Martins formation to form as a passive fill. This highstand may correlate with cycles LZA-4.5 of Haq *et al.* (Fig. 4), yet the lower sequence boundary on seismic sections (Fig. 14) appears to be at the base of the Tojeira member, the base of which is situated in the Upper Oxfordian, and so would correlate better with cycle LZA-4.4. Resolution of this problem must await more detailed biostratigraphical studies. It is tempting to suggest that the drowning of buildups and the appearance of resedimented carbonate breccias was synchronous throughout most of the basin and was caused by a combination of latest Oxfordian and earliest Kimmeridgian eustatic and tectonic effects. Throughout much of the Basin, this event was followed by the southwards progradation of siliciclastics of the Abadia formation, but shallow-water carbonate sedimentation continued on salt- and fault-controlled highs.

AN ATLANTIC PERSPECTIVE

The purpose of the final section of the paper is to examine the nature of the carbonate buildups comprising the 'Mesozoic reef trend' beneath the Atlantic continental shelf of North America in the light of conclusions made in Portugal. Despite the very different sizes of the two regions, far more is known

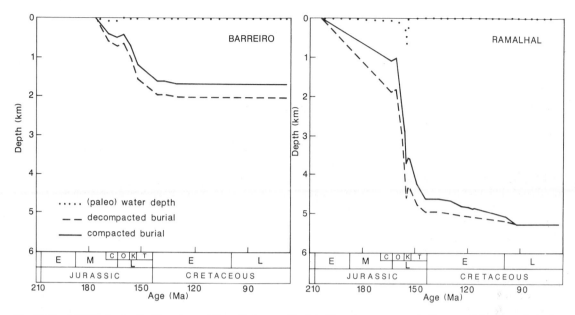

Fig. 22. Burial histories for the Barreiro and Ramalhal areas. That for Barreiro was computed from well data from Barreiro #1, but the well data for Ramalhal was augmented using seismic data below the Upper Jurassic down to the top-Dagorda level, and above the Lourinhã formation using thicknesses from nearby measured sections. Palaeowater depths were estimated from sedimentological and palaeoecological studies.

about the lithological content and lateral facies variation of the Portuguese buildups.

Buildup geometries and settings

A number of carbonate buildups from the North American Atlantic margin have similar cross-sectional geometries and settings to those in Portugal. Fault-controlled shelves analogous to Montejunto and Ota, for example, were identified from the Grand Banks Le Have platform by Jansa (1981). The carbonate ramp of the Mem Martins formation displays similar lithofacies and cross-sectional geometry to sequences cored by Petro-Canada Shell Penobscot L-30 on the Scotian Shelf (Eliuk, 1981; Ellis, 1984: although breccias were not recorded from this Canadian example), while the Amaral Formation has a similar stratigraphic setting to the 'O' marker carbonate of the Grand Banks (Jansa, 1981), for both units overly prograding slope sediments. The majority of the eastern North American Upper Jurassic buildups and those taken to represent the main 'reef trend' (Schlee & Grow, 1982), were located over a rapidly subsiding continental margin and facing the newly-opening ocean. Though this represents a very different setting from

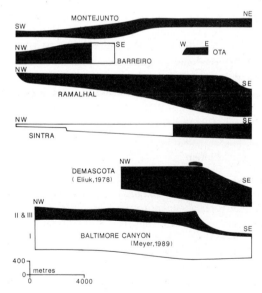

Fig. 23. Comparative cross-sectional geometries of Portuguese and North American buildups discussed in the text. The white areas in the Portuguese examples represent speculative extensions of the buildings not seen at outcrop or penetrated by wells. The white area in the Baltimore Canyon example comprises siliciclastic sediments to the north west, which pass basinwards into carbonates and then into presumed slope deposits.

the Portuguese buildups, they developed comparable thicknesses and geometries to the salt-influenced Ramalhal buildup.

Meyer (1989) recorded three stages in the development of North American Atlantic margin carbonate buildups. The first of these (Stage I) consists of prograding carbonate clinoforms which developed into aggradational carbonate shelves (Stage II). A thin cap of deeper-water carbonates records a drowning event (Stage III). Explanations for the two main styles of buildup growth (Meyer's Stages I and II) vary widely from author to author. Meyer suggested that buildups receiving siliciclastic sediments from the hinterland prograde (Stage I), while those that do not, aggrade (Stage II). Mattick (1982), however, takes exactly the opposite view! Erlich *et al.* (1987), in contrast, postulated a eustatic control, whereby rapid shelf-margin progradation was caused by a sudden short-lived sea-level fall, whereas aggradation was caused by a rapid rise.

Both aggradational and progradational shelves occur within the Upper Jurassic of the Lusitanian Basin as well as a number of examples of drowned buildups, equivalent to Meyer's Stage III and containing similar facies (Ellis *et al.*, 1985). Clastic input played no significant part in the development of these Portuguese buildups, though this does not rule it out as a factor influencing either the position or geometry of the American buildups. In Portugal, aggradational and progradational buildup types developed during the same period of relative sea-level rise. The two buildup geometries in the Lusitanian Basin are linked to local tectonic conditions, rather than differing eustatic sea-level changes.

Facies variation

The nature of the shelf-edge along the North American Atlantic margin has been described as largely reefal, similar to modern reefs (Schlee & Grow, 1982), reefal only in aggradational phases (Mattick, 1982; Erlich *et al.*, 1987; Meyer, 1989), or as a carbonate bank system in which 'oolitic shoals were present near the edge, and skeletal, peloidal wackestones and biomicrites were deposited in the inner part of the platform', where 'coral−stromatoporoid and sponge bioherms were only rare constituents' (Jansa, 1981).

The Nova Scotian shelf has yielded the greatest amount of lithofacies data based on core samples. Eliuk (1978, 1979, 1981) described shelf-break sediments in this region to consist largely of non-skeletal oncolitic grainstones and packstones, with some well-developed coral−stromatoporoid reefs, such as those cored in Shell Demascota G-32 (Fig. 24). Like the Barreiro and Ramalhal buildups from Portugal, this sequence shows a general shallowing-upwards trend, capped by deeper water facies following a drowning event. Eliuk (1978), Jansa *et al.* (1983) and Ellis (1984) agreed that the cored intervals within this well exhibit a true reefal framework, but there was less agreement about the significance of the shallow-water facies in cores 2 and 3.

Deeper-water thrombolitic boundstones, containing the characteristic microfossil *Thartharella*, form much of the lowest core in Demascota. Eliuk envisaged that this passed up into a persistent coral-dominated reefal structure, analogous to modern shelf-edge reefs, building up from the deep-water bioherms into a semi-exposed reef-flat environment. He considered that the presence of red algae (*Solenopora*) also gave a 'modern aspect' to the reef. However, on the basis of electric log data, Jansa *et al.* (1983) and Ellis (1984) suggested that the reefal intervals represented by cores 2 and 3 (Fig. 24) were no more than 15−20 m thick (consisting of coral−chaetitid framestone), and bounded above and below by wackestones and shales. They compared them to Wilson's (1979) 'knoll reefs', which formed in quiet-water conditions at or just below, wave base.

A modern, reef-flat analogy for the Demascota buildup is challenged by two other observations. Firstly, *Solenopora* identified by Eliuk (1978) and likened to red algae in modern reef flats, was re-identified as the milleporid *Milleporidium remesi* by Ellis (1984), who also remarked at the conspicuous lack of binding and encrusting algae within the framework. In Portugal, *Solenopora* only occurs as individual heads within grainstones and packstones and is extremely rare in knoll-reef frameworks. Secondly, the breccias within Demascota core 2 which were considered by Jansa *et al.* (1983) to be reef-flat debris, were interpreted by Ellis (1984) as a diagenetic breccia.

Recent Shell−Amoco−Sun wells drilled in the Baltimore Canyon Trough area also indicate a grainstone and packstone-dominated shelf edge (Erlich *et al.*, 1987) similar to those of the salt-controlled buildup in Portugal. Cores from an eastward prograding clinoform interval penetrated by 0337 Civet (equivalent to Meyer's Stage II buildup) show a similar shallowing-upwards sequence (Fig. 24), from deeper-water wackestones to coral−stromatoporoid

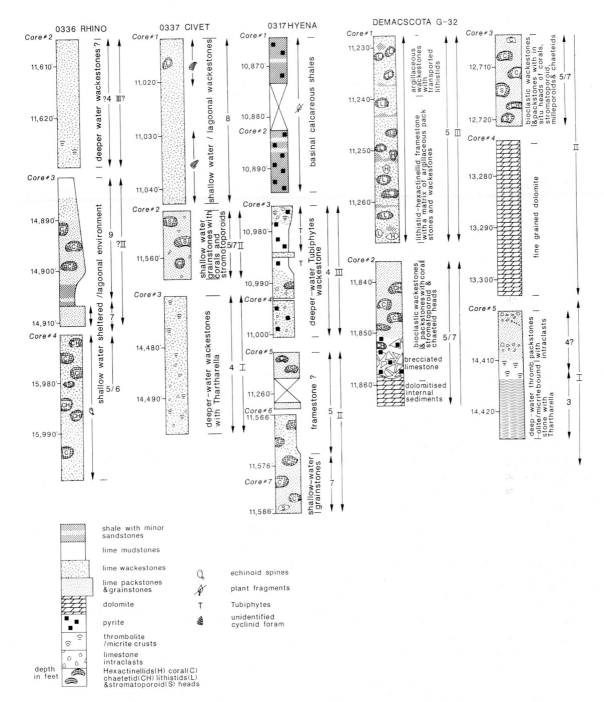

Fig. 24. Summary logs of cores taken in wells drilled into the Upper Jurassic buildups on the North American margin of the North Atlantic. The numbers 1–9 refer to facies associations described in Table 1, and I–III indicate Meyer's (1989) stages of building development.

grainstones. The grainstones exhibit no framestone structure and, in this case, do not have a deeper-water cap. A more argillaceous lagoonal facies (similar to that probably occurring southeast of Barreiro) occurs shelfward from the Civet well in 0336 Rhino (Fig. 24). Only one well was drilled into an aggrading reefal structure (equivalent to Meyer's Stage II). This yielded coral–stromotoporoid grainstones, with a possible true high-energy framestone interval (cores 5 and 6, Fig. 24) and was capped with slightly deeper-water *Tubiphytes* wackestones.

Though core data from the American buildups are sparse, they tend to support Jansa's (1981) idea of a largely grainstone-dominated carbonate shelf edge, with significant shelf-break reefal structures being a relatively rare component. Comparison with the Portuguese buildups supports this conclusion for reefal frameworks are not an important influence on sedimentation, other than the Ota Reef, which was an unusual structure being strongly influenced by a shallow basement fault. In the larger buildup developments of the American seaboard, shelf-break reefs, when they are present, may largely be of the 'knoll-reef' type envisaged for the Demascota buildup. Jansa (1981) suggested that many of the steep carbonate platform edges known from the area might have formed by early submarine lithification and may not be reefal in origin.

The general lack of well-developed coral–stromatoporoid reefs within the high-energy environments of the North Atlantic basin during the Jurassic is surprising in view of James's (1984) conclusion that this was a period of significant reefal growth. It appears that at least in the Atlantic Basin, corals and stromatoporoids were not able to outpace other types of carbonate sedimentation to produce major reef structures. Except where carbonate deposition was strongly controlled by faulting (such as Ota), the evidence available suggests that corals and stromatoporoids were only able to construct smaller-scale knoll-reefs in lower, energy environments such as those in the Portuguese Montejunto buildup and the Canadian Demascota example. Scott (1984) envisaged a similar habitat for coral–stromatoporoid bioherms in the Lower Cretaceous of North America, leaving higher-energy environments free for colonization by rudists.

ACKNOWLEDGEMENTS

Peter Ellis acknowledges the generous financial assistance of a Royal Society Grant (through the award of a European Postdoctoral Research Fellowship) and funding from Amoco (Tulsa) and Texaco (Houston). Reinhold Leinfelder acknowledges financial support from the Deutsche Forschungsgemeinschaft (Project Le 580/1). The assistance of staff of the exploration department of Petrogal and the Serviços Geológicos de Portugal in Lisbon, and Sceptre Resources in Calgary in providing well and seismic data is gratefully acknowledged. We thank Miguel Ramalho of the Serviços Geológicos for his assistance whilst working in Portugal. Felix Gradstein computed the burial curves depicted in Fig. 21, using data supplied by us. Subsurface data in this paper are published by kind permission of the Director General of the Gabinete para a Pesquisa e Exploração de Petroleo in Lisbon. The stratigraphic summary chart shown in Fig. 2 was devised in collaboration with Peter Ellwood, Graham Hill, Richard Hiscott, David Kitson, Miguel Ramalho and Peter Skelton. The authors are grateful to Beryl West and Carol Whale for patiently typing the manuscript and to John Taylor and Andrew Lloyd for drafting the figures. The authors wish to thank reviewers Paul Crevello, Franz Meyer and Wolfgang Schlager for the helpful comments which stimulated the extensive revision of the first draft of the paper.

REFERENCES

ATROPS, F. & MARQUES, B. (1986) Mise en évidence de la zone à Platynota (Kimmeridgien Infèrieur) dans le Massif de Montejunto (Portugal): conséquences statigraphiques et paléontologiques. *Geobios* **19**, 537–547.

CREVELLO, P.D. & HARRIS, P.M. (1982) Depositional models and reef building organisms, Upper Jurassic reefs of the Smackover Formation. *Gulf Coast Sect. Soc. Econ. Paleont. Mineral., Foundation Res. Conf.* 1, pp. 25–28.

ELIUK, L.S. (1978) The Abenaki Formation, Nova Scotia shelf, Canada – a depositional and diagenetic model for a Mesozoic carbonate platform. *Bull. Can. Petrol. Geol.* **26**, 424–514.

ELIUK, L.S. (1979) Abenaki update: variations along a Mesozoic carbonate shelf edge, Nova Scotia shelf, Canada (Abstract). Canadian Society of Petroleum Geology, Baillie Carbonate Symposium, *Recent Advances in Carbonate Sedimentology*, Sept. 20–21 1979, Calgary, Alberta, Canada.

ELIUK, L.S. (1981) Abenaki update: variations along a Mesozoic carbonate shelf, Nova Scotia Shelf, Canada. In: *Can. Soc. Petrol. Geol. Annual Core. Field Sample Conf. January 15th 1981* (Compiler, Stokes, F.A.S.), pp. 15–19.

ELIUK, L.S. (1982) Nova Scotia Shelf Mesozoic analogy to the south. *Bull. Assoc. Petrol. Geol.* **66**, 567.

ELLIS, P.M. (1984) *Upper Jurassic carbonates from the Lusitanian Basin, Portugal, and their subsurface counterparts in the Nova Scotian Shelf.* Unpublished PhD Thesis, Open University, p. 193.

ELLIS, P.M., CREVELLO, P.D. & ELIUK, L.S. (1985) Upper Jurassic and Lower Cretaceous deep-water buildups, Abenaki Formation, Nova Scotia Shelf. In: *Deep Water Carbonates* (Eds Crevello, P.D. & Harris, P.M.) Soc. Econ. Paleont. Mineral. Core Workshop 6, pp. 212–248.

ELLIS, P.M., ELLWOOD, P.M. & WILSON, R.C.L. (1987) Kimmeridgian siliciclastic and carbonate slope deposits, in Excursion B: Structural control of sedimentation during Upper Jurassic in Meridional region of the Lusitanian Basin. *2nd Int. Symp. Jurassic Stratigraphy*, Lisbon, 3, pp. 1–34.

ELLWOOD, P.M. (1987) *Sedimentology of the Upper Jurassic Abadia Formation and its equivalents, Lusitanian Basin, Portugal.* PhD Thesis, Open University, Milton Keynes, UK, p. 337.

ENOS, P. & PERKINS, R. (1979) Evolution of Florida bay from island stratigraphy. *Geol. Soc. Am. Bull.* **90**, 59–83.

ERLICH, R.N., MAHER, K.P., HUMMEL, G.A., BENSON, D.G., KASTRITIS, G.J., LINDER, H.D., HOAR, R.S. & NEELEY, D.H. (1987) Baltimore Canyon Trough, Mid-Atlantic O.C.S: Seismic stratigraphy of Shell/Amoco/Sun Wells. In: *Atlas of Seismic Stratigraphy*, Vol. 2 (Ed. Bally, B.W.). Am. Assoc. Petrol. Geo. Studies in Geology 27, 2, pp. 5–65.

GUERY, F. (1984) *Evolution sédimentaire et dynamique du bassin marginal ouest portugais au Jurassique: Province Estémadure secteur de Caldas da Rainha-Montejunto.* Thesis, Université de Lyon, p. 477.

HAQ, B.U., HARDENBOL, J., VAIL, P.R. & ERLICH, R.N. (1987) Chronology of fluctuating sea levels since the Triassic. *Science* **235**, 1156–1166.

JAMES, N.P. (1984) Reefs. In: *Facies Models* (Ed. Walker, R.G.) Geoscience Canada, Reprint Series 1 (2nd edn) p. 229–244.

JANSA, L.F. (1981) Mesozoic carbonate platforms and banks of the eastern North American margin. *Marine Geol.* **44**, 97–117.

JANSA, L.F. (1986) Paleooceanography and evolution of the North Atlantic Ocean basin during the Jurassic. In: *The Western North Atlantic Region* (Eds Vogt, P.R. & Tucholke, B.E.) (The Geology of North America, volume M). Geol. Soc. Am., pp. 603–616.

JANSA, L.F. & WEIDMANN, J. (1982) Mesozoic–Cenozoic development of the Eastern North American and Northwest African continental margins: a comparison. In: *Geology of the Northwestern African Continental Margin* (Eds von Rad, U., Hinz, K., Sarnthein, M. & Seibold E.) Springer Verlag, pp. 215–269.

JANSA, L.F., TERMIER, G. & TERMIER, H. (1983) Les bioherms à algues, spongiares et coreaux des series carbonateés de la flexure bordiere du 'paleoshelf' au marge du Canada oriental. *Rev. Micropaleont.* **25**, 181–219.

KENT, D.V. & GRADSTEIN, F.M. (1985) A Cretaceous and Jurassic geochronology. *Geol. Soc. Am. Bull.* **96**, 1419–1427.

LANCELOT, Y. & WINTERER, E.L. (1980) Evolution of the Moroccan oceanic basin and adjacent continental margin – a synthesis. *Init. Rep. Deep-Sea Drilling Project* **50**, 801–821.

LEINFELDER, R.R. (1985) Cyanophyte calcification morphotypes and depositional environments (Alenquer Oncolite, Upper Kimmeridgian, Portugal). *Facies* **12**, 253–274.

LEINFELDER, R.R. (1986) Facies, stratigraphy and paleogeographic analysis of Upper? Kimmeridgian to Upper Portlandian sediments in the environ of Arruda dos Vinhos, Estrémadura, Portugal. *Munchner Geowiss. Abh. (A)*, **7**, 1–216.

LEINFELDER, R.R. (1987) Formation and significance of black pebbles from the Ota Limestone (Upper Jurassic, Portugal). *Facies* **17**, 159–170.

LEINFELDER, R.R. & RAMALHO, M.M. (1987) Age and general facies development of the Ota Limestone (Estremadura, Portugal) (Abstracts). *2nd Int. Symp. Jurassic Stratigraphy*, Universidade Nova Lisboa, Lisbon.

LEINFELDER, R.R. & WILSON, R.C.L. (1989) Seismic and sedimentologic features of Oxfordian–Kimmeridgian syn-rift sediments on the eastern margin of the Lusitanian Basin, Portugal. *Geol. Runds.* (in press).

MANUPPELLA, G. & BALACO MOREIRA, J.C. (1984) Calcários da Serra de Ota, Materias, Primas Minerais Não Metálicas. *Estudeos Notas Trabalho Servicos Fomento Minero, Porto*, **26**, 29–52

MASSON, D.G. & MILES, P.R. (1986) Development and hydrocarbon potential of Mesozoic sedimentary basins around margins of North Atlantic. *Am. Assoc. Petrol. Geol. Bull.* **70**, 721–729.

MASSON, D.G. & ROBERTS, D.G. (1981) Late Jurassic–early Cretaceous reef trends on the continental margin SW of the British Isles. *J. Geol. Soc. Lond.* **138**, 437–443.

MATTICK, R.E. (1982) Significance of the Mesozoic carbonate bank-reef sequence for the petroleum geology of the Georges Bank Basin. In: *Geological Studies of the COST nos. G-1 and G-2 Wells, United States North Atlantic Shelf* (Eds Scholle, P.A. & Wenkam, C.R.) US Geological Survey Circular 861, pp. 93–104.

MAUFFRET, A., MOUGENOT, D., MILES, P.R. & MALOD, J. (1989) Results from multichannel reflection profiling of the Tagus abyssal plain (Portugal): comparison with the Canadian margin. In: *Extensional Tectonics and Stratigraphy of the North Atlantic Margins* (Eds Tankard, A.J. & Balkwill, H.R.) Am. Assoc. Petrol. Geol. Mem. 46 (in press).

McIVER, N.L. (1972) Cenozoic and Mesozoic stratigraphy of the Nova Scotian Shelf. *Can. J. Earth Sci.* **9**, 54–70.

MEMPEL, G. (1955) Zur palaogeographie der Oberen Jura in Mittel-Portugal. *Zeit. Deutsche Geol. Gessellchaft* **105**, 106–126.

MEYER, F.O. (1989) Siliclastic influence on mesozoic platform development; Baltimore Canyon Trough, Western Atlantic. In: *Controls on Carbonate Platform and Basin Development* (Ed. Crevello, P.D., Wilson J.L., Sarge, J.F. & Read, J.F.). Spec. Publ. Soc. Econ. Paleont. Mineral. 44; 213–232.

MOUTERDE, R., ROCHA, R.B. & RUGET, C. (1971) Le Lias moyen et superieur de la region de Tomar. *Comm. Serv. Geol. Portugal* **55**, 50–80.

RAMALHO, M.M. (1971) Contribution à l'etude micro-palaeontologique et stratigraphique du Jurassique superieur et du Cretace inferieur des environs de Lisbonne (Portugal). *Mem. Serv. Geol. Portugal* **19**, 212.

RAMALHO, M.M. (1981) Note préliminaire sur le microfacies du Jurassique Supérieur Portugais. *Comm. Serv. Geol. Portugal* **67**, 41−45.

READ, J.F. (1980) Carbonate ramp-to-basin transition and foreland basin evolution, Middle Ordovician, Virginia Appalachians. *Am. Assoc. Petrol. Geol. Bull.* **64**, 1575−1612.

READ, J.F. (1982) Carbonate platforms of passive (extensional) continental margins: types, characteristics and evolution. *Tectonophys.* **81**, 195−212.

REY, J. (1972) Récherches geologiques sur le Cretacé Inferieur de l'Estremadura (Portugal). *Mem. Serv. Geol. Portugal* **21**, 477.

RIBEIRO, A., ANTUNES, M.T., FERREIRA, M.P., ROCHA, R.B., SOARES, A.F., ZBYSZEWSKI, G., MOINTINHO DE ALMEIDA, F., DE CARVALHO, D. & MOTEIRO, J.H. (1979) *Introduction à la géologie générale de Portugal.* Serv. Geol. Portugal, Lisboa, p. 114.

RUGET-PERROT, C. (1961) Études stratigraphiques sur le Dogger et le Malm inférieur au Nord du Tage. *Mem. Serv. Geol. Portugal* **7**, 197.

SCHLEE, J.S. & GROW, J.A. (1982) Buried carbonate shelf beneath the Atlantic continental slope. *Oil Gas J.* Feb. 25, 148−159.

SCOTT, R.W. (1984) Evolution of early Cretaceous reefs in the Gulf of Mexico. *Paleontograph. Am.* **54**, 406−412.

STAM, B. (1985) *Quantitative analysis of middle and late Jurassic foraminifera from Portugal and its implications for the Grand Banks of Newfoundland.* Unpublished PhD Thesis, Dalhousie University, Halifax, Nova Scotia, p. 212

TUCKER, M.E. (1985) Shallow-marine carbonate facies and facies models. In: *Sedimentology: Recent Developments and Applied Aspects* (Eds Brenchley, P.J. & Williams B.P.J.) Spec. Publ. Geol. Soc. Lond., **18**, 147−169.

VOGT, P.R. & TUCHOLKE, B.E. (Eds) (1986) The Western

North Atlantic Region. *The Geology of North America*, Vol. M. Geol. Soc. Am., p. 696.

WESTERHAUSEN, H. (1956) *Seismic reflection and refraction investigations carried out in the North Tajo Basin.* Unpublished report (available on open file at the Servicos Geológicos de Portugal, Lisbon) for Companhia dos Petróleos de Portugal and Mobil Exploration Inc. by Prakla Gesellschaft für praktische Layerstättenforschung G. mb. H, Hanover.

WILSON, J.L. (1975) *Carbonate Facies in Geologic History.* Springer-Verlag, New York, p. 471.

WILSON, R.C.L. (1979) A reconnaissance study of Upper Jurassic sediments of the Lusitanian Basin. *Ciências da Terra* (Univ. Nova de Lisboa) **25**, 53−84.

WILSON, R.C.L. (1988) Mesozoic development of the Lusitanian Basin, Portugal. *Rev. Soc. Geol. Espanha*, **1**, 394−407.

WILSON, R.C.L., HISCOTT, R.N., WILLIS, M.G. & GRADSTEIN, F.M. (1989) Lusitanian Basin of west Central Portugal: Mesozoic and Tertiary tectonic, stratigraphic and subsidence history. In: *Extensional Tectonics and Stratigraphy of the North Atlantic Margins* (Eds Tankard, A.J. & Balkwill, H.R.) Am. Assoc. Petrol. Geol. Mem. 46 (in press).

WRIGHT, V.P. (1985) Algal marsh deposits from the Upper Jurassic of Portugal. In: *Palaeogeology* (Eds Toomey, D.F. & Niteck, M.H.) Springer Verlag, Berlin, pp. 330−341.

WRIGHT, V.P. & WILSON, R.C.L. (1985) Lacustrine carbonates and source rocks from the Upper Jurassic of Portugal. *Int. Assoc. Sed. European Meeting Abstracts*, pp. 487−490.

WRIGHT, V.P. & WILSON, R.C.L. (1987) A Terra Rossa-like palaeosol complex from the Upper Jurassic of Portugal. *Sedimentology* **34**, 259−273.

ZBYSZEWSKI, G., VEIGA FERREIRA, O.D, MANUPPELLA, G. & TORRE DE ASSUNCAO, C. (1966) *Carta Geológica de Portugal na escala 1:50 000 Noticia Explicativa da Folha 30-B, Bombarral.* Servicos Geológicos de Portugal, p. 90.

Spec. Publs int. Ass. Sediment. (1990) **9**, 203–233

Hauterivian to Lower Aptian carbonate shelf sedimentation and sequence stratigraphy in the Jura and northern Subalpine chains (southeastern France and Swiss Jura)

A. ARNAUD-VANNEAU *and* H. ARNAUD

Laboratoire de Géologie Alpine associé au CNRS (URA 69). Université Joseph Fourier, Institut Dolomieu, Rue Maurice Gignoux, 38031-Grenoble cedex, France.

ABSTRACT

In the Jura and the northern sub-alpine chains (SE France), sedimentological and palaeontological analysis of Hauterivian–early Aptian platform carbonates shows that platform development was affected by both the morphological inheritance of Liassic rifting and variations in sea-level. Three major variations in relative sea-level can be observed: (1) a sudden drop in relative sea-level at the beginning of the Barremian, allowing the Jura platform to become emergent and a lowstand wedge to develop along the previous Hauterivian slope; (2) a rise in sea-level in the late Barremian, bringing about the submergence of the Jura platform and the deposition throughout the region of Urgonian limestone characterized by the development of rudist facies over a vast inner-shelf; (3) an emergence, preceding the transgression at the end of the early Aptian, characterized by the carving of incised valleys filled with marls of the upper *Orbitolina* beds. Each stage of this evolution is characterized by a proper model of carbonate platform with shallowing-upward sequences.

In the western Alps and the Franco-Swiss Jura (Fig. 1), the late Hauterivian–early Aptian carbonates were deposited in generally shallow-marine environments contained within a vast system of shelves and ramps, the Urgonian platform *sensu lato*, which developed on the European passive margin during the opening stage of the Ligurian Tethys. In southeast France, the Urgonian platform surrounds a small, deeper basin, the western appendage of the Alpine sea called the Dauphinois basin (Jurassic and Lower Cretaceous) or the Vocontian Basin (Barremian–Aptian). During certain periods, this platform was along continental areas: the Paris Basin to the north and the Corsican–Sardinian continent to the south. The evolution of the Urgonian platform *sensu lato* is complex, affected by both a palaeogeography inherited from the Liassic rifting and variations in sea-level. A previous study (Arnaud-Vanneau, 1980; Arnaud, 1981) proposed continuous sedimentation in the northern Subalpine chains. The acquisition of new data, especially in the Jura, and the reinterpretation of some earlier observations now lead us to propose a new dynamic evolutionary scheme for this platform. After briefly

summarizing the effects of Liassic rifting, the main stages of development linked to variations in sea-level are listed. In spite of the uncertainty resulting from the effects of alpine tectonism and frequently mediocre outcrops, the reconstructions obtained can be usefully compared to the sedimentary sequence models of Haq *et al.* (1987) and Vail *et al.* (1977, 1987). They can also be compared to the platform models of Kendall & Schlager (1981), Schlager (1981), Scholle *et al.* (1983), Read (1985) and Tucker (1985).

INTRODUCTION

The western Alps are a classic mountain chain constructed during the Alpine (i.e. late Cretaceous and Tertiary) compressional phases. They resulted from a collision between two continental domains which were originally separated by an oceanic area whose crust is now represented by overthrust ophiolitic remnants. This palaeo-ocean was the Ligurian Tethys, a part of the Mesozoic Tethys. It was situated between the European continent to the northwest

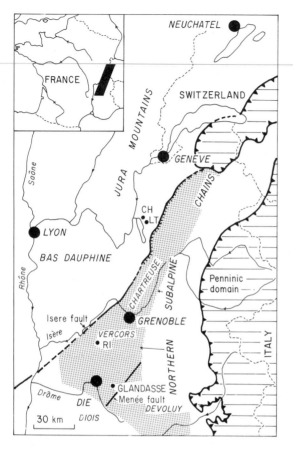

Fig. 1. Map of France shows the position of the area studied (in black). Larger map shows the position of the northern Subalpine chains (dotted area) and the Jura mountains: CH, La Chambotte; LT, La Tailla 1 borehole; RI, Les Rimets.

and the Apulian−Adriatic continental block to the southeast (Lemoine, 1984; Lemoine *et al.*, 1986).

Considering the main sedimentological features of the European continental margin, to which the Lower Cretaceous deposits described here belong, four main stages may be distinguished: prerift Triassic platform, Liassic and Middle Jurassic rifting, late Jurassic−early Cretaceous subsidence of the passive margin and onset of the seafloor spreading, and late Cretaceous−Tertiary convergence of margins and continental collision. In the external domain of the Alps and in the Jura, the Jurassic and Cretaceous shelves were derived from the break-up of the Triassic platform during the rifting period.

During the Trias and particularly in the Lias, the rift structures were all normal faults, some of which

may have a certain strike−slip component (Lemoine, 1984). The interplay of two groups of fault (namely the N−S to NE−SW faults) allows the delineation of three large domains in southeastern France (Fig. 2): (1) the Jura−Bas Dauphiné platform situated to the northwest of the Cévennes faults and the Isère fault (Vialon, 1974); (2) the Haute−Provence platform located to the southeast of the Durance fault. These were rather shallow-marine domains where thin deposits show few lateral variations. Known liassic palaeofaults have small throws; (3) Dauphinois or Vocontian Basin and Languedoc trench located in the centre represents a deeper-marine domain where thick and laterally variable basinal sediments occur. Palaeofaults are found wherever Liassic rocks crop out. They have throws which are often considerable and which delimit a series of tilted blocks (Lemoine, 1984).

From the Dogger to the late Jurassic (onset of the spreading stage), the Dauphinois Basin (central domain) was subsiding rapidly and a thick sequence of deep-marine facies was deposited there. The more stable Jura platform and Haute−Provence platform were subjected to reduced subsidence, and shallow-marine environments extended over large areas. On the platform margins, sedimentation was strongly affected by sea-level variations.

From the end of the Jurassic to the early Aptian (Figs 3 & 7), many palaeogeographical changes took place, including progradation of the platform facies into the NW part of the unstable and rapidly subsiding Dauphinois Basin (Arnaud, 1988). The shelf margin corresponded locally to a bypass margin (Read, 1985), with an accumulation of breccia at the foot. During the Berriasian, the margin evolved into a rimmed shelf (Read, 1985) with reefs. The deposition of thick hemipelagic sequences in the Dauphinois Basin, along the Isère fault scarp, brought about a gradual filling of this marine domain and decreasing of the outer slope of the shelf. From the early Valanginian to the early Hauterivian, the shelf was gradually drowned. The gradual filling of the NW part of the Dauphinois Basin (northern Subalpine chains) by hemipelagic sequences continued and differences in seafloor relief were removed. The former shelf margin was obliterated and a ramp developed in the Hauterivian (Read, 1985; Tucker, 1985) sloping steadily towards the SE. In the late Hauterivian, the outer edge of this hemipelagic sediment ramp was located 40 km to the SE of the Isére fault, along the Menée fault (Fig. 1), another earlier major structure which runs in the

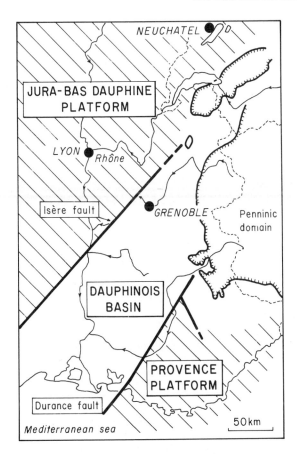

Fig. 2. Location of the stable and subsident areas. The Jura-Bas Dauphiné platform and the Provence platform correspond to stable areas, the Dauphinois Basin was an unstable domain.

same NE−SW direction (Arnaud, 1981). It was largely through this palaeotopography that Urgonian limestone was deposited during the late Barremian over both the former Jura platform and the northwest part of the early Dauphinois Basin.

HAUTERIVIAN, BARREMIAN AND EARLY APTIAN CARBONATE FORMATIONS

Four carbonate platform formations, characterized by associations of benthic foraminifera, occur in the northern Subalpine chains and Jura massif region (Arnaud-Vanneau, 1980; Arnaud, 1981; Remane,

1982; Viéban, 1983; Zweidler, 1985). These are (Figs 4 & 5):

1 the Pierre Jaune de Neuchâtel Formation (bioclastic and oolitic packstone and grainstone) belonging to the Hauterivian (mostly the early Hauterivian), found only on the Jura platform;

2 the Borne bioclastic limestone formation deposited close to the Hauterivian−Barremian boundary in the eastern Diois, at the foot of the outer hemipelagic slope;

3 the Glandasse bioclastic limestone formation deposited in the southern Vercors during the early Barremian and at the beginning of the late Barremian;

4 the Urgonian limestone formation (usually wackestone−packstone with rudists) deposited towards the top of the Barremian and in the lowest Aptian throughout the Jura and the northern Subalpine chains. The Urgonian limestone rests upon the Pierre Jaune de Neuchâtel in the Jura and on the Glandasse bioclastic limestone in the southern Vercors.

Pierre Jaune de Neuchâtel

This formation exists only on the Jura platform (Fig. 4), occurring above the *Acanthodiscus radiatus* marls (Marne d'Hauterive; Remane, 1982) of the basal Hauterivian. It consists of:

1 at the base, varied bio- and oo-bioclastic limestones dated by ammonites as early Hauterivian (Nodosoplicatum and, possibly, Cruasense Zones; Busnardo & Thieuloy 1989);

2 at the top, several meters of marls, bioclastic limestone and locally *Pachytraga* limestone, attributed to the late Hauterivian. The benthic microfauna, poorly preserved and still relatively little-known, is characterized at the top of the formation by the association of *Citaella? favrei*, *Cribellopsis* aff. *elongata* and *Urgonina alpillensis-protuberans* (small-sized forms). This association (Fig. 5) has not been identified farther to the south in the Barremian bioclastic limestone of the Chartreuse, Vercors and Diois.

Borne bioclastic limestone

This formation has been found in the eastern Diois and the western part of the Dévoluy (Figs 3 & 4). Consisting of grain-flow deposits in the Vocontian Basin at the foot of the hemipelagic slope, 120−200 m thick, it is dated as the basal Barremian

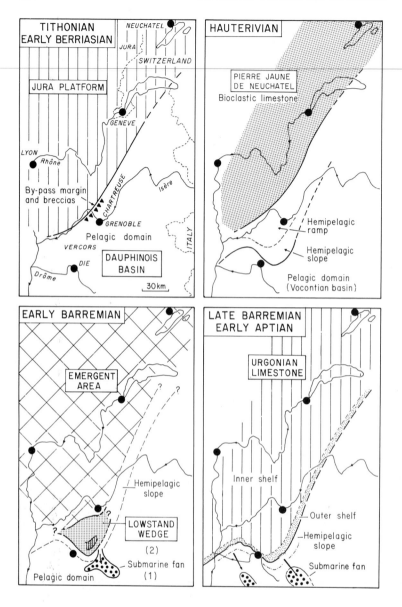

Fig. 3. Palaeogeographic maps corresponding to four stages in the development of the platform during the early Cretaceous. Tithonian–early Berriasian: the southeastern edge of the Jura platform is located near the Isère Fault (Fig. 2); this edge was a bypass margin corresponding to the boundary between the pelagic domain of the Dauphinois Basin and the inner domain of the Jura platform. Hauterivian (middle to late Hauterivian): the 'Pierre Jaune de Neuchâtel' bioclastic limestone is a drowned platform sediment; hemipelagic sediments were well developed in the northern Subalpine chains (Fig. 1). Early Barremian: the Jura platform emerged during a major drop in sea-level; (1) submarine fan (Borne bioclastic limestone formation) of Lowermost early Barremian age, located in the pelagic domain of the Dauphinois Basin; (2) lowstand wedge (Glandasse bioclastic limestone formation) above the previous Hauterivian hemipelagic ramp and slope (within dotted area, the small hatched area corresponds to the first appearance of the lowstand wedge). Late Barremian–early Aptian, deposition of the Urgonian limestone.

(probable Hugii Zone). It lies above an unconformity surface, which is gullied, cutting Hauterivian or even Valanginian pelagic limestone and marl. Reworked microfaunas are extremely rich in the gravitational deposits, and are characterized by (Fig. 5): at the base, the association of *Cribellopsis* aff. *elongata* and *Urgonina alpillensis–protuberans* (large-sized forms), and at the top, by the appearance of *Cribellopsis thieuloyi*, *Orbitolinopsis debelmasi* and *Paracoskinolina? jourdanensis*, species typical

of the early Barremian.

Two observations may be stressed. (1) Only one of these foraminifera has been found in the carbonate formations of the northern Subalpine chains and of the Jura (*C.* aff. *elongata*, observed at the top of the Pierre Jaune de Neuchâtel). (2) The presence of large-sized forms of *Urgonina alpillensis–protuberans* clearly indicates that the basal association of the Borne bioclastic limestone is stratigraphically located above the topmost levels of the

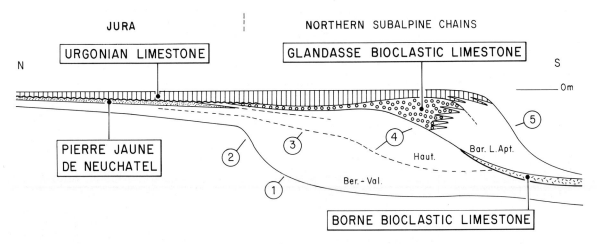

Fig. 4. Arrangement of the Hauterivian−Lower Aptian carbonate formations (Jura and northern sub-alpine chains). (1) Jurassic−Cretaceous boundary; (2) Bypass margin of the Jura platform (Tithonian and early Berriasian); (3) Upper Valanginian and Lower Hauterivian outer shelf; (4) Hauterivian hemipelagic slope; (5) Lower Aptian hemipelagic slope.

Pierre Jaune de Neuchâtel, levels which contain only small forms of this species.

Barremian of the southern Vercors

In this region (Fig. 5), the Glandasse bioclastic limestone and the Urgonian limestone are separated by the *Matheronites limentinus* marly level (Arnaud-Vanneau et al., 1976; Thieuloy, 1979) of the late Barremian (Feraudi Zone at the base, Astieri Zone at the top) which contains a large number of ammonites. In the carbonate facies, which are exceptionally rich in benthic foraminifera, two groups of useful stratigraphic species have been identified. The first, observed in the Glandasse bioclastic limestone, has been found only below the *M. limentinus* marls. It includes *Cribellopsis thieuloyi*, *Orbitolinopsis debelmasi*, *Orbitolinopsis flandrini* and *Paracoskinolina? jourdanensis*. The second group contains *Paracoskinolina reicheli*, *Eopalorbitolina charollaisi*, *Valserina bronnimanni* and *Palorbitolina lenticularis*. This association of the four species characterizes the top of the late Barremian and the Urgonian limestone, although the first two appear at the top of the Glandasse bioclastic limestone whereas the last, appearing in the late Barremian, is especially common in the early Aptian.

Barremian of the northern Vercors, the Chartreuse and the Jura

In this region, two fundamental observations can be made. (1) The foraminifera typical of the early Barremian are not found from the central Vercors to the Swiss Jura. In the Vercors, the absence of this microfauna can be explained by a very substantial reduction in the thickness of the early Barremian which decreases from ≈ 2000 m in the southeast to just a few metres to the northeast of the massif. (2) The foraminifera which are typical of the late Barremian are found all through the Urgonian limestone, from the Vercors to the Neuchâtel area (Swiss Jura). Throughout this region, the Urgonian limestone always begins in the late Barremian, including the Jura where it is directly above the Pierre Jaune de Neuchâtel dated to the Hauterivian. In the Jura, a part of the late Hauterivian, the early Barremian and, probably, a part of the late Barremian are missing between these two formations. Their absence is highlighted by a clear lithological boundary, as confirmed by the examination of wire-line logs (Fig. 6).

In conclusion, between the Hauterivian and the early Aptian, there is no steady progradation of carbonate platform facies towards the Vocontian

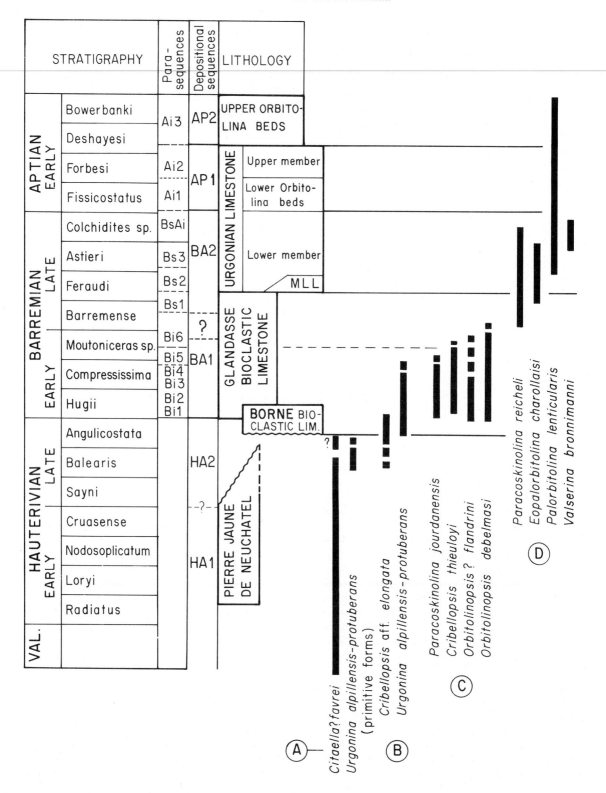

Basin (as interpreted by Clavel *et al.*, 1987), nor any steady sedimentation during this period on the Jura platform or its edges. On the contrary, the absence of latest Hauterivian−early Barremian strata, on the one hand, and the unexpected geometrical distribution, on the other, suggest a complex history marked by a substantial drop in sea-level in the early Barremian and by the emergence of the Jura platform.

SEA-LEVEL VARIATIONS IN THE HAUTERIVIAN AND EARLY BARREMIAN

Hauterivian depositional sequence

On the Jura platform, the basal Hauterivian corresponds to the ultimate transgression stage during which the platform was drowned and covered with

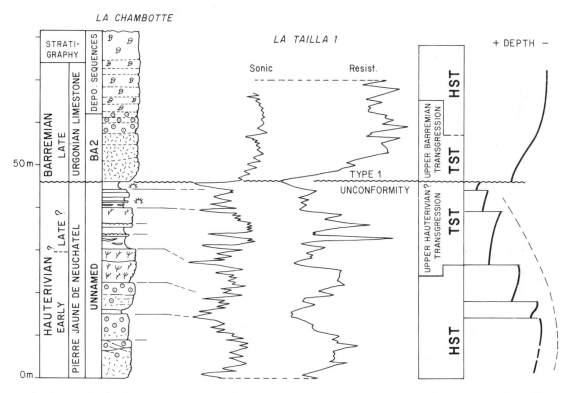

Fig. 6. Comparison between La Chambotte section and La Tailla 1 borehole showing: the top of the Lower Hauterivian depositional sequence (bioclastic and oolitic limestone of the Pierre Jaune de Neuchâtel Formation); the probable Upper Hauterivian transgressive systems tract; the type 1 unconformity at the base of the Urgonian limestone; and the initial sequence of the late Barremian Urgonian limestone.

Fig. 5. [*Opposite*] Main stratigraphic units and distribution of the more important benthic foraminifera in the Jura and the northern Subalpine chains. Aptian zonation according to Casey (1961). Barremian zonation according to Busnardo (1984). The microfaunas A, B, C and D characterize the Pierre Jaune de Neuchâtel Formation, the Borne bioclastic limestone formation, the Glandasse bioclastic limestone formation and the Urgonian limestone formation, respectively. MLL, *Matheronites limentinus* level. The Aptian foraminifera are not represented.

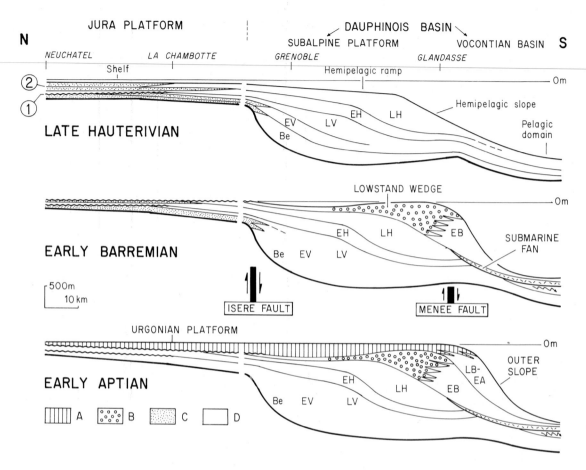

Fig. 7. Schematic palaeogeographic sections between Neuchâtel and the southeastern Vercors. (A) Urgonian limestone formation; (B) Glandasse bioclastic limestone formation; (C) Berriasian–Hauterivian platform facies; (D) basinal hemipelagic to pelagic facies. (1) Erosion surface of the Valanginian deposits in the Neuchâtel area (Swiss Jura); (2) Pierre Jaune de Neuchâtel Formation (bioclastic limestone). Be = Berriasian; EV = early Valanginian; LV = late Valanginian; EH = early Hauterivian; LH = late Hauterivian; EB = early Barremian; LB–EA = late Barremian–early Aptian.

cephalopod-bearing marls (Marne d'Hauterive Formation, with *Acanthodiscus radiatus*). These marls are overlain by highstand deposits (Pierre Jaune de Neuchâtel) which are arranged in shallowing-upward sequences consisting of blue marly limestone with irregular echinoids (*Toxaster*) passing up into yellowish bioclastic limestone and oobioclastic limestone. The late Hauterivian parasequence is locally capped by *Pachytraga* rudist limestone or white oobioclastic limestone. This sequence is commonly truncated at the top or even missing entirely due to the Barremian emergence.

The southern boundary of this outer shelf bioclastic limestone is situated along the Isère fault. Farther to the southeast, in the Dauphinois Basin,

thick hemipelagic sequences were deposited in the northern Subalpine chains. They correspond to the highstand wedges that prograded towards the southeast, changing laterally in the Diois to the pelagic marl-limestone alternations of the Vocontian Basin. The lengthwise profile of the seafloor during the late Hauterivian shows the existence of three distinct parts: a sub-horizontal submarine surface in the Jura Domain, a ramp sloping slightly towards the southeast in the northern Subalpine chains ($\approx 1°$ slope) and a hemipelagic slope joining the deep pelagic part of the Vocontian Basin (5–10° slope). The first part corresponds to the area of bioclastic limestone and the last two correspond to the hemipelagic wedge (Figs 7 & 8).

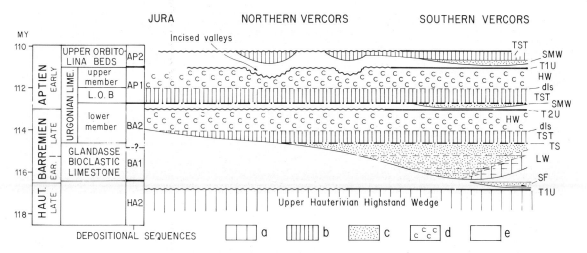

Fig. 8. Sequence stratigraphy of the Barremian−early Aptian: in geological time (time scale according to Haq *et al.*, 1987). dls = downlap surface; HW = highstand wedge; LW = lowstand wedge; SF = submarine fan; SMW = shelf-margin wedge; T1U = type 1 sequence boundary; T2U = type 2 sequence boundary; TS = transgressive surface; TST = transgressive systems tract (abbreviations according to Vail *et al.*, 1987). (a) Hemipelagic facies; (b) transgressive systems tract; (c) outer-shelf facies; (d) inner-shelf facies; (e) omission surface. LOB = lower *Orbitolina* beds, early Aptian (Bedoulian)

Early Barremian−basal late Barremian depositional sequence (sequence BA1)

This sequence, which is usually very thick (700−2000 m), corresponds to the Glandasse bioclastic limestone formation and its lateral hemipelagic equivalents (Figs 4 & 5). Palaeogeographical, morphological, sedimentological and stratigraphical reasons indicate that it is most probably a lowstand wedge deposited during a period of substantial drop in sea-level. (1) From a palaeogeographical point of view, it is located in the Dauphinois (or Vocontian) Basin, on the late Hauterivian hemipelagic wedge, i.e. 15−40 km outside the edge of the Jura platform (Figs 3 & 7). (2) From a morphological point of view, the lower parasequences of this depositional sequence are located along the hemipelagic slope (downcurrent part of the late Hauterivian hemipelagic wedge). Only the two highest parasequences (Bi5 and Bi6) were deposited on the top part of the Hauterivian wedge (Figs 7 & 8). (3) From a sedimentological point of view, this sequence is made up of outer-shelf bioclastic facies and by hemipelagic facies which occur above, without any transition, either hemipelagic facies in the first case (La Montagnette section, Glandasse plateau) or pelagic facies in the second (Col de Rousset road section). Furthermore, the bioclastic sequences disappear

laterally both towards the southeast in the direction of the basin (downlap, Fig. 9) and towards the NW in the direction of the platform (onlap, Fig. 10). (4) From a stratigraphical point of view, the sequence contains an early Barremian microfauna which is abundant and diversified but totally unknown farther to the north where this period has been shown to be missing on the Jura platform. To summarize, this sedimentary body was deposited beyond and below the Jura platform which was emergent during the early Barremian.

It is difficult to date accurately the sudden drop in sea-level. It is, however, very close to the Hauterivian−Barremian boundary, for three main reasons. (1) In the Col de Rousset section, as throughout this part of the southern Vercors, it is at the top of the terminal Hauterivian Angulicostata Zone that the vertical change can be observed, without any transition, between Hauterivian pelagic facies and the hemipelagic facies dated a few metres above by early Barremian ammonites. (2) In the Montagnette region (northern Glandasse plateau), it is also at this level, or very slightly above, precisely in the first Barremian beds (Pas de l'Essaure section, Arnaud, 1981), that coarse bioclastic facies can be observed on a gullied surface on the hemipelagic facies. (3) In the Jura, marine sedimentation continued during most of the late Hauterivian since the

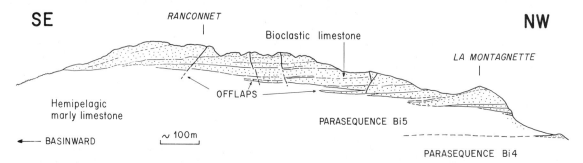

Fig. 9. Lower Barremian lowstand prograding wedge (or late highstand wedge) of the Montagnette (northern Glandasse plateau). Outcrop view offlap arrangement of bioclastic limestone (Bi4 and Bi5 parasequences, part of the Glandasse bioclastic limestone).

highest levels of this sequence contain a microfauna older than the basal Barremian.

DEPOSITIONAL DYNAMICS

Several stages can be identified (Figs 7 & 8). The first stage corresponds to the sudden drop, probably of several decameters, in sea-level at the top of the Hauterivian or at the base of the Barremian. This drop in depth had two consequences. The first, noticeable in the Jura, was the emergence of the inner domain and the edge of the Hauterivian platform (Jura platform), followed by the erosion of most or all of the late Hauterivian deposits. This erosion, varying in importance from one area to another, provides an understanding of the particularly rapid and often unexplained variations in thickness and facies at this level in the Vaudois and Neuchâtelois Jura. The second consequence of the Substantial drop in sea-level is in the northern Subalpine chains, where outer-platform bioclastic sedimentation took place along the late Hauterivian

hemipelagic slope. Located on a slope, but not seen in outcrops, these deposits must have accumulated along a narrow, more or less unstable strip that was the source for the grain flows that led to the Borne bioclastic limestone formation.

The second stage corresponds to the deposition of a submarine fan in the eastern Diois and the Dévoluy (northern Subalpine chains, Fig. 3), at the bottom of the Hauterivian hemipelagic slope. In this region, at the Hauterivian–Barremian boundary, the Borne bioclastic limestone corresponds to an accumulation of grain flows located geometrically to the southeast and below the initial levels of the Glandasse bioclastic limestone. It is found above a large truncation surface responsible for the synsedimentary sliding of the Hauterivian (Fig. 7), and even locally of a part of the Valanginian, in the $\approx 30\,km^2$ region between the Col de Menée and the Col de la Croix-Haute (Arnaud, 1981). This submarine fan can thus be interpreted as being the first level of the early Barremian depositional sequence (BAl sequence, Fig. 8). It was fed by shallow-subtidal sediments, the exact location of which are unknown, but which

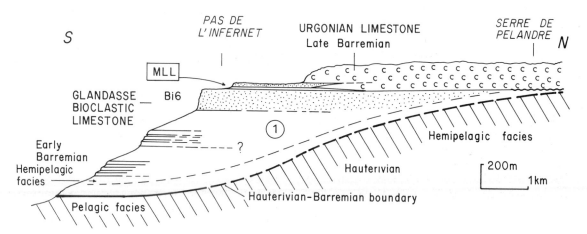

Fig. 10. Schematic section through the Lente Forest (southwestern Vercors) showing the probable onlap arrangement of the Lower Barremian deposits above the Hauterivian hemipelagic slope. In the Serre de Pélandré Mountain, the top of the hemipelagic facies observed below the Urgonian limestone unconformity is of basal early Barremian age (according to Clavel *et al.*, 1987). It then seems that, as in the 'La Montagnette' section, the maximum sea-level fall took place in the basal Barremian. MLL: *Matheronites limentinus* level. 1: Unknown, but possible position of a bioclastic lowstand wedge in this region.

must have been located to the north of the Glandasse plateau, not far from the first visible outcrops of the Glandasse bioclastic limestone formation.

The third stage, which is now only identified with certainty in the Vercors, is distinguished by the deposition of a general lowstand wedge made up of Glandasse bioclastic limestone and its lateral hemipelagic equivalents. This lowstand wedge corresponds to a relatively narrow strip, several kilometres wide, occurring on and basinwards of the late Hauterivian wedge. It was formed by a parasequence set (depositional sequence BA1) made up of six groups of shallowing-upward sequences (sequences Bi1 to Bi6, Arnaud-Vanneau *et al.*, 1976) showing: at the base, a thin level of hemipelagic marly limestone and marl, and at the top, bioclastic, oolitic or coral limestone. Downslope (towards the southeast), these sequences change laterally to hemipelagic cycles which are usually marly at the base and more calcareous at the top (Arnaud, 1981; Ferry & Rubino, 1988). The thickness of parasequences Bi2 to Bi5 decreases rapidly towards the southeast, displaying an offlap arrangement on the Borne bioclastic limestone (Fig. 9; Arnaud, 1981, fig. 59).

Upslope (to the northwest), the onlap arrangement of bioclastic parasequences on the late Hauterivian hemipelagic wedge is very rarely visible because of the vegetation, but it can be deduced indirectly from field data. It definitely exists on the

northern slope of the Tête Chevalière (northern Glandasse plateau), to the north of the Deux Soeurs mountain (for location, see Fig. 13) and near the Grand Goulets (central Vercors) where the disappearance of several bioclastic toplaps is observed towards the northwest. It probably also occurs throughout the southern Vercors where ≈ 1000 m of hemipelagic marly limestone and 200–300 m of bioclastic limestone (Bi6 sequence) disappear towards the north within < 7 km (Fig. 10). The late Barremian Urgonian limestone of the southern Vercors above the Bi6 bioclastic limestone is then observed directly above the hemipelagic marly limestone of the Hauterivian–Barremian boundary (Serre de Pélandré, central Vercors, Fig. 10).

Given present erosion and the abundance of vegetation, data remain sketchy; there is, however, sufficient to prove that the early Barremian deposits correspond to a wedge ≈ 10 km in width, the total thickness of which reaches 2000 m below the Glandasse plateau. Except for its location outside the Hauterivian hemipelagic wedge, the early Barremian wedge is outstanding: (1) because of the very clear sedimentological break at the base (ranging from region to region one observes Barremian bioclastic or hemipelagic limestone, respectively, on Hauterivian hemipelagic or pelagic facies; (2) because of the substantial development of bioclastic facies on the shallowest part of the lowstand wedge; (3) because of sedimentation in a highly subsident area (up to 2000 m of early Barremian, including $\approx$ 800 m of shallow-subtidal bioclastic facies). After the sudden drop in the relative sea-level at the Hauterivian–Barremian boundary, sedimentary productivity and very high sedimentation rates compensated for both the very rapid subsidence and the probable variations in sea-level (Arnaud, 1988).

In this rapidly subsiding area, it is difficult to list in detail the variations in sea-level for the early Barremian. Nevertheless, it does appear that there was a rise in the early Barremian (Hugii *pro parte* and Compressissima Zones) and a drop in relative sea-level at the early–late Barremian boundary (Moutoniceras, Barremense and Feraudi *pro parte* Zones). The former corresponds to a deepening of environments, displayed by the temporary retrogradation of bioclastic facies of the Bi2 to Bi4 parasequences and by marls in the hemipelagic series (Fontaine Graillère marls, Arnaud-Vannheau *et al.*, 1976). Within this high sea-level context, it is conceivable that at a given time, there could have been a change from a lowstand wedge to a highstand wedge then a shelf-margin wedge since parasequence Bi5 and Bi6 were no longer confined to the outer slope of the previous Hauterivian wedge, but spread widely towards the north above this wedge. The relative drop of sea-level of the early–late Barremian boundary would then have resulted in the rapid and substantial progradation towards the basin of the bioclastic facies of parasequences Bi5 (Fig. 9) and Bi6.

Late Barremian–early Aptian depositional sequences: the Urgonian platform (sequences BA2 and AP1)

The Urgonian limestone consists of two depositional sequences (BA2 and AP1), the first dating to the upper part of the late Barremian (Feraudi *pro parte*, Astieri and *Colchidites* sp. zones, parasequences Bs2, Bs3 and BsAi), the second to the lower part of the early Aptian (parasequences Ai1 and Ai2). These depositional sequences, which have been described in detail (Arnaud-Vanneau, 1980; Arnaud, 1981), begin by a transgressive systems tract and end with a highstand deposit (Figs 8 & 9). Only the main features of the transgressive systems tracts and the highstand deposits are presented here.

Trangressive systems tracts

The transgressive systems tract of the late Barremian (Fig. 8), dated by Feraudi and Astieri Zone ammonites, is located at the base of the first Urgonian limestone depositional sequence (BA2 depositional sequence). In the southern Vercors, on the outer edge of the Urgonian platform, three stages can be distinguished (Fig. 11): (1) The deepening of environments begins in the Bs1 parasequence, identified only in the southern Vercors where it corresponds lithologically to the top of the Glandasse bioclastic limestone formation (for this reason, this thin parasequence is put in the depositional sequence BA1 while it is really at the base of the depositional sequence BA2). This Bs1 parasequence is characterized by a clear retrogradation of bioclastic facies at the top of the early Barremian lowstand wedge, by the presence of small coral reefs surrounded by fine bioclastic facies, by the appearance of facies with large rudists and corals and by the presence of transgressive facies (Arnaud-Vanneau *et al.*, 1987). (2) The maximum deepening corresponds to the discontinuity surface covered with ammonites at the top of the Bs1 parasequence and to the overlying marls at the base of the Bs2 parasequence (maximum flooding surface, Haq *et al.*, 1987). This horizon, which is extremely rich in ammonites and locally contains glauconite, probably corresponds to a condensed section highlighting the ultimate stage of the transgression (*Matheronites limentinus* level). (3) The very high sea-level period ends in the deposition of Bs2 parasequence and the base of Bs3. These thin sequences are characterized, at the base, by the existence of marls and, at the top, by hemipelagic marly limestone or fine bioclastic facies which change laterally, over a distance of several tens to several hundreds of metres, into rudist wackestone–packstone (Fig. 11). To the southeast, in the Vocontian basin, this transgressive systems tract is represented by the *Heteroceras* marls.

To the northwest (northern Vercors, Chartreuse, Jura), the transgressive systems tract is always represented at the base of the Urgonian limestone (Figs 19 & 20), by thin transgressive facies overlain by coarse bioclastic or oolitic facies (initiation facies or 'facies d'installation', Arnaud-Vanneau *et al.*, 1987). From the north of the Vercors to the Jura, the transgressive systems tract lies directly upon Hauterivian beds; available palaeontological data do not make it clear whether the transgression was synchronous over the region or whether there is an

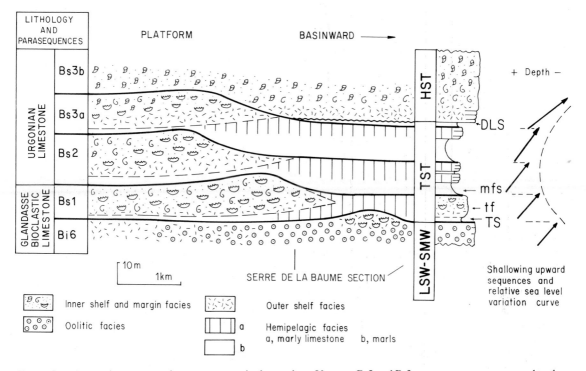

Fig. 11. Late Barremian transgressive systems tract in the southern Vercors. Bs2 and Bs3a parasequences correspond to the *Matheronites limentinus* level (*Hemihoplites feraudi* and *Heteroceras astieri* zones, Arnaud-Vanneau *et al.*, 1976). DLS = downlap surface; mfs = maximum flooding surface (with glauconite and ammonites); tf = transgressive facies; TS = transgressive surface; HST = highstand systems tract; LSW = lowstand wedge; SMW = shelf-margin wedge; TST = transgressive systems tract (nomenclature according to Haq *et al.*, 1987). For organism platform legend, see Fig. 18.

onlap arrangement of the first Urgonian limestone parasequences.

To summarize, two conclusions may be drawn from the study of the most complete sections of the transgressive systems tract: (1) the start of the deepening is highlighted by carbonaceous transgressive facies below the major discontinuity surface; and (2) the ultimate transgressive stage is distinguished by marly sedimentation both on the platform edge and in the adjoining basin.

The basal Aptian transgressive systems tract (Fig. 12) can be clearly observed in the inner part of the Urgonian platform where it is represented by the lower *Orbitolina* beds (parasequence Ai1 *pro parte*) Here again, several stages can be distinguished above the emergence surface at the top of the BsAi parasequence, the top of the BA2 depositional sequence (Fig. 12). (1) The rise in sea-level is recorded in two superposed parasequences. The first (parasequence Ai1a) is made up of several metres of transgressive

facies characterized at the base by a relative abundance of detrital material (quartz and clay) ending with a hardground. The second parasequence (Ai1b) is characterized by a greater abundance of detrital grains, a wide variety of facies (including micro-oolite and algal nodule facies) and, above all, by the presence of numerous channels filled by *Palorbitolina lenticularis* marls containing a very large number of benthic foraminifera (tidal channels?). The deepening of environments remains slight, however, since the benthic foraminifera show no fundamental change in populations compared to those of the carbonate facies of the inner Urgonian platform. Because of this, the facies of parasequence Ai1b reflect more the relative abundance of detrital minerals than a major deepening. (2) The ultimate deepening stage (maximum flooding stage) corresponds to the base of the Ai1c parasequence which begins practically throughout the northern part of the Vercors and the Chartreuse with a marly level

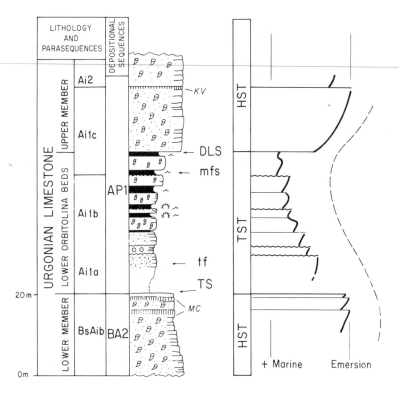

Fig. 12. Early Aptian transgressive systems tract in the northern Vercors (Ai1 parasequence). This transgressive systems tract corresponds to the lower *Orbitolina* beds. Legend, see Fig. 18. KV = beach-level with keystone vugs; MC = storm deposit level with vadose (microstalactitic) cement. The downlap surface (DLS) corresponds to a clear lithologic boundary between the transgressive systems tract (marls and argillaceous limestones) and the highstand deposits (upper Urgonian limestone).

which is locally several meters thick and particularly rich in coarse grains of detrital quartz. The rarity of ammonites in the marls clearly highlights the maximum deepening of this inner platform. No condensed level has been observed. (3) The high sea-level period, or even the beginning of the sea-level fall, is distinguished by the disappearance of detrital material and by a return to exclusively carbonate sedimentation. The lower boundary of these highstand deposits is a flat lithological surface.

In conclusion, the study of this transgressive systems tract shows, once again, that the rise in sea-level preceded the maximum deepening period and was accompanied by the deposition of detrital terrigenous material even though the latter are usually unknown in inner-platform type environments of this region.

Highstand deposits

In terms of thickness, the highstand deposits make up most of the Urgonian limestone formation. The Bs3 and BsAi parasequences of the late Barremian BA2 depositional sequence are composed of inner-platform facies (dominantly rudistid facies) and

end in emergence levels of various types (beach grainstones, fenestral micrites, storm deposits with microstalactitic cements). The emergence horizons appear to be localized and do not pass into distinct erosional surfaces. The Ai2 parasequence, making up the highstand deposit of the basal Aptian depositional sequence (A1 depositional sequence) is very similar. It does differ though in the fact that at its upper boundary, there is a major erosion surface highlighting the emergence of the Urgonian platform in the early Aptian.

Early Aptian emergence and the transgression of the middle−early Aptian

The emergence at the top of the BsAi parasequence, located very near the Barremian−Aptian boundary, appears to be minor. It is distinguished either by the rather unremarkable gullying of underlying limestone or by a very rarely preserved thin level of *Characae* micrite above the erosion surface (Chartreuse) (Arnaud-Vanneau, 1980).

The emergence at the top of the Ai2 parasequence, on the other hand, is quite substantial. It is marked first of all by diagenetic changes in the sediment at

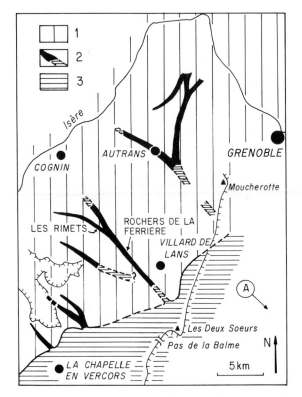

Fig. 13. Schematic map of the early Aptian incised valleys in the northern Vercors. (1) Erosion surface of the Urgonian limestone; (2) incised valleys with transgressive systems tract of the upper part of the early Aptian (upper *Orbitolina* beds); (3) Glauconitic sandy marls which were deposited above the Urgonian platform during the upper early Aptian. These marls may be considered as the lateral equivalent of the upper *Orbitolina* beds; they are observed only on the outer part of the previous inner-shelf of the Urgonian platform. (A) Direction of the Urgonian platform margin.

the top of the Ai2 parasequence (shell dissolution and vadose cementation), and then by distinct cavities of various forms and dimensions. Some of these cavities, up to 1 m across, are rootmolds, while others are karstic cavities. In both cases, they are filled first by limestone containing crinoids, and then by *Palorbitolina* limestone similar to that of the upper *Orbitolina* beds. Finally, large extensive depressions, tens of kilometres in length, hundreds of metres wide and tens of metres deep cut into the surface of the Urgonian platform in the Vercors. During the trangression of the upper early Aptian, these depressions served as tidal channels, and then

they were later filled by upper *Orbitolina* beds. The channels are very deep, especially downcurrent, at least 30 m deep in the case of the Rochers de la Ferrière channel north of the Gorges de la Bourne (Arnaud, 1981, fig. 281). This depth would appear to be incompatible with simple erosion by tidal currents since this region was affected by early lithification and by a probably small tidal range, given the dearth of intertidal facies in the Urgonian limestone. It therefore appears more likely that these valleys were cut in a continental environment. The maximum depth of the valleys implies that there was at least a 30–40 m fall in sea-level. In this event, the shoreline must have dropped over the edge of the Urgonian platform, i.e. to the southeast of the northern Subalpine chains.

In the Vercors, numerous outcrops have made it possible to reconstruct the network of palaeovalleys (Fig. 13). The valleys mostly run perpendicularly to the edge of the Urgonian platform. Their depth decreases rapidly towards the northwest and they disappear within 20 km, probably explaining their absence in the Subalpine chains farther north.

At the end of this period, the rapid rise in sea-level brought about the submergence of the area. The valleys were first tidal channels, and they were then filled by a variety of sediments making up the depositional sequence AP2 (parasequence Ai3) at the top part of the early Aptian (*deshayesi* and *bowerbanki* ammonite zones, zonation according to Casey, 1961). The transgressive systems tract exists solely in the palaeovalleys and on their edges, corresponding to upper *Orbitolina* beds. These marls are very rich in quartz, clay and heavy minerals, as is usually the case in transgressive sediments.

One of these valleys, near Les Rimets farm (northern Vercors), which is remarkably well preserved, shows three main stages of fill (Fig. 14). (1) At the beginning of the platform drowning, the sea must have been fairly shallow. Tidal currents were rapid and they were channelled by the valleys, in the same way as those presently observed on the Pelagian platform to the east of the Kerkennah Islands, Tunisia (Burollet *et al.*, 1979). A condensed section covers the bottom and sides of this valley and consists of a thin, iron-rich centimetre–decimetre-thick level which was lithified early (numerous boring organisms). It locally contains numerous corals and large rudistids on the edge of the valley. (2) At the highest point of the sea-level rise, the currents were less strong, allowing the filling of the tidal channels by marls which are very

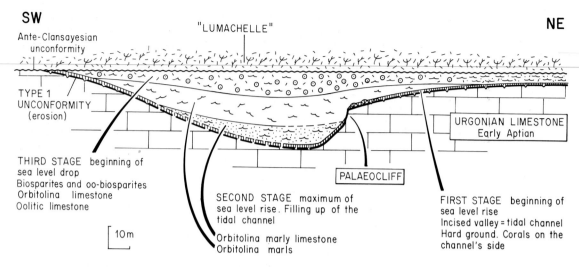

Fig. 14. Schematic section through the incised valley of 'Les Rimets' (northern Vercors). This section shows: (1) the type 1 unconformity above the Urgonian limestone (which was eroded during the early Aptian emergence); (2) the three main stages of filling by the upper *Orbitolina* beds during the sea-level rise of the upper part of the early Aptian (the valley was then used as a tidal channel); and (3) the onlap arrangement of the different levels of the upper *Orbitolina* beds. Legend: see Fig. 18.

rich in *Palorbitolina lenticularis*. The channel running through Les Rimets and continuing on to the eastern edge of the Vercors (south of Villard-de-Lans) shows a longitudinal zonation of the fill from the edge towards the inner platform. Seaward, the fill is coarse outer-shelf type grainstone whereas towards the inner platform (Rochers de la Ferrière), this bioclastic limestone is overlain by *Palorbitolina* marls. Finally, farther to the northwest, in Les Rimets, the multi-phased filling consists of several distinct layers of marl and marly *Palorbitolina* limestone resting against the slopes of the valley. These observations confirm the onlap arrangement of the various sedimentary bodies on the erosion surface, both longitudinally and transversely. (3) The beginning of the sea-level fall corresponds to the accumulation in a subtidal, shallow-marine environment of bioclastic (*Palorbitolina* grainstone) and oo-bioclastic sands, either on the previous transgressive systems tract or, laterally, directly upon the Urgonian limestone. This highstand deposit is dated either in the upper part of the Bedoulian (early Aptian) or the basal Gargasian (middle Aptian) according to whether the *bowerbanki* zone is assigned to the first (Casey, 1961; Arnaud-Vanneau *et al.*, 1976) or the second (Kemper, 1982) of these sub-stages. At the most a few metres thick, the base of this highstand deposit is separated from the Upper Aptian bryozoan and

crinoidal grainstone (referred to as 'Lumachelle' by French authors) by an unconformity which corresponds to the Gargasian. It is therefore impossible to identify sea-level variations in this region during the Gargasian. It is not possible to determine whether this hiatus is due to an emergence or not.

MAJOR FACIES OF THE HAUTERIVIAN–EARLY APTIAN DEPOSITS

The many carbonate facies in this series are most commonly arranged in shallowing-upward sequences, the nature of which varies according to their palaeogeographical location and their position relative to sea-level. For the sake of simplicity, three palaeogeographical regions characterized by distinct facies and sequence types are distinguished. In the first region are the early Barremian lowstand wedge facies and the fairly similar ones on the edge and outer domain of the Urgonian platform. These facies consist of deposits overlying the hemipelagic and pelagic facies of the Vocontian Basin. The second region includes the facies corresponding to deposits situated in the inner domain and outer-shelf of the previous Upper Jurassic–Valanginian Jura platform. These facies consist of Hauterivian sediment deposited after the maximum deepening of the late Valanginian—early Hauterivian transgression and, above, of the first Urgonian limestone sequence deposited during the late Barremian transgression. The third region is represented by inner Urgonian platform facies deposited on both the Jura domain and in the northwest part of the Dauphinois Basin (northern Subalpine chains). A fourth group of facies, independent of the palaeogeography, may also be added, grouping together the facies linked to the two extremes of the sea-level variation curve: on the one hand, low-level and emergence facies, and on the other, high-level and transgressive and/or submergence facies.

Lowstand wedge located in the Vocontian Basin and Urgonian platform margin

Vocontian Basin pelagic and hemipelagic facies

The pelagic facies are represented by regular decimetre to metre-thick alternations of argillaceous limestone and marl, with high clay content, silted quartz, glauconite, radiolaria and ammonites.

The hemipelagic facies are represented either by irregular alternations of marl and marly limestone, or marly limestone. Two microfacies can be distinguished: (1) mudstone—wackestone with sponge spicules and small foraminifera, and (2) peloidal wackestone—packstone with clay, silted quartz and sparse sponge spicules. Glauconite is present, but the pelagic fauna has practically disappeared (few ammonites).

Lowstand wedge bioclastic facies and outer-shelf facies of the Urgonian platform

These consist of bedded bioclastic limestone with coarse debris and low clay content (87–100% carbonate). Free of quartz and glauconite, they are blue or ochre in colour due to iron oxides. These facies are arranged in shallowing-upward sequences (Figs 15 & 16) grouped together in parasequence sets. When complete, they are arranged in the following order above the hemipelagic facies: (1) argillaceous wackestone with 'ball-shaped' bryozoans and crinoids; (2) poorly-sorted packstone—grainstone with little or no clay and dominant deep-subtidal skeletal grains; (3) grainstone with coarse or very coarse debris (corals, large foraminifera and dasyclad algae) arranged in well-sorted millimetric beds, alternately fine and coarse (influence of

the waves?); and (4) oolitic grainstone with cross-bedding. This top level is locally overlain by small patch-reefs. The nature and vertical distribution of bioclasts suggests a change from deep-subtidal at the base to shallow-subtidal at the top. The sedimentological features indicate deposition above fairweather wave-base.

Too few examples of bioclastic lowstand wedges are presently known for comparisons to be made. Nevertheless, the southern Vercors early Barremian lowstand wedge, that of the Chartreuse late Berriasian (unpublished) and that of the early Aptian which probably exists at the Belle—Motte Mountain (northeastern Diois) permit three fundamental features to be distinguished: (1) very rapid lateral seaward passage from oolitic shallow-subtidal facies and coral reefs to hemipelagic facies (in places < 1 km); (2) very rapid vertical passage, commonly with gullying and without any transition in each parasequence, from hemipelagic facies to coarse bioclastic facies; and (3) presence of bioclastic facies made up of very coarse debris locally mixed with hemipelagic mud. These features indicate a morphology containing many contrasts: small reefs and narrow oolitic shoals on the edge of the lowstand wedge and steep outer slope upon which heavy storms (cyclones?) brought about several metres of accumulation of very poorly-sorted coarse sands (as in the Little Bahamas Bank; Hine & Neumann, 1977).

The outer regions of the Urgonian platform, on the other hand, are characterized by finer bioclastic facies and, above all, by far less rapid and more gradual vertical and lateral passages (generally 4–5 km) between shallow-subtidal bioclastic facies and hemipelagic facies. This is usually associated with a more gentle slope.

Hauterivian facies and Urgonian limestone transgressive sequence, Jura platform

Hauterivian facies

These facies (Figs 17 & 18), more-or-less argillaceous and always containing glauconite, differ according to their palaeogeographical position in relation to the earlier Jura platform.

On the slope of the outer shelf of the Jura platform (northern Vercors), marls alternate with argillaceous limestone in wavy beds (nodular limestone, the French 'calcaires à miches') containing

Fig. 15. Lowstand wedge shallowing-upward sequence. (A) Outcrop view showing the basal part of a shallowing-upward sequence (10–30 m thick) starting above hemipelagic argillaceous limestone. (B to D) Selected microfacies in thin sections from base B to top D of the sequence; (B) hemipelagic facies corresponding to sponge spicule peloidal mudstone; (C) coarse skeletal fragment wackestone with bryozoans (B) and crinoids; (D) very coarse skeletal fragment grainstone with corals (c) and larger foraminifera (F). (Scale bar: 1 mm)

mud-dwelling benthic fauna such as irregular echinoids (Spatangidae, particularly *Toxaster retusus*) and bivalvia (*Pinna, Panopea*). The nodularity is perhaps due to deposition on a slope. The microfacies is an argillaceous mudstone–wackestone with sponge spicules, small agglutinated foraminifera, silt-size quartz and glauconite. These 'calcaires à miches' are associated with other facies in shallowing-upward sequences 7–15 m thick consisting, from the bottom upwards, of: (1) nodular marly limestone ('calcaires à miches') with irregular echinoids; (2) more calcareous beds, arranged in centimetric to decimetric layers of packstone with peloids and small foraminifera; and (3) the sequence may end in a bed of packstone–grainstone with bryozoans and crinoids, locally corresponding to storm deposits.

Near the previous margin of the Jura platform (southern Jura), the dominant facies is glauconitic packstone with bryozoans and crinoids. Shallowing-upward sequences, 5–10 m thick, are usually composed of muddy limestones with irregular echinoids at the bottom and grainstones with oolites and cross-bedding at the top.

Above the earlier inner-shelf of the Jura platform, the dominant facies is grainstone with oolites and cross-bedding. Shallowing-upward sequences, 15–20 m thick, usually begin with packstone with bryozoans and crinoids and end in grainstones (bio-oosparites) which may contain varying amounts of coral debris. Sequences of this type are observed throughout the Pierre Jaune de Neuchâtel Formation.

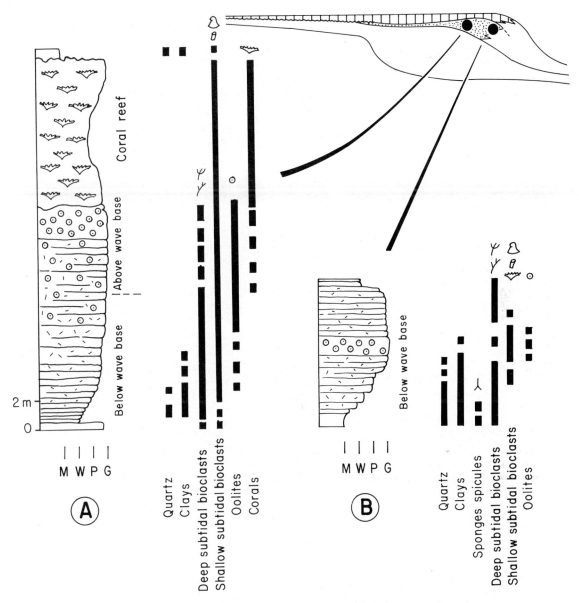

Fig. 16. Sketches of lowstand wedge sequences. (A) Section of the shallowing-upward sequence from the inner part of the lowstand wedge. From base to top, wackestone with debris of 'ball-shaped' bryozoans and crinoids, poorly-sorted bioclastic packstone–grainstone, well-sorted oolitic grainstone, coral reef. (B) Section of the cyclic sequence from the outer part of the lowstand wedge. The deepest facies corresponds to hemipelagic facies with sponge spicules, the shallowest, to oolitic grainstone. M. mudstone; W, wackestone; P, packstone; G, grainstones.

In conclusion, the geographical distribution of Hauterivian facies and their very gradual lateral changes testify to their deposition on a ramp, deepening slightly towards the southeast. The shallowest part of this ramp, located in the Franco–Swiss Jura, was affected by currents and, at least locally, by waves. The deepest part of the ramp, found in the southeast of the northern sub-alpine chains, was located well below fairweather wave-base.

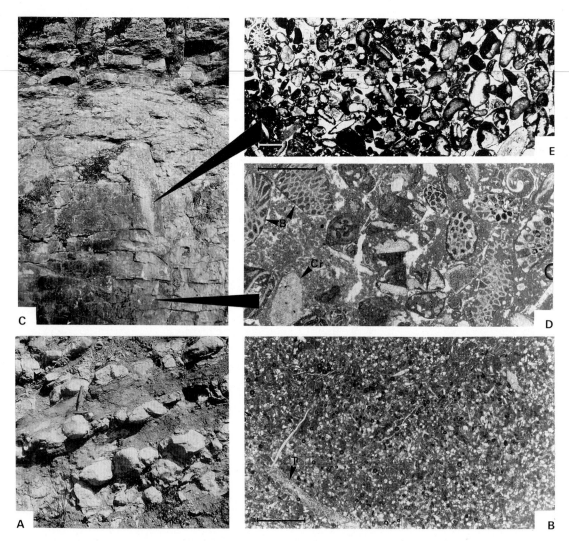

Fig. 17. Hauterivian sequence. (A) From the Vercors, outcrop view of the base of Hauterivian sequence showing interbedded nodular argillaceous limestone ('calcaire à miches') and marls, both with irregular echinoids (Spatangidae, *Toxaster*); note wavy stratification and irregular thickness of strata; (B) photomicrograph of glauconitic argillaceous wackestone with sections of echinoid tests (*Toxaster* test, T); (C) from the southern Jura, outcrop view of the top of the Hauterivian sequence (8–20 m thick) showing irregular bed thickness and wavy stratification. Cherts may appear in this part of the sequence; (D) photomicrograph of crinoid (Cr) and bryozoan (B) wackestone with well-preserved organisms; (E) photomicrograph from the top of the sequence: medium-grained, rounded bioclast grainstone. (Scale bar: 1 mm)

Facies of the Urgonian limestone basal sequence

These facies (Figs 19 & 20) again differ according to their location in relation to the earlier margin of the Jura platform. They correspond throughout to a shallowing-upward sequence, the base of which is transgressive and discordant on the Hauterivian beds of varying ages. In this case, there is commonly a very clear-cut lithological boundary, easily seen on wire-line logs (Fig. 6), between the Urgonian limestone transgressive calcareous sequence and the marly limestone or marl of the underlying Hauterivian. The presence or absence of clay and detrital quartz usually allows the muddy limestones of the Hauterivian to be distinguished from the

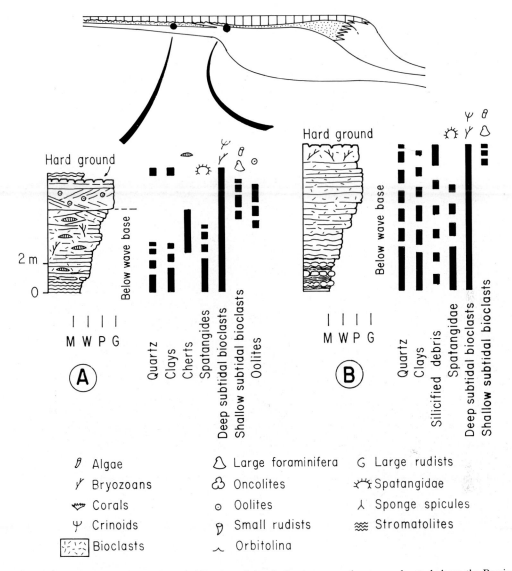

Fig. 18. Sketches of Hauterivian sequences. (A) Section of the shallowing-upward sequence located above the Berriasian platform (southern Jura). From base to top, nodular argillaceous limestone (mudstone–packstone) containing irregular echinoids, crinoid–bryozoan wackestone–packstone with glauconitic grains and cherts, rounded bioclast–oolitic grainstone with cross-bedding and hardground. (B) Section of the shallowing-upward sequence located above the slope of the Berriasian platform (Vercors). From base to top, interbedded nodular argillaceous limestone and marl with irregular echinoids, crinoid bryozoan packestone and hardground.

pure limestones of the late Barremian (Urgonian platform).

Near the edge of the outer slope of the earlier shelf, the sequences, ≈ 20 m thick, display features linking them to both the Hauterivian sequences (from which they differ through the absence of glauconite) and those of lowstand wedges. These Urgonian platform initiation sequences ('séquence d'installation'; Arnaud-Vanneau *et al.*, 1987) display, from the bottom upwards: (1) marls, commonly with irregular echinoids; (2) wackestone–packstone with small foraminifera, in decimetric wavy beds; and (3) grainstone with bioclasts and/or oolites above which there occur cross-bedded oolites in the Chartreuse and southern Jura) and, in the northern Vercors, first coral reefs, and then, still farther towards the southeast, grainstone with corals debris. This sequence may end locally in the first wackestone–packstone beds with rudists.

Above the inner-shelf of the earlier Jura platform, the initial sequence is almost totally missing; in the Neuchâtel area, for example, wackestone–packstone with rudists can be observed almost at the base of the Urgonian limestone formation. In this case, there is either a lateral passage between the initial platform facies and the rudists facies or else an onlap arrangement of basal parasequences onto the transgressive surface. The latter hypothesis has

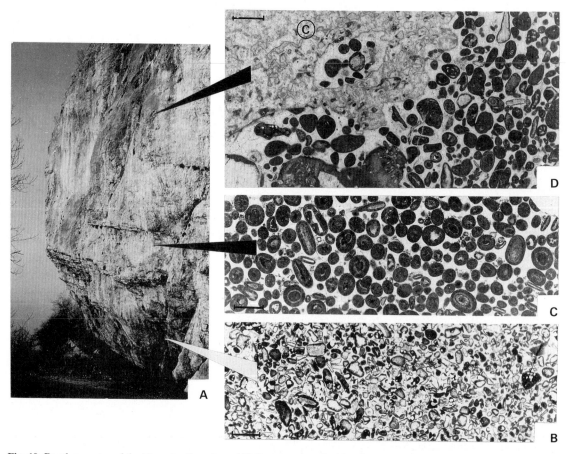

Fig. 19. Basal sequence of the Urgonian limestone. (A) Outcrop view showing the toe of the Urgonian cliff (30 m high) and the basal part of the sequence. Note thin, wavy stratification in sub-horizontal sets at the base. (B to D) Selected photomicrographs from base B to top D of the sequence; (B) well-sorted, fine-grained peloidal grainstone; (C) well-sorted medium-grained oolitic grainstone; (D) coarse, rounded skeletal debris grainstone with coral fragments (c). (Scale bar: 1 mm)

Fig. 20. [*Opposite*] Sketches of basal shallowing-upward sequences of the Urgonian limestone. The sequences are partly similar to those of the Hauterivian sequences from the Neuchâtel region, but they are devoid of glauconite. (A) Section located above the Berriasian platform (Chartreuse, Le Frou section). At the base, wavy argillaceous limestone (wackestone) containing irregular echinoids, at the top, oolitic grainstone. (B) Section located above the Hauterivian hemipelagic ramp (northern Vercors, Gorges du Nant section). This section ends with coral reef and the first appearance of rudist limestone (partly dolomitized wackestone). At the top, beach facies with keystone vugs.

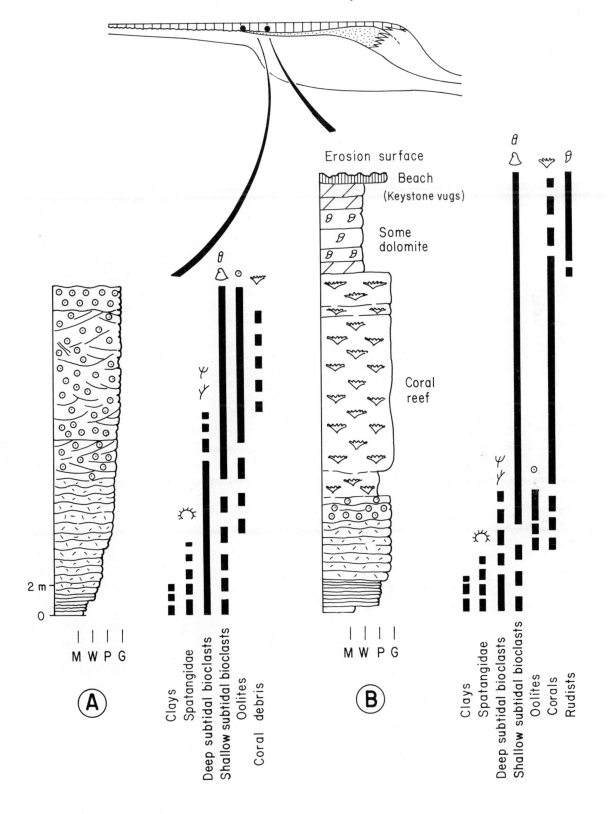

Erosion surface

Beach (Keystone vugs)

Some dolomite

Coral reef

M W P G

A

Clays
Spatangidae
Deep subtidal bioclasts
Shallow subtidal bioclasts
Oolites
Coral debris

M W P G

B

Clays
Spatangidae
Deep subtidal bioclasts
Shallow subtidal bioclasts
Oolites
Corals
Rudists

2 m
0

not been proved, but it is feasible since, in this region, certain foraminifera from the top of the Barremian (*Valserina bronnimanni*, *Palorbitolina lenticularis*) are found just a few metres above the base of the Urgonian limestone.

Inner-platform facies (Urgonian platform)

Throughout the Jura and the northern Subalpine chains (Figs 21 & 22), inner-platform facies are observed above the basal transgression sequence. These are white, exclusively carbonate facies characterized by the micritization of bioclasts due to the activity of endolithic micro-organisms (Cyanophycae, bacteria, fungi) and by the presence of algal oncolites. These sequences are from 4 to 10 m thick. Always of the shallowing-upward type, they display from the bottom upwards: (1) packstone with large and small rudists, large benthic foraminifera, dasyclad algae, miliolidae and locally, a few corals; and (2) packstone–wackestone with small rudists and oncolites which are locally very common. Among the foraminifera, only miliolidae are abundant. The sequence may end in two different ways.

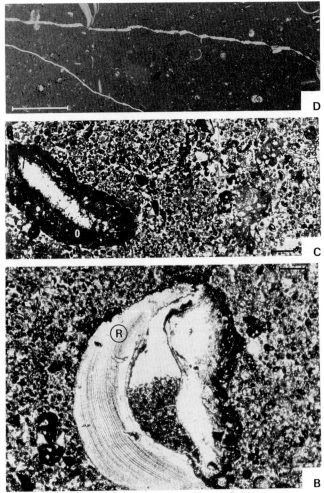

Fig. 21. Inner-platform shallowing-upward sequence (Urgonian platform). (A) General view of the Urgonian cliff. The arrow marks the location of an inner-platform shallowing-upward sequence (S). Thickness of the cliff ≈ 300 m, thickness of sequence: 4–10 m. (B to D) Selected microfacies in thin sections from base B to top D of the sequence; (B) micritized rounded bioclast–foraminiferal packstone with rudists (R); (C) micritized rounded bioclast–miliolid wackestone with oncolites (O); (D) *Pseudotriloculina* mudstone (restricted environment). (Scale bar: 1 mm)

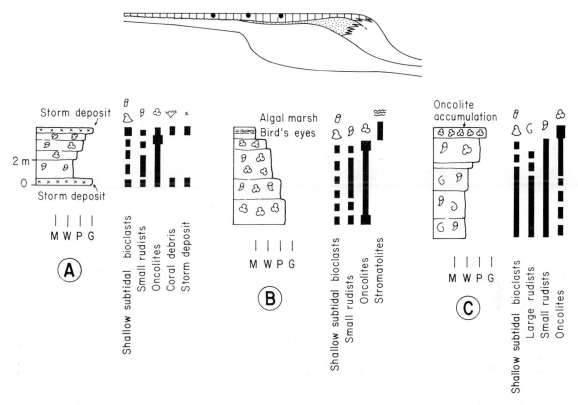

Fig. 22. Sketches of inner-platform shallowing-upward sequences (Urgonian platform). These sequences may be observed in different places on the Urgonian platform. Nevertheless, sequences with large rudists are more common in the outer part of the inner platform, and sequences with oncolites and small rudists in the inner part. Sequences may be interrupted by coarse storm grainstone, but, usually, they end with either *Pseudotriculina* mudstone, or mudstone with bird's eyes, or mudstone with stromatolites.

In one case, there is a grainstone of storm deposits made up of either oncolites (Fig. 22C), numerous rudist shells or coral fragments (Figs 22A & 23). In the other case, the top of the sequence is composed of a more restricted facies: first wackestone–mudstone with oncolites and then mudstone with restricted microfauna (one or two types of miliolidae), bird's eyes or stromatolites (Figs 22B & 24).

Very low or very high sea-level facies

Low sea-level facies on the Hauterivian ramp and Urgonian platform

On the Hauterivian ramp, low sea-level periods involved emergence and thus erosion, the record of which varies according to the calcareous or muddy calcareous nature of the facies. In the oolitic facies, emergence involved an early dissolution of aragonite ooid cortexes. In the muddy limestones, evidence of emergence is far more difficult to find and is deduced in part from the examination of overlying deposits. It is detected either by the presence of small millimetre–centimetre-sized lithoclasts at the base of the overlying sequence, by the presence of an irregular hardened surface encrusted by bryozoans, serpulids and foraminifera, or else indirectly through the recognition of a stratigraphic hiatus (the hiatus must be equal to at least a half-stage in order to be detected by palaeontological means).

On the Urgonian platform, the emergences linked to low sea-level periods are locally characterized by fresh-water deposits (*Chara* limestone) in the Vercors and the Chartreuse. Most of the time, the emergences are highlighted by the erosion of underlying beds or by particular diagenetic fabrics: shell

Fig. 23. Storm deposit capping the inner platform shallowing-upward sequence. (A) Outcrop showing a lower bed rich in large gastropods and an upper bed with coral fragments. (B) Very coarse skeletal fragment grainstone with branching coral fragments displaying microstalactitic cement (ac). (Scale bar: 1 mm)

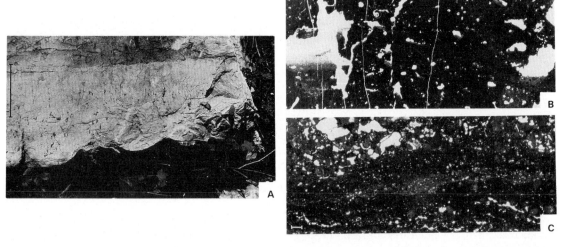

Fig. 24. Normal facies of the top of the inner-platform shallowing-upward sequence. (A) Outcrop view showing calcite-filled vertical tubes in limestone (vertical scale bar: 10 cm). (B) Photomicrograph of bird's eyes in mudstone. (C) Planar stromatolite including fine fenestral fabric (horizontal scale bar : 1 mm).

dissolution, dolomite crystal dissolution and subsequent vadose silt, vadose cement... The erosion corresponds either to the start of karstification (cavities penetrating down to a metre or more below the surface) or, more rarely, to the cutting of incised valleys up to > 30 m in depth.

High sea-level facies on the Hauterivian ramp and Urgonian platform: transgressive facies

The horizon of maximum depth is always preceded by a particular type of transgressive facies (Arnaud-Vanneau, 1980; Arnaud-Vanneau *et al.*, 1987).

Whatever the location or the stratigraphic level, transgressive facies always display four fundamental features (Figs 12 & 25). (1) Appearance or notable increase in clay, quartz and iron oxide content, resulting in a highly recognizable bluish, reddish or yellowish colour. (2) Rounded lithoclasts in a fine-grained sediment, or a mixture of bioclasts of different sizes or a mixture of two facies types. (3) Substantial reworking and mixing of fauna and flora: brackish−fresh-water organisms (characae) occur with deep-subtidal marine organisms (small agglutinated foraminifera), or even pelagic organisms (ammonites). (4) Skeletal grains are commonly

Fig. 25. Transgressive sequence. (A) Outcrop photograph showing argillaceous limestones and marls of the transgressive sequence Ail (lower *Orbitolina* beds) interbedded between the lower and the upper member of the massive Urgonian limestone. Thickness of the transgressive sequence: ≈ 25 m. (B to D) Selected microfacies in thin section from base B to top D of the sequence; (B) bimodal, medium–very coarse fossil fragments, rounded intraclasts and limonitized oncoids with fine peloidal–foraminifera wackestone; (C) well-sorted limonitized peloidal grainstone from hardground; (D) poorly-sorted wackestone with large foraminifera (*Orbitolina*, centre), algae and gastropods. (Scale bar : 1 mm)

surrounded by a coating of iron oxides, locally cemented together in a ferruginous hardground (Fig. 25).

Transgressive facies are observed either directly on an emergence surface or at the top of a shallowing-upward sequence characterized by the gradual vertical deepening of the environment from one parasequence to the next. The interval between the transgressive facies and the maximum transgression level is usually taken up (late Barremian transgression in the southern Vercors, lower *Orbitolina* beds in the northern Vercors and the Chartreuse) by one or more shallowing-upward sequences composed of argillaceous facies rich in irregular echinoids, *Palorbitolina*, benthic foraminifera and dasyclad algae. The maximum transgression stage generally corresponds to marls which, especially on the outer-shelf, may include glauconitic beds that are locally rich in ammonites (likelihood of condensed sections).

COMPARISONS WITH HAQ *ET AL*.'S EUSTATIC SEA-LEVEL CURVE

Two major observations may be drawn from the comparison between the sea-level variation curve established in the region studied and the eustatic sea-level curve of Haq *et al.* (1987). The two curves (Fig. 26) are similar for the Hauterivian and the late Barremian–early Aptian. The slight differences which still exist at these levels are due to the difficulties encountered in accurately locating events on a timescale rather than any true differences in interpretation. On the other hand, the two curves are very different for the early Barremian. Because of this, the variations in depth recorded in the region studied may be of eustatic origin in the first case, whereas they would appear to arise from tectonic causes in second case. This last hypothesis is probable since an analysis of subsidence curves, not only in the region studied (Arnaud, 1988) but also farther north (Funk, 1985), shows the existence of a Barremian crisis beginning in the vicinity of the Hauterivian–Barremian boundary which is marked by wide regional variations in subsidence rates and by the probable effect of NE−SW basement faults. Of the latter, vertical movements on the Isère fault are shown by the emergence of the northern region (Jura domain) and the rapid sinking of the southern region (Vercors) where bioclastic sediments accumulated, forming a lowstand wedge. The significance of this Barremian crisis is questionable, but conceivably it is related to the opening of the North Atlantic (Arnaud, 1988) and, in particular, to the events which, on the Galicia margin are responsible for the unconformity of Barremian strata on the Neocomian (Boillot *et al.*, 1986) and for the intrusion of peridotites (Feraud *et al.*, 1988). It could also be linked to the beginning of the subduction in the eastern Alps (Tollmann, 1987).

CONCLUSION

Sedimentological and palaeontological data collected in the Jura and northern Subalpine chains allow a large-scale interpretation of the development of the Hauterivian to early Aptian platforms. Of these regions, the Vercors is the most interesting for two reasons: little tectonism, and outcrops of the transition between platform and basin.

Three major observations can be made. (1) The deposits of the early Barremian (Glandasse bioclastic limestone formation), which are very thick in the southern Vercors, disappear very rapidly towards the north. (2) The lower part of the Urgonian limestone has been dated as late Barremian in both the Jura and the northern Subalpine chains. (3) The early Barremian and part of the late Hauterivian are absent on the Jura platform between the Pierre Jaune de Neuchâtel Formation and the Urgonian limestone formation.

These observations can be explained by substantial variations in relative sea-level, namely two major low sea-level stands separated by two high sea-level stands. (1) The very low sea-level period at the beginning of the early Barremian is characterized by the emergence and erosion of at least part of the late Hauterivian sequence in the Jura and by the development of a lowstand wedge in the southern Vercors (Fig. 3). (2) The very high sea-level stand in the late Barremian (Feraudi and

Astieri Zones) is responsible for the deposition of the first transgressive sequences of the Urgonian limestone formation on the previously emergent Jura platform (Fig. 7). (3) The low sea-level stand localized towards the middle of the early Aptian is marked in the northern Vercors by the cutting of a system of incised valleys, tens of kilometres in length and locally up to 30 m in depth (Fig. 8). (4) The high sea-level stand towards the end of the early Aptian is characterized by the final drowning of the Urgonian platform. The corresponding depositional sequence is composed mainly of marly layers that filled the valleys (upper *Orbitolina* beds).

From a sedimentological point of view, emphasis has been placed on the lowstand wedge features of the early Barremian and the sequence stratigraphy. The lowstand wedge of the early Barremian crops out in the southern Vercors (Glandasse plateau). It is characterized by its width (up to 10 km), by its palaeogeographical position on the hemipelagic slope of the late Hauterivian, i.e. below the Jura platform margin, and by the outstanding development of coarse bioclastic facies. The latter, whose overall thickness reaches 800 m, displays all facies from deep-subtidal to shallow-subtidal. The sequences usually end in oosparites and/or small coral reefs. The development of these reefs and oolitic shoals on the outer part of the lowstand wedge indicates that the wedge's upper surface was not flat, but corresponded to a shoal that sloped steeply towards the basin in the southeast and passed laterally towards the northwest to less agitated environments.

The sequence stratigraphy is relatively complex. Four major types of sequence have been distinguished. (1) Lowstand wedge and Urgonian platform margin sequences, characterized by the preponderance of bioclastic facies, the presence of small reefs, the abundance of benthic foraminifera and the presence of storm beds, locally gullying the outer slope. (2) Drowned platform sequences that were well developed on the Jura platform during the Hauterivian, characterized by the preponderance of bryozoan and crinoid facies overlying the spantangid facies, by the local abundance of oolites and by the presence of glauconite and detrital minerals. (3) Sequences of the inner part of the Urgonian platform are characterized by the preponderance of rudist facies and the existence of various restricted facies deposited in shallow-subtidal to supratidal environments. (4) High sea-level sequences (first Urgonian limestone parasequence, lower *Orbitolina* beds)

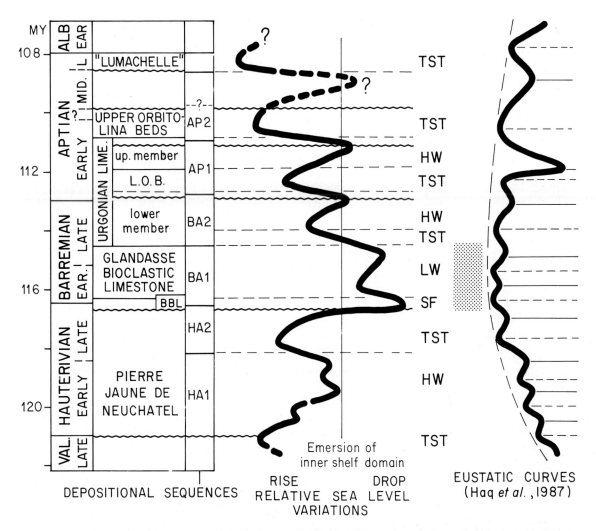

Fig. 26. Comparison between the eustatic curves of Haq *et al.* (1987) and the relative sea-level variations in the Jura and Subalpine domains. Dotted interval corresponds to a large difference between curves. Abbreviations: see Fig. 8. LOB, lower *Orbitolina* beds; BBL, Borne bioclastic limestone formation.

characterized at the base by transgressive facies and, at the top, by the deposition of oolites, coral debris facies and coral reefs (initiation facies).

From a morphological point of view, the lowstand wedge of the early Barremian can be considered as a shoal which prepared the way for the development of the Urgonian platform ('haut-fond précurseur'). The rapid sediment accumulation on the hemipelagic slope, i.e. outside the platform, in a part of the Vocontian basin that remained deep until the end of the Hauterivian, did in fact enable the inner part of the Urgonian platform to extend very rapidly more than 30 km to the southeast of the former edge (along the Isère fault) of the Neocomian platform (Berriasian to Hauterivian platforms) (Fig. 3).

It should be emphasized finally that the Urgonian limestone is a transgressive formation deposited over a relatively short span of time. The particularly high sedimentation rates are thus a reflection of both the substantial rise in sea-level at this time and a likely acceleration of subsidence after the Barremian exposure.

ACKNOWLEDGEMENTS

The authors wish to thank K. Lewis for the English translation, P. Crevello, M. Gidon and K. Shanley for their help, and M. E. Tucker who kindly reviewed the manuscript and provided many valuable suggestions.

REFERENCES

ARNAUD-VANNEAU, A. (1980) Micropaléontologie, paléo-écologie et sédimentologie d'une plate-forme carbonatée de la marge passive de la Téthys: l'Urgonien du Vercors septentrional et de la Chartreuse (Alpes occidentales). *Géologie Alpine*, mém. 10, 3 Vols, 874 p.

ARNAUD-VANNEAU, A., ARNAUD, H., & THIEULOY, J.-P. (1976) Bases nouvelles pour la stratigraphie des calcaires urgoniens du Vercors. *Newsl. Stratig.* **5**(2−3), 143−149.

ARNAUD-VANNEAU, A., ARNAUD, H., MEUNIER, A.-R. & SEGUIN, J.-C. (1987) Caractères des transgressions du Crétacé inférieur sur les marges de l'océan ligure (Sud-Est de la France et Italie centrale). *Mém. Géol. Université de Dijon* **11**, 167−182.

ARNAUD, H. (1979) Caractères sédimentologiques et paléo-géographiques du haut-fond du Vercors méridional (S.E. de la France), le problème des corrélations séquentielles haut-fond − bassin. *Geobios*, Lyon Mém. 3, 103−119.

ARNAUD, H. (1981) De la plate-forme urgonienne au bassin vocontien: le Barrémo-Bédoulien des Alpes occidentales entre Isère et Buëch (Vercors méridional, Diois oriental et Dévoluy). *Géologie Alpine*, mém. 11, 3 Vols, 804 p.

ARNAUD, H. (1988) Subsidence in certain domains of Southeastern France during the Ligurian Tethys opening and spreading stages. *Bull. Soc. Géol. France* **IV**(8), 725−732.

ARNAUD, H. & ARNAUD-VANNEAU, A. (1989) Séquences de dépôt et variations du niveau relatif de la mer au Barrémien et à l'Aptien inférieur dans les massifs subalpins septentrionaux et le Jura (SE de la France). *Bull. Soc. Géol. France* **8**, 5, no. 3, 651−660.

BOILLOT, G., RECQ, M., WINTERER, E., MEYER, A.W., APPLEGATE, J., BALTUCK, M., BERGEN, J.A., COMAS, M.C., DAVIES, T.A., DUNHAM, K., EVANS, C.A., GIRARDEAU, J., GOLBERG, D.G., HAGGERTY, J., JANSA, L.F., JOHNSON, J.A., KASAHARA, J., LOREAU, J.P., LUNA SIERRA, E., MOULLADE, M., OGG, J., SARTI, M., THUROW J. & WILLIAMSON, M.A. (1986) Amincissement de la croûte continentale et dénudation tectonique du manteau supérieur sous les marges stables: A la recherche d'un modèle − L'exemple de la marge occidentale de la Galice (Espagne). *Bull. Centres Recherch. Explor.-Prod. Elf-Aquitaine* **10**(1), 95−104.

BUROLLET, P.F., CLAIREFOND, P. & WINNOCK, E. (1979) La Mer Pélagienne. *Rev. Geol. Méditerranéenne*, **6**(1), 345.

BUSNARDO, R. (1984) Echelles stratigraphiques du Crétacé inférieur, Ammonites, In *Synthèse géologique du Sud-Est de la France*. (Eds Debrand-Passard S. *et al.*) Mém. Bull. Rech. Géol. Min. 125, pp. 291−293.

BUSNARDO, R. & THIEULOY, J.-P. (1989) Les Ammonites de l'Hauterivien jurassien. Révision des faunes de la région du stratotype historique de l'étage hauterivien. *Mem. Soc. Neuchâteloise Sci. Nat.* **11**, 101−147.

CASEY, R. (1961) The stratigraphical Palaeontology of the Lower Greensand. *Palaeontology* **3**, 487−621.

CLAVEL, B., CHAROLLAIS, J. & BUSNARDO, R. (1987) Données biostratigraphiques nouvelles sur l'apparition des faciès urgoniens du Jura au Vercors. *Eclogae Geol. Helv.* **80**(1), 59−68.

CONRAD, M.-A. (1969) Les calcaires urgoniens dans la région entourant Genève. *Eclogae Geol. Helv.* **62**(1), 1−79.

FERAUD, G., GIRARDEAU, J., BESLIER, M.-O. & BOILLOT, G. (1988) Datation 39Ar−40Ar de la mise en place des péridotites bordant la marge de la Galice (Espagñe). *Comp. Rend. Acad Sci.* **307**, serie II, 49−55.

FERRY, S. & RUBINO, J.-L. (1988) Eustatisme et séquences de dépôt dans le Crétacé du Sud-Est de la France. Livret-guide de l'excursion du Groupe Français du Crétacé en fosse vocontienne (25−27 Mai 1988). *Geotrope*, Lyon **1**, 131.

FUNK, H. (1985) Mesozoische Subsidenzgeschichte im Helvetischen Schelf der Ostschweiz. *Eclogae Geol. Helv.* **78**(2), 249−272.

HAQ, B.U., HARDENBOL, J. & VAIL, P.R. (1987) The chronology of fluctuating sea level since the Triassic. *Science* **235**, 1156−1167.

HINE, A.C. & NEUMANN, A.C. (1977) Shallow carbonate bank. Margin growth and structure, Little Bahama Bank, Bahamas. *Am. Assoc. Petrol. Geol. Bull.* **61**(3), 376−406.

KENDALL, C.G. ST. & SCHLAGER, W. (1981) Carbonates and relative changes in sea level. *Marine Geol.* **44**, 181−212.

KEMPER, E. (1982) Zur Gliederung der Schichtenfolge Apt-Unter-Alb. *Geol. Jb.* **65**, 21−33.

LEMOINE, M. (1984) La marge occidentale de la Tethys ligure et les Alpes occidentales. In: *Les Marges Continentales Actuelles et Fossiles autour de la France* (Ed. G. BOILLOT.) Masson edn., Paris, pp. 155−248.

LEMOINE, M., BAS, T., ARNAUD-VANNEAU, A., ARNAUD, H., DUMONT, T., GIDON, M., BOURBON, M., DE GRACIANSKY, P.-C., RUDKIEWICZ, J.-L., MEGARD-GALLI, J. & TRICART, P. (1986) The continental margin of the Mesozoic Tethys in the Western Alps. *Marine Petrol. Geol.* **3**, 179−199.

READ, J.F. (1985) Carbonate platform facies models. *Am. Assoc. Petrol. Geol. Bull.* **69**(1), 1−21.

REMANE, J. (1982) Die Kreide des Neuenburger Juras. *Jber. Mitt. Oberrhein. Geol. Ver, N.F.* **64**, 47−59.

SCHLAGER, W. (1981) The paradox of drowned reefs and carbonate platforms. *Geol. Soc. Am. Bull.* **92**(1), 197−211.

SCHOLLE, P.A., BEBOUT, D.G. & MOORE, C.H. (1983) Carbonate Depositional Environments. *Am. Assoc. Petrol. Geol. Mém.* **33**, 708.

THIEULOY, J.-P. (1979) *Matheronites limentinus* n. sp. (Ammonoidea) Espèce type d'un horizon-repère Barrémien supérieur du Vercors méridional (massif subalpin français). *Geobios*, Mém. 3, 305−317.

TOLLMANN, A. (1987) Geodynamic concepts of the Evolution of the Eastern Alps. The Alpidic Evolution of the Eastern Alps. In: *Geodynamics of the Eastern Alps, Vienne* (Eds Flugel, H. & Faupl, P.) pp. 361–378.

TUCKER, M.E. (1985) Shallow-marine carbonate facies and facies models. In: *Sedimentology, Recent Developments and Applied Aspects* (Eds Brenchley & Williams). pp. 147–169.

VAIL, P.R., COLIN, J.-P., Jan du CHENE, R., KUCHLY, J., MEDIAVILLA, F. & TRIFILIEFF, V. (1987) La stratigraphie séquentielle et son application aux corrélations chronostratigraphiques dans le Jurassique du Bassin de Paris. *Bull. Soc. Geol. France* **III**(8), 1301–1321.

VAIL, P.R., MITCHUM, R.M. Jr., TODD, R.G., WIDMERI, J.W., THOMPSON, S., SANGREE, J.B., BUBB, J.N. & HATELID, W.G. (1977) Seismic stratigraphy and global changes of sea level. In: *Seismic stratigraphy. Application to Hydrocarbon Exploration*. Am. Assoc. Petrol. Geol. Mém. 26, pp. 49–212.

VIALON, P. (1974) Les déformations "synchisteuses" superposées en Dauphiné. Leur place dans le collision des éléments du Socle Préalpin. Conséquences paléostructurales. *Schweiz. Mineral. Petrogr. Mitt.* **54**(2/3), 663–690.

VIEBAN, F. (1983) *Installation et évolution de la plate-forme urgonienne (Hauterivien à Bédoulien) du Jura méridional aux chaînes subalpines (Ain, Savoie, Haute-Savoie)*. These 3e Cycle, Université Grenoble, 291 p.

ZWEIDLER, D. (1985) *Genèse des gisements d'asphalte des formations de la Pierre Jaune de Neuchâtel et des calcaires Urgoniens du Jura (Jura neuchâtelois et nord-vaudois, Suisse)*. These Université Neuchâtel, 107 p.

Spec. Publs int. Ass. Sediment. (1990) **9**, 235–255

Basement structural controls on Mesozoic carbonate facies in northeastern Mexico—a review

J. L. WILSON

1316 Patio Drive, New Braunfels, Texas 78130, USA

ABSTRACT

Clear marine tropical water and the proper oceanographic environment cause formation of carbonate ramps and rimmed platforms but their trends and orientations are controlled partly by tectonic basement framework. Elongate buildups may form parallel to subsiding passive cratonal margins, or platforms may develop over and around fault blocks along such borders. In many instances, individual isolated and steep buildups rise from earlier formed wide platforms of low relief. Narrow platform rims, with a widely recognized spectrum of carbonate facies, may evolve around major subsiding basins.

With the opening of the Gulf of Mexico in early Mesozoic time, left-lateral, northwest-directed rifting occurred through eastern Mexico, cutting across a previously emplaced Permo–Triassic orogenic belt bordered on its east by Triassic granodiorite batholiths. This orogenic belt may have a western provenance and may have been moved east against a coastward continuation of the Quachita–Marathon metamorphic rocks, or it simply may be a dislocated western continuation of this belt. A prominent series of blocks and intervening basins developed in the region during the succeeding Liassic rifting. Maps are presented along with discussion of the complex basement.

Topographic relief produced variations in sedimentation during the irregular but continued transgression during Jurassic and Cretaceous times. The rift-produced grabens were filled with redbeds and arkose, followed by evaporites in middle Jurassic time, and at the beginning of late Jurassic, basinal evaporites and oolitic grainstones surrounded some uplifts. The positive tectonic blocks partly controlled development of spectacular rimmed platforms in the late Mesozoic, responding to renewed subsidence and to development of organic framework potential of both corals and large bivalves.

A Lower Cretaceous carbonate platform developed around the Coahuila block and across the mouth of Sabinas Basin to the Sligo reef trend of Texas. A reef trend also formed along a N–S directed tectonic ridge of Precambrian gneiss and late Palaeozoic schist on the east side of the Valles platform; perhaps it encircled the area to form a large atoll with a central evaporite basin. Gulfward subsidence of the Tamaulipas arch, Golden Lane and Cordova basement prevented shallow-water carbonate development here during early Cretaceous time. In middle Cretaceous the Sabinas Basin was encircled by reefy development around the Coahuila block and along the west flank of the Burro–Salado uplift. This continued up the Gulf Coast of Texas as the Stuart City (Deep Edwards) reef trend. Middle Cretaceous Valles and Golden Lane platforms kept up with subsidence, grew to heights approaching 1000 m, and furnished debris into the Chicontepec Basin separating the platforms. The smaller El Doctor and Toliman banks and the narrow Actopan extension result from basement block fragmentation along the Transverse Mexican Neovolcanic belt which must have been a major lineament separating northern from southern Mexico. The major platforms continued development into Turonian time despite general sea-level lowerings during the middle Cretaceous.

Jurassic oolite, Cretaceous reefs and forereef debris furnish excellent reservoir rock and provide large oil fields in central and southern Mexico. The Mesozoic of Mexico can be used as a model for predicting trends of carbonate reservoir development in both North Africa and the Middle East.

INTRODUCTION

It is generally considered that northeastern Mexico consists of early Mesozoic fault blocks which transect one or more Palaeozoic orogenic belts. In turn these fault blocks partly controlled late Jurassic and Cretaceous facies as well as Tertiary structural patterns resulting from Laramide compressive folding. This has been well recorded by Mexican and North American geologists (Alfonso-Zwanzigger, 1978; Gonzales-Garcia, 1984; Padilla y Sanchez, 1982; Lopez-Ramos, 1972, 1981, 1982; DeCserna, 1970; Belcher, 1979; Mixon, 1963; Charleston, 1981; Wilson et al., 1984). In this paper the geological history of this part of Mexico is briefly reviewed. An attempt is then made to show how the areas of Palaeozoic metasedimentary rocks, Permo-Triassic granite intrusions, and Mesozoic-rifted fault blocks influenced late Jurassic and Cretaceous carbonate facies and platform construction.

PERMO-TRIASSIC OROGENY AND EARLY MESOZOIC RIFTING IN NORTHEAST AND CENTRAL MEXICO

There exists scattered evidence of an important fold belt whose age of deformation and metamorphism is Permo-Carboniferous and which is distributed in central Mexico from Chihuahua to the Trans-Mexico Neovolcanic belt ($\approx$ 2000 km N–S). This area of deformation lies west of and along the front of the eastern Sierra Madre, its outcrops brought to the surface by major Laramide anticlines. It includes clastic flyschoid sediments generally of Permian and Carboniferous age (Guacamaya and Plomosas Formations). In the Perigrina Canyon these Wolfcampian sediments are in tectonic contact with metasediments of the same age (Granjeno schist). There exist two outcrop areas of Grenville age gneiss in Huayacocotla and Perigrina Canyons (1200 Ma to 900 Ma) as well as some sub-surface reports of this gneiss in the Poza Rica area. These gneisses have the same late Proterozoic age as those in southern Chihuahua. It is conjectured that this gneiss ridge and attendant Palaeozoic metasediments, have been moved eastward against the Quachita–Marathon trend by Permo-Triassic orogeny and perhaps also by left-lateral early Jurassic strike–slip faulting.

To the north in Coahuila 4000 m of fossiliferous wildflysch, volcanoclastic and calcareous sediments

with olistostromes, occur in an area west of Nueva Delicias, Coahuila. They range in age from middle Pennsylvanian through to the Permian and represent an island arc facies now exposed as part of the Coahuila block basement (McKee et al., 1988). They have been intruded by Triassic granites.

A better known late Palaeozoic fold belt, the Quachita–Marathon trend (Fig. 1), curves southward out of the Big Bend region of Texas into northern Mexico where, after 100 km, a gravity anomaly indicates that it abruptly stops as if transected by faulting (Handschy et al., 1987). In Mexico its interior portion consists of clastic metasediments (low metamorphic grade schist) of Siluro-Devonian to Pennsylvanian age inclusive. The belt is conjectured to underlie the Burro–Peyotes uplift and may trend south to two wells north of Tampico where Wolfcampian unmetamorphosed terrigenous clastics overlie metamorphic rocks reheated in Triassic or later time. The metasediments of the Lampazos–Picacho highs are like the above but no radiometric dates have been published. They are known only to be of pre-Huizachal (early Mesozoic) age. Later, important left-lateral faulting, including the conjectured Mojave–Sonora megashear of Anderson & Schmidt (1983), may have displaced the first described orogenic belt of central Mexico to the southeast. Originally this may have connected to the west-trending Quachita–Marathon belt some hundreds of kilometres to the north. Table 1 gives a summary of the pertinent localities where Palaeozoic rocks are seen in outcrops or reached in wells.

In Triassic time, south and east of the central orogenic belt, there exist extensive granodiorite intrusions whose K/Ar and R/Sr dates range from Permian to middle Jurassic (Fig. 2). The intrusives stretch from the Coahuila block at Nueva Delicias, southeast to Vera Cruz. The granodiorite is seen in many wells along the Tamaulipas Arch, the Arenque block offshore of Tampico, the Golden Lane Tuxpan massif and the more southerly Tezuitlan massif. The granodiorites are closely interspersed with metasedimentary and terrigenous clastic Wolfcampian–Pennsylvanian rocks all along their trend. Rhyolites and shallow-emplaced granitic rocks are known at a few localities near the Laramide front of the eastern Sierra Madre but apart from these no Triassic granites are known to intrude the central Palaeozoic orogenic belt. It is assumed that this granitic belt represents the roots of an island arc. Rhyolitic–

Table 1. Areas of folded late Palaeozoic strata in northern Mexico

Locality	Underlying folded and metamorphosed strata	Overlying strata
1 Nuevo Casas Grandes & Janos, Chih.	Leonardian flysch	Tertiary volcanics
2 Samalayuca, Chih.	Metamorphosed Palaeozoic terrigenous clastics	Late Jurassic
3 Sierra de Cuervo Aldama, Chih.	Early Permian−Leonardian flysch with Precambrian tectonic slices	Early Cretaceous
4 Sierra de Mojina	Reworked boulders from nearby uplift contain Middle Permian rhyolite (246 Ma Rb/Sr) and Permian metasediments. (231−266 Ma, K/Ar)	From boulders in basal Cretaceous 'Glance Conglomerate'
5 Minas Plomosas−Carrizalillo	Lower Permian and Leonardian flysch with Leonardian age rhyolite. Carrizalillo has tightly folded strata	Late Jurassic LaCasita Formation
6 Las Delicias−Pènn. Acatita Valley, Coahuila	Complete Pènn. and Permian section. Youngest Permian in N. America. Volcaniclastic flysch with olistoliths intruded by late Triassic (?) granite	Early to middle Cretaceous limestone
7 Sta. Maria del Oro	Pescadito schist whose metamorphism is dated as Mississippian (326 Ma)	Liassic dated on palynomorphs (fault contact)
8 Catorce, SLP	Varied metasediments, schists, sandstones, ophiolite (?) Some late Mississippian−early Pennsylvanian spores	
9 Caopas−Rodeo or Apizolaya area, Zacatecas	Very thick, metasedimentary sequence, and metavolcanics. Maybe metamorphosed Nazas. In a nearby area is the Taray Fm. a cherty argillilte & quartzite. Has Penn. fusulinids (*in situ?*)	Nazas redbeds and volcanoclastics of probable Callovian−Oxfordian age
10 Aramberi, N. Leon.	Granjeno-type schists & some gneiss (Novillo?) Schists dated 270−290 Ma (K/Ar)	Huizachal or Nazas Formations. Late Triassic−mid-Jurassic
11 Miquihuana, Tamaulipas	Talcose sericite schists	Huizachal or Nazas Fm (late Triassic)
12 Perigrina Canyon C. Victoria, Tamaulipas	Precambrian gneiss (Novillo Fm.) and shelf-upper slope Mid Palaeozoic carbonate, Penn−lower Permian flysch sequence. Overthrust (?) by metamorphozed Permian Pennsylvanian Granjeno schist. Dates: 262−271 Ma, 294−315 Ma, (K/Ar), 315 Ma, Rb/Sr	Huizachal Formation (late Triassic)
13 Huiznopala, Hidalgo Huayacocotla anticline	Pre-Cambrian gneiss overlain by thick (2000 m) of Guacamaya Wolfcampian−Leonardian flysch	Late Triassic? Huizachal and Liassic marine sediments
14 Zacatecas City	Panuco Fm. Sericite schists with quartzite, limestone breccia	Late Triassic marine schists. Structural contact?

andesitic volcaniclastic sediments with redbeds (Nazas Formation) lie to the south and west of these batholiths and also above them in places. The granites are overlain unconformably in some places by late Triassic–Liassic strata as young as late Jurassic with a basal weathered zone or breccia. (See Table 2 for detailed list of localities for these rocks).

Rifting began after the granitic emplacements and typical rift-valley sediments are widely known south of Chihuahua and the Coahuila block in basins between horst blocks. These redbeds are grouped as the Huizachal–Nazas Unit and are typical taphrogenic deposits. Their provenance and alluvial, fandelta and lacustrine environments are well known

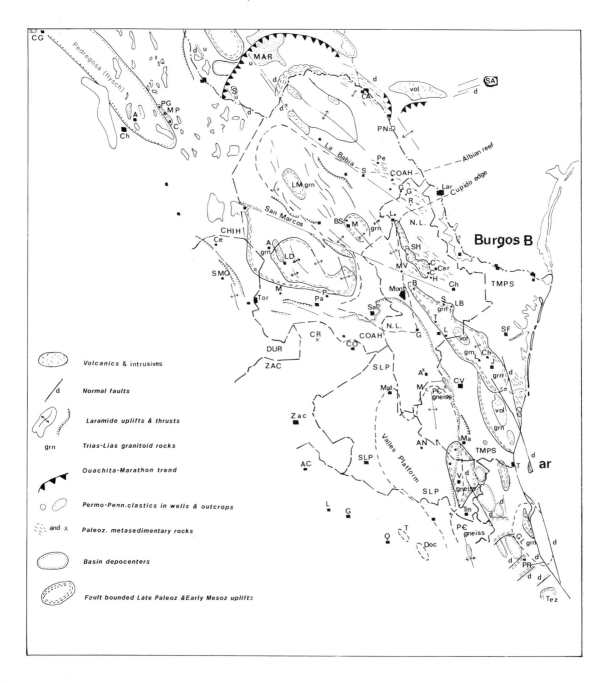

Volcanics & intrusives

Normal faults

Laramide uplifts & thrusts

grn   Trias-Lias granitoid rocks

Ouachita-Marathon trend

Permo-Penn. clastics in wells & outcrops

and x   Paleoz. metasedimentary rocks

Basin depocenters

Fault bounded Late Paleoz & Early Mesoz uplifts

(Belcher, 1979; Mixon, 1963; Blickwede, 1981). They are replete with rhyolite to andesitic shallow intrusions as sills and dykes. The Huizachal consists of two stratigraphic units, an older and thicker one of late Triassic−Liassic age (La Boca Formation), and a probably middle Jurassic basal conglomerate and rift valley fill, La Joya Formation, which persistently grades upward to Oxfordian carbonates or clean quartz sandstones. Both units occur in north trending areas, apparently fault basins along the front of the eastern Sierra Madres. Another basin lies south and west of Torreon and Saltillo, and is filled with red volcaniclastic sediments and igneous intrusive rocks and tuffs (Nazas Formation). The exact relation between the Nazas and Huizachal is not known but they are assumed to be approximately the same age. Table 3 gives details of various localities.

Fig. 1. Late Palaeozoic−early Mesozoic tectonic framework of northern Mexico.

Index to locations' west to east (across) and secondarily north to south

CG	Casas Grandes, Chih.	Tor	Torreon, Coahuila
MAR	Marathon uplift (Tertiary)	M	Mayran well
S	Solitario uplift (Tertiary igneous intrusion)	Pa	Parras, Coahuila
		P	Paila well
Ch	Chihuahua, Chih.	Mont	Monterrey, Nuevo León
A	Aldama, Chih.	Sal	Saltillo, Coahuila
PG	Placer de Guadalupe, Palaeozoic outcrops	S	Suarez well
		LB	Las Blancas well
MP	Minas Plomosa, Palaeozoic outcrops	CR	Coapas Rodeo, Palaeozoic metasedimentary outcrops
C	Carrizalillo, Palaeozoic outcrops	CO	Concepción del Oro, Zac.
CA	Cd. Acuna, Coahuila	G	Galeana, Nuevo León
SA	San Antonio, Texas	T	Trincheras well
PN	Piedras Negras, Coahuila	L	Linares, Nuevo León and Linares well
La Babia	basement fault		
Pe	Peyotes well	SF	San Fernando, Tamps.
S	Sabinas, Coahuila	CH	Chaneque well
GG	Garza wells	L	Lantrisco well
R	Ramones well	J	Jimenez, Tamaulipas
Lar	Laredo, Texas	Ar	Arenque offshore wells and structure
LM	La Mula Uplift, granite exposure		
San Marcos	basement fault	Zac	Zacatecas, Zac.
BS	Buena Suerte well	SLP	San Luis Potosí
M	Monclova, Coahuila	Mat	Matehuala, San Luis Potosí
L	Lampazos, city and well	A	Aramberri outcrops of Palaeozoic schists
Ce	Ceballos well		
A	Acatita valley, Palaeozoic and Triassic outcrop	M	Miquihuana outcrops of Palaeozoic schists
LD	Las Delicias area of Palaeozoic outcrop	CV	Cd. Victoria, Tamps.
		AC	Aquascalientes
SH	Sabinas Hidalgo, NL	AN	Agua Nueva well
MV	Minas Viejas well	Ma	Cd. Mante, Tamaulipas
C	Cerralvo well	T	Tampico
C	Carbajal well	L	León
Cer	Cerralvo, Nuevo León	G	Guanajuato
H	Herreras	Q	Queretaro
B	Benemerito well	T	Toliman Cretaceous platform
Ch	China, Nuevo León	V	Valles, San Luis Potosí
SMO	Santa Maria del Oro, outcrops of Palaeozoic metamorphics and Jurassic island arc sediments	Tm	Tamazunchale, SLP
		GL	Golden Lane/Cretaceous platform and basement high
		PR	Poza Rica
		Tex	Tezuitlan Uplift

Table 2. Areas of presumed Triassic granitic intrusions

Locality	Petrography	Age and evidence, (method of dating indicated where known)
1 Las Delicias—Acatita	Granodiorite, Las Delicias pluton is tonalite	208 Ma K/Ar (Denison et al. 1970) Late Trias, cuts Penn. rocks. Possibility of Permo-Penn. intrusions during deposition also exists. Canon Rosillo volcanic sediments are Leonardian
2 Sierra De Mojado	Granite boulders in conglomerate	225 Ma Permo-Trias (McKee et al., 1984, 1988)
3 La Mula	Granite—Granodiorite	204 Ma Shell; 211—213 Trias (Jones et al., 1984)
4 Menchaca well	Granite	162 Ma (mid-Jur.)
5 Monclova 5 well	Intrusive igneous rock on strike with La Mula	Pre-Huizachal (Wilson et al., 1984)
6 Anahuac field wells	Altered granodiorite	Pre-Huizachal (Wilson et al., 1984)
7 Pecten 1 well	Granite basement	Alfonso-Zwanziger (1978), Wilson et al. (1984)
8 Benmerito well	Granite	139±9 Ma Rb/Sr (late Jur.) Rivera (1976 in Padilla y Sanchez, 1982)
9 Teran well	Granite	184±14 Rb/Sr (Lias?) above reference
10 Linares well	Granodiorite	234 Ma (late Permian)
11 Arenque field	Granite	Eocene date for granite which is below Up. Jur. (Lopez-Ramos, 1972)
12 Muleto (Block E)	Granite	212 Ma±5 (Permo-Trias)
13 Papaya 1–A northern Block E	Granite	320 Ma above reference (Mississippian)
14 Juana Ramirez Tezuitlan block	Gneiss	192±3 (late Triassic) above reference
15 Arroyo Viejo Tezuitlan block	Granodiorite	273±5 (Permo-Carboniferous) above reference
16 Punta Jerez 5 Block E	Granite	183 Ma (Liassic?) above reference.
17 San Rafael 2 north of Block E	Granite	183 Ma (Liassic?) above reference
18 Barreta 2 SW of Sabinas Hidalgo	'Granite basement'?	Under 137 m of Upper Jurassic redbeds
19 Camotal 1 and 2 Sierra de Tamaulipas	Granite	Under Upper Jurassic
20 Tepehauje 2	Fossiliferous Permian clastics over schist	210 Ma (late Trias.) Date is caused by heating owing to Triassic batholith intrusion
21 Ojital 101 south of Poza Rica	Schist	237±5 (Permian) Denison et al. (1969)

Table 2. (*Cont.*)

Locality	Petrography	Age and evidence, (method of dating indicated where known)
22 Ebano−Panuco fields Nucleus H	Tonalite & granite surrounding metamorphic rocks	Lopez-Ramos (1972), no radiometric dates
23 Tuxpan 3 Golden Lane	Granite−syenite	Above reference
24 Frijolillo Golden Lane	Tonalite	Above reference
25 Muro 2 Golden Lane	Tonalite	Above reference
26 Salto 1 Golden Lane	Tonalite	Above reference
27 Sebastian 101 Golden Lane	Granite-derived arkose	Above reference
28 Canyon Caballero, Tamaulipas	Rhyolite and rhyodacites (Assadero Rhyolite)	Gursky & Ramirez-R. (1986)

After granite pluton emplacement, but probably during redbed deposition, the whole area of northeastern Mexico was affected by NW−SE strike−slip and normal faults which subdivided pre−late Jurassic topography into high and low areas. Proceeding in a southeasterly direction from the Texas−Mexican border, these trends are as follows: Chittim anticline−Devil's River uplift, the Burro−Peyotes high, La Babia Fault, the Sabinas Basin, the San Marcos fault, the Coahuila block and the Torreon−Monterrey lineament. In general, the blocks have individually distinctive basement rocks. For example, the Burro−Peyotes high is underlain principally by metasedimentary rocks, the Coahuila block by granite−granodiorite intruding an extensive, folded but little metamorphozed late Palaeozoic flysch. A minor isolated positive area between the Tamaulipas and Burro−Peyotes block, beneath the Sierras de Picacho−Lampazos, has a metasedimentary basement. The metasedimentary belt of terrigenous clastics continues south to the coast at Tampico, lying east of a large granite batholith along the Laramide Tamaulipas arch. These blocks must represent sialic crust close to the cratonal edge in early Mesozoic time.

The Torreon−Monterrey lineament separates the granitic Coahuila block from the area of metasediments to its south and is the favoured pathway down the Mojave−Sonora megashear of Silver & Anderson (1974) and Anderson & Schmidt (1983).

The eastern Sierra Madre salient at Monterrey is thrust over late−middle Jurassic evaporites in a depression along this megashear. Two other fault lineaments trending more NNW are postulated to extend down toward the northern Mexico Gulf coast. It is not clear what these lineaments really represent. The western lineament separates the granitic Tamaulipas arch from the central metasedimentary province and follows the Laramide Magiscatzin syncline to the Tampico−Misantla embayment. The eastern NNW fault is reflected in pre−late Jurassic topography; it follows the eastern edge of the Tamaulipas arch (San Juan de Las Rusias homocline) past Arenque and the east side of the Tuxpan−Golden Lane uplift. It is a major fault downthrown to the east.

The postulated strike−slip faults, La Babia and San Marcos, seem to converge slightly into the Burgos Basin which may have formed as the Yucatan block separated from the western Gulf area. This area subsided dramatically in the ensuing Cretaceous and early Tertiary. The Burgos Basin lies in the projected path of the megashear of Anderson & Schmidt (1983) which might trend into the Gulf south of Brownsville, forming the break along which Yucatan is proposed to have moved out of the western Gulf area. An optional path for the megashear is a southward projection down the axis of the Magiscatzin Basin just in front of the Laramide eastern Sierra Madre.

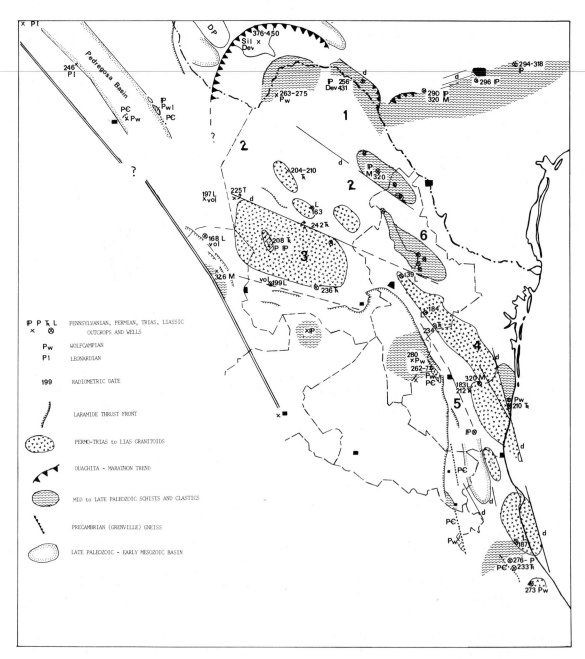

Fig. 2. Late Palaeozoic−early Mesozoic tectonic framework of northern Mexico showing areas of pre−late Jurassic granitic and metasedimentary basement and their radiometric dates.

(1) Burro−Peyotes high; (2) Sabinas Basin; (3) Coahuila block; (4) Tamaulipas arch; (5) Magiscatzin Basin; and (6) Lampazos−Picacho highs.

Table 3. Huizachal–Nazas localities

Southern Chihuahua and Coahuila: no Huizachal at Las Delicias (on Coahuila block)

Ceballo-1: date of 168 Ma (K/Ar) in Jurassic metasediments. Nazas?

Sierra de Diablo (McKee *et al.*, 1988): Welded rhyolite tuffs and ash-fall deposits. Pre-Cretaceous. Nazas Fm? Or older volcanics. 197 Ma (Rb/Sr) late Triassic (Halpern)

Paila-1: igneous intrusive basement 236 Ma Permo-Triassic. Nazas or older

Mayran-1: volcanic–metamorphic complex. 199 Ma (Rb/Sr) Nazas or older

Villa Juarez, Durango, south of Torreon: Coahuila Nazas Fm ≈ 50% volcanics, varying from 35 to 1100 m, not much conglomerate

San Pedro del Gallo; only top of Nazas Fm, 35 m of fan and channel deposits

Catorce, San Luis Potosi: 200 m of silt, shale, slate, reddish and grey–green colour, contains an igneous flow. Lies above metamorphic rocks; arietitids (Arnioceras and Vermiceras) of Sinemurian age reported by Erben (1956)

Matehuala (west of city): 100 m of conglomeratic and non-conglomeratic fan and channel deposits

Caopas–Rodeo (Grunidora) area, Sierra de San Julian: ≈ 500 m of conglomerate, shale, red–green silt of the La Joya Formation. Channels and fans cut into dacite and rhyodacite volcanics. A source area from the north and northeast is indicated

Agua Nueva 1 well: light to dark grey calcareous cemented sandstone, rare grey to reddish shales, sandy with some light grey to red sandstones, bentonites

Guaxcama 1 well: 800 m of medium-grained reddish sandstone and red–green shale

Tampico Area wells: (Imlay *et al.*, 1948 p. 1757, fig. 1), San Manuel 82; Chocoy-1; Altamira, 11; Chijel, 1012 (45 km west of Tampico) has 674 m under Upper Jurassic; Comales 102 (85 km SW of Tampico) has 182 m of redbeds over 395 m of dark shales, silts, with Liassic plants

Galeana area (Rancho Alamar and San Pablo): sandstone and siltstones, distal fan channels and fan plains, 300 m of La Joya Fm., with transport direction from the northwest. Igneous sill of trachyandesite (Michalzik, 1986)

Sierra de la Ventura, Coahuila, Imlay, 1938: Miquihuana area: Huizachal Fm., 300 m (undivided), overlain by Neocomian marly limestone

Ciudad Victoria area: Peregrina Canyon has Huizachal Fm (La Boca Member), 100–130 m fan and channel deposits. Caballero and Novillo Canyons: 150 m of green and grey, red mudstone, sandstone and conglomerate. Late Triassic plants and a skull of Triadon (at University of Chicago)

Huizachal Canyon: Huizachal Fm (La Boca Member) 400+ m, non-conglomeratic channel and inter-fluve deposits (20 km SW of Ciudad Victoria)

San Marcos Canyon: 2150 m of conglomerate and sandstone, forming a fan against faulted side of graben

Guayabas Canyon: 400 m of fan and channel deposits, both conglomerate and non-conglomeratic sandstone

Mesquital area (Rio Blanco) (Aramberri area): up to 125 m thick. Rhyolites and tuffs overlain by red lacustrine sediments and polymictic breccia (Meiburg *et al.*, 1987)

Sabinas Basin: wells with suspected Lias–Triassic sediments. Ines-1, Monclova 5, Pecten-1, conglomerate and sandstone at least 1000 m thick (Gonzales-Garcia, 1984). El Gato-1 has 350 m of basal conglomerate

La Joya Formation: the upper member of the Huizachal Fm is thinner and separated from La Boca Member by an important unconformity. La Joya is considered to be middle–upper Jurassic.

  Huizachal Canyon: 18–50 m, conglomeratic fan deposited into a lake, overlain by a second fan deposit

  Peregrina Canyon: 60–100 m of coarse sandstone

  Novillo Canyon: 70 m debris flow and channel deposits

  Caballero Canyon: 270 m, proximal fan and channel deposits

  Arroyo Seco: 115 m

  Galeana: 200 m, debris flows and channels, braided stream deposits; Pablillo area, 350 m

All these faults are Liassic to middle Jurassic in age. At La Mula, Coahuila, this faulting cuts across a batholith which yields K/Ar isotope ratios indicating 208 Ma and in wells near Monterrey cooling dates on the granites are between 139 and 234 Ma. Faulting appears to cut across solidified crust forming eroded and weathered granite batholiths of varying ages. The intrusive rocks and resulting higher topography are overlapped by sediments from Callovian to Tithonian age, although lower areas preserve older Jurassic to late Triassic redbeds, as for example in the Sabinas Basin, south of the Coahuila block and within and west of the Tampico–Misantla Basin.

BASEMENT CONTROLS ON JURASSIC SEDIMENTATION

General Jurassic stratigraphy of northeast Mexico has been reviewed by Salvador (1987) whose correlation chart is given in Fig. 3. The formations from bottom to top are as follows:

1 Late Triassic–Liassic redbeds and basalts to andesites and rhyolites (Huizachal Formation) formed during inception of rift, are of variable thickness because of deposition and preservation in grabens. They may range up to 2 500 m thick (see above discussion and Table 3). These feldspathic lithic arenites filled troughs south of the Coahuila and Burro–Peyotes uplifts, such as the Sabinas Basin. Some of these sediments should occur on the east flank of the Tamaulipas arch as well.

2 A major rift graben shown by Salvador (1987) ends to the south in a marine channel or aulacogen-like embayment leading to the Pacific, filled with ≈ 1500 m of fine terrigenous clastics of euxinic or deeper-water environment, with ammonites and plant fragments, the Huayacocotla Formation.

3 The top of the Huizachal sequence, the Cahuasas or La Joya Formation, may be the updip redbed equivalent of the Middle Jurassic salt (the Norphlet of Louisiana.) Its base is unconformable over the lower Huizachal (La Boca Formation).

4 Extensive Callovian (Tepexic) to Oxfordian (Santiago–Zuloaga–Smackover) and Kimmeridgian (San Andres Olvido–Haynesville) carbonate facies complexes overlie the salt and redbeds all around the Gulf of Mexico and are > 1000 m thick. The Kimmeridgian basinal mudstone facies is known as Taman and the equivalent San Andres represents an oolitic shoal facies. These transgressive units cover most of the horst blocks caused by

earlier Mesozoic rifting. Facies are strongly influenced by numerous islands which protruded above sea-level (Figs 1 and 5).

5 The Jurassic ends with a blanket of fine-terrigenous clastics equivalent to the Cotton Valley of the northern Gulf coast. These grade up into Lower Cretaceous deposits. The Coahuila block furnished arkosic sandstones of late Jurassic and early Cretaceous age to areas in the Sabinas Basin and to the south in what was to become the Parras Basin. Likewise the Burro–Peyotes uplift provided arkosic sands from its northeast flank. The silt–shale facies grades gulfward to carbonate exactly as the Cotton Valley does in Louisiana. The Kimmeridgian–Tithonian beds in northern Mexico are known as the La Casita or La Caja Formation and in southern Mexico they are termed Pimienta.

Jurassic facies are influenced in several ways by basement high areas:

1 Eroded islands may be surrounded by arkosic debris and redbed strata of various Jurassic ages.

2 Evaporites. Both middle Jurassic (Louann of northern Gulf Coast) and late Jurassic (Buckner equivalent) may exist in northeast and central Mexico. Their distribution helps outline palaeotectonic elements. During times of aridity, evaporites form both at the updip lagoonal margins of carbonate complexes, commonly mixed with fine clastics, but also in tectonically subsident areas, filling basins. In the latter situation, updip equivalents are usually redbeds of arkosic sandstones and fluvial deposits.

Salvador (1987) has outlined the distribution of the earliest Jurassic evaporites (Callovian) in the Gulf of Mexico and discussed the correlations in central and northern Mexico. Callovian evaporites occur in rifted basins within the Gulf of Mexico and as scattered salt domes in the Burgos Basin or Rio Grande embayment. The marine waters which furnished the evaporites were apparently derived from the Pacific and entered the Gulf via the Huayacocotla channel where a thick fossiliferous Liassic sequence exists. The earliest updip equivalent to the Gulf evaporites is the Huehuetepec Formation which overlies the middle Jurassic or late Liassic Cahuasas redbeds.

An unpublished stratigraphic study of Jurassic in eastern Mexico by Sandstrom (1982) has indicated a Kimmeridgian age for carbonates and evaporites, the Olvido Formation, exposed around Galeana and C. Victoria (see also Michalzik, 1988).

The thick Minas Viejas evaporites which form a pocket around Monterrey may also be all or in part

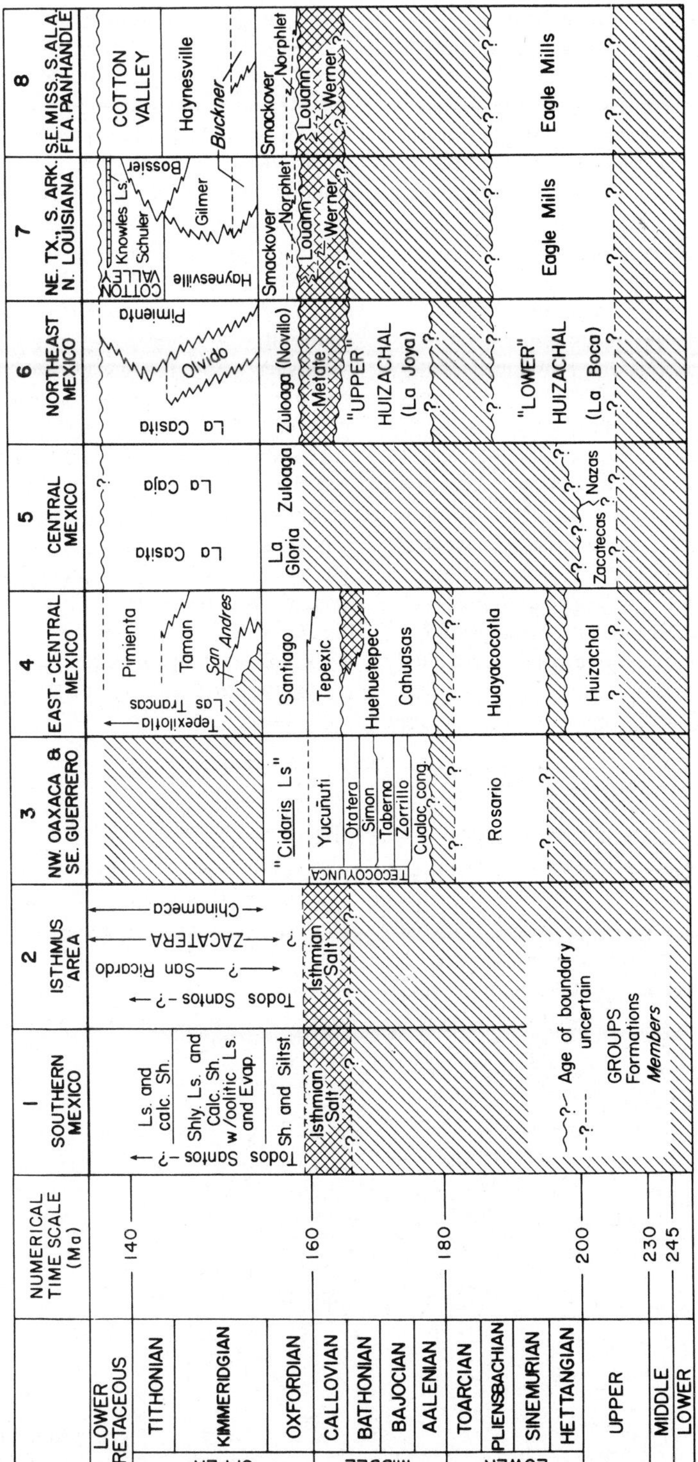

JURASSIC / TRIASSIC	Stage	NUMERICAL TIME SCALE (Ma)	1 SOUTHERN MEXICO	2 ISTHMUS AREA	3 NW OAXACA & SE. GUERRERO	4 EAST-CENTRAL MEXICO	5 CENTRAL MEXICO	6 NORTHEAST MEXICO	7 NE. TX., S. ARK., N. LOUISIANA	8 S.E. MISS., S. ALA., FLA. PANHANDLE
LOWER CRETACEOUS			Ls and calc. Sh.		"Cidaris Ls"	Tepexilotla / Las Trancas	La Casita	La Casita	COTTON VALLEY / Haynesville	COTTON VALLEY / Haynesville
UPPER	TITHONIAN	140	Shly. Ls. and Calc. Sh. w/oolitic Ls. and Evap.	Todos Santos –? Chinameca		Pimienta	La Caja	Pimienta / Olvido	Knowles Ls / Bossier / Gilmer / Schuler	Buckner
UPPER	KIMMERIDGIAN		Sh. and Siltst.	Zacatera / San Ricardo –?	Yucuñuti / Otatera / Simon / Taberna / Zorrillo	Santiago / Taman / San Andres	Zuloaga	Zuloaga (Novillo) / Metate	Smackover / Norphlet	Smackover / Norphlet / Louann / Werner
MIDDLE	OXFORDIAN	160	Isthmian Salt	Isthmian Salt / TECOCOYUNCA	Cualac cong. / TECOCOYUNCA	Tepexic	La Gloria	"UPPER" HUIZACHAL (La Joya)	Louann / Werner	Eagle Mills
MIDDLE	CALLOVIAN				Rosario	Huehuetepec				
MIDDLE	BATHONIAN					Cahuasas	Zacatecas / Nazas			
MIDDLE	BAJOCIAN					Huayacocotla			Eagle Mills	Eagle Mills
MIDDLE	AALENIAN	180						"LOWER" HUIZACHAL (La Boca)		
LOWER	TOARCIAN					Huizachal				
LOWER	PLIENSBACHIAN									
LOWER	SINEMURIAN	200								
LOWER	HETTANGIAN									
TRIASSIC	UPPER	230								
TRIASSIC	MIDDLE	245								
TRIASSIC	LOWER									

GROUPS
Formations
Members
–?– Age of boundary uncertain

Fig. 3. Jurassic correlation chart. From Salvador (1987, fig. 2).

246 *J.L. Wilson*

Kimmeridgian but this is not now determined. The anhydrite fringe of these deposits trends northwest up the Sabinas Basin and the halite facies lies in the basin's southwestern projection, probably filling a trough caused by the San Marcos and the Saltillo−Monterrey lineaments. Salt lies in the vicinity of Monterrey but west of the Lampazos−Picacho high. It may be distinct from, and younger than, the Callovian salt lying in the Burgos Basin to the east. These 'interior' evaporites may be Buckner equivalent and may lie shelfward of extensive oolite shoals completely along the margin of the Gulf.

3 Jurassic carbonate facies are also influenced locally by basement palaeotopography. The earliest shoal calcarenite is the Tepexic Formation which is of Callovian age and occurs locally in central Mexico in the Chicontepec basin. Oxfordian strata consist of Smackover−Zuloaga−Santiago−Novillo Formations which are calcarenitic. In northeastern Mexico a regional study by Oivanki (1974) of the Zuloaga limestone indicated that the widespread Smackover oolite facies is developed around the southern edge of the Coahuila block and to the north of the Valles platform (Fig. 4)

It is not yet clear whether the Zuloaga around the Saltillo−Monterrey area is Oxfordian as traditionally dated or may be in part Kimmeridgian. To the south of Galeana, around Aramberri, two limestones may be seen down the flanks of a local high area separated by redbeds and gypsum. The lower limestone (Meiburg *et al.*, 1987) is thinner, more restricted marine and tidal flat in character, and is presumed to be Oxfordian. The thick upper unit is more oolitic and may be Kimmeridgian. It is overlain by identifiable La Caja−La Casita shales with Kimmeridgian ammonites.

Farther south Kimmeridgian beds (Taman−San Andres−Haynesville) form a facies complex around the Gulf of Mexico in which shoal oolite, calcarenites or patch-reefs occur in shallow water over high areas and dark argillaceous or purely micritic limestone occurs in basinal areas in between complexes of islands and shoal areas (Sansores & Girard, 1969). The southern part of the Tamaulipas arch breaks up into an irregular mosaic of island topography which occurs south to the Tuxpan−Golden Lane high. These islands may be surrounded by oolite (Fig. 5).

On the northern edge of the Tezuitlan massif in

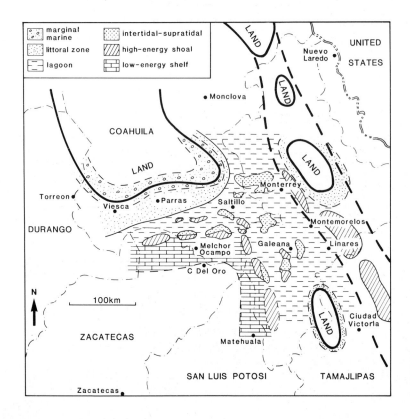

Fig. 4. Oxfordian−Kimmeridgian facies map of Zuloaga Formation in northern Mexico. From Oivanki (1974, fig. 10).

Vera Cruz State and on the southern end of the Golden Lane−Tuxpan block, high topography is particularly reflected in oolite developed in banks of the Kimmeridgian San Andres Formation (see Figs 5 & 6; Gonzales-Garcia, 1969). Such high-energy sediments also occur at the southeastern edge of the Valles platform (Tavitas-Galvan & Solano-Maya, 1984) and on subsidiary ridges in the Chicontepec Basin as illustrated by Pedrazini & Basañez, 1978). These strata have been studied in some detail because of the calcarenitic San Andres reservoir developed in the Constituciones−Tamaulipas and San Andres fields. A major oil discovery, made in the 1960s in the Arenque block offshore of Tampico, contains reservoir rock consisting of richly fossiliferous Kimmeridgian limestone developed over a high area, part of which furnished arkosic sand as well as the development of oolite.

BASEMENT CONTROL OF CRETACEOUS CARBONATE PLATFORM DEVELOPMENT

This discussion draws heavily on a palaeogeographic study by Enos (1983) and earlier work reported by Carrillo-Bravo (1971) and Lopez-Ramos (1972, 1981, 1982). (Refer to figs 4 & 5, Enos, 1983 and Figs 7 & 8 of this paper).

Lower Cretaceous

The Coahuila block controls the trend of the Cupido−Sligo Reef which surrounds it but to the south of the Coahuila block the reef is displaced northward many kilometres by Laramide folding to form a 'dog leg' in the Sierra Madre Oriental (see Figs 9 & 10; Conklin & Moore, 1977; Wilson & Pialli, 1977).

The Cupido−Sligo shelf margin cuts across the mouth of the Sabinas Basin apparently not influenced by Jurassic positive blocks of the Lampazos−Tamaulipas arches. The Sabinas Basin interior to the west continued to subside once the shelf margin was established. It was filled early in the Cretaceous with deltaic sediments derived from the elevated northeast edge of the Coahuila block. Later, during Barremian time the basin was filled with gypsum−anhydrite, the La Virgen Formation.

Gulfward subsidence lowered the Jurassic Tamaulipas−Lampazos arch; from the beginning of Cretaceous time it was no longer a high area. The

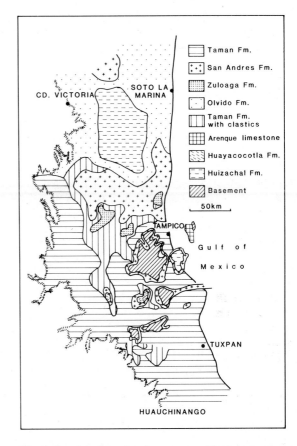

Fig. 5. Pre−late Jurassic sub-crop map with superposed facies of Oxfordian−Kimmeridgian strata in the northern district of Petroleos Mexicanos, southern Tamaulipas arch to Tuxpan−Golden Lane platform. From Sansores & Girard (1969, fig. 5). Shows numerous 'islands' or positive areas with cores of Triassic redbeds (Huizachal) or 'basement rocks' surrounded by Kimmeridgian-restricted marine sediments (Olvido) and grainstone (San Andres). The basinal Taman Formation occurs off the palaeohigh areas.

eastern side of the Tamaulipas−Lampazos arch is underlain by relatively non-resistant schistose metasedimentary strata. This erosion may also have played a role in wearing down the arch in early Cretaceous time. The effect of the arch is seen, however, in a slight thinning of the late Aptian La Peña shale which marks the top of the Lower Cretaceous and buries the Cupido−Taraises carbonate complex.

The Valles platform of early Cretaceous time is poorly known because of sparse control. Pre-reefal

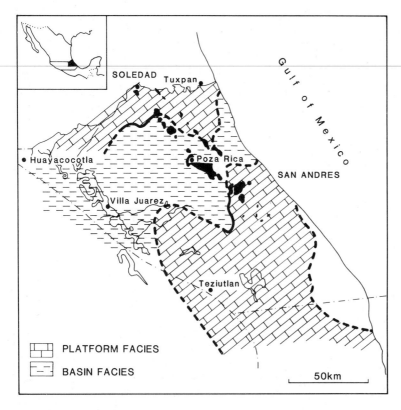

Fig. 6. Upper Jurassic facies in the Poza Rica district of Petroleos Mexicanos showing platform facies of San Andres Formation and basinal Taman facies developed between the Tezuitlan massif on south and Tuxpan-Golden Lane block on the north. From Gonzales-Garcia (1969, fig. 6).

rocks are known in outcrops of Neocomian–Aptian age on the west side of the platform in Sierra de Alvarez, northeast side in the arc of Nahola, and northeast of Jalpan, Queretaro (Suter, 1984, 1987). Tavitas-Galvan & Solano-Maya (1984) indicated three possible Jurassic basement ridges which might exist under the platform; these are now anticlines folded in the Tertiary. The N–S Tanchipa ridge, the most easterly of these, caused formation of some Aptian–Barremian dolomite and limestone which is also seen on outcrops to the north at Miquihuana. This carbonate ridge, and the underlying block-faulted uplift, formed a barrier west of which Aptian–Barremian evaporites were deposited (Fig. 11).

The following is evidence that the Valles platform originated in early Cretaceous time along a N–S line on what is now its eastern margin and in time grew westward and southward.

1 The Miquihuana high ridge and its southern projection along the Tanchipa block correspond with a pre-Cambrian gneiss and Palaeozoic schist basement high which trends along the present front of the

Sierra Madre Oriental (Lopez-Ramos, 1972). This lies west of the Triassic granite of the Tamaulipas arch.

2 The base of the El Abra across the platform is Barremian to the north and Albian to the south and west, according to Suter (1984, 1987).

3 At the eastern margin of the platform the top of the El Abra remained high and was subjected to subaerial exposure until post-Turonian. It was buried only in Campanian–Maastrichtian. Presumably, the block was progressively down-tilted to the west during the late Cretaceous (Smith, 1987).

It is unclear whether the early Cretaceous shallow-water carbonate, cut off a basin to the west and induced evaporite deposition, or whether Lower Cretaceous carbonate development encircled the present position of the Valles platform and formed an atoll whose centre was filled with evaporites. The Guaxcama and Agua Nueva wells contain Lower Cretaceous evaporites thickened by diapirism (Fig. 11; Carillo-Bravo, 1971). The age of the inception of the smaller platforms, El Doctor and Toliman, is not known. Likewise, the early Cretaceous history of

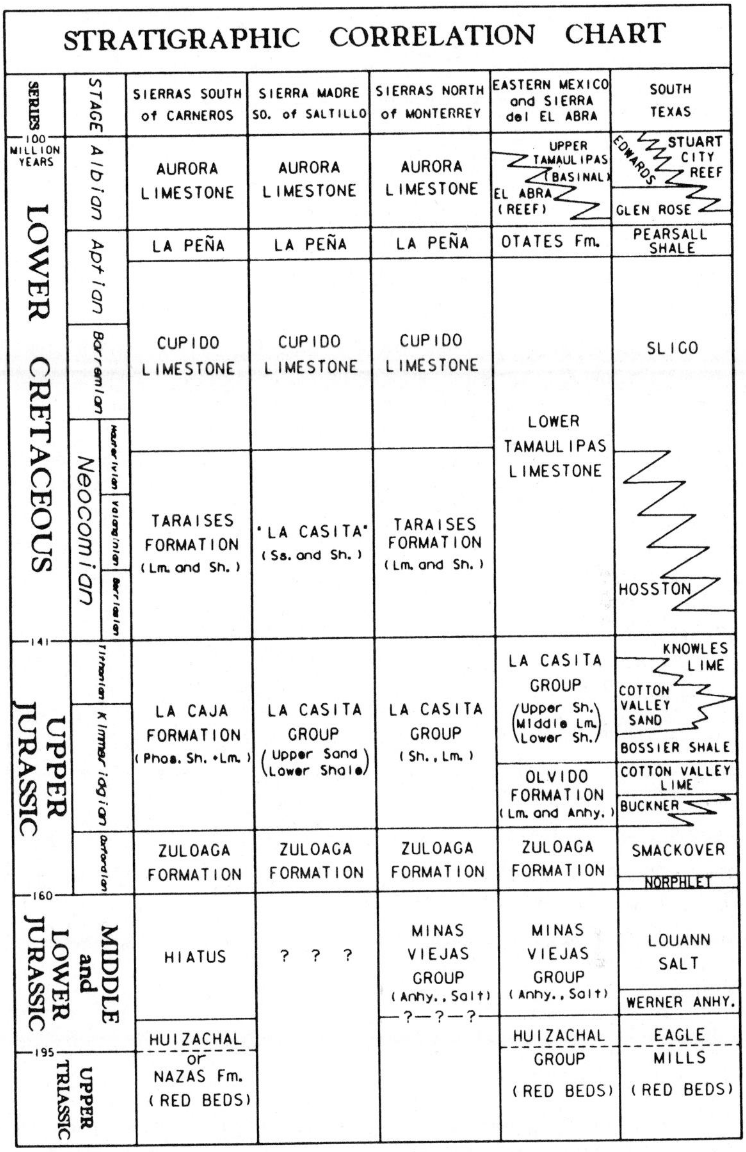

STRATIGRAPHIC CORRELATION CHART

SERIES	STAGE	SIERRAS SOUTH of CARNEROS	SIERRA MADRE SO. of SALTILLO	SIERRAS NORTH of MONTERREY	EASTERN MEXICO and SIERRA del EL ABRA	SOUTH TEXAS
LOWER CRETACEOUS	Albian	AURORA LIMESTONE	AURORA LIMESTONE	AURORA LIMESTONE	UPPER TAMAULIPAS (BASINAL) / EL ABRA (REEF)	EDWARDS / STUART CITY REEF / GLEN ROSE
LOWER CRETACEOUS	Aptian	LA PEÑA	LA PEÑA	LA PEÑA	OTATES Fm.	PEARSALL SHALE
LOWER CRETACEOUS	Barremian	CUPIDO LIMESTONE	CUPIDO LIMESTONE	CUPIDO LIMESTONE	LOWER TAMAULIPAS LIMESTONE	SLIGO
LOWER CRETACEOUS	Neocomian (Hauterivian / Valanginian / Berriasian)	TARAISES FORMATION (Lm. and Sh.)	"LA CASITA" (Ss. and Sh.)	TARAISES FORMATION (Lm. and Sh.)	LOWER TAMAULIPAS LIMESTONE	HOSSTON
UPPER JURASSIC	Tithonian / Kimmeridgian	LA CAJA FORMATION (Phos. Sh. + Lm.)	LA CASITA GROUP (Upper Sand / Lower Shale)	LA CASITA GROUP (Sh., Lm.)	LA CASITA GROUP (Upper Sh. / Middle Lm. / Lower Sh.)	KNOWLES LIME / COTTON VALLEY SAND / BOSSIER SHALE
UPPER JURASSIC					OLVIDO FORMATION (Lm. and Anhy.)	COTTON VALLEY LIME / BUCKNER
UPPER JURASSIC	Oxfordian	ZULOAGA FORMATION	ZULOAGA FORMATION	ZULOAGA FORMATION	ZULOAGA FORMATION	SMACKOVER / NORPHLET
MIDDLE and LOWER JURASSIC	Callovian	HIATUS	? ? ?	MINAS VIEJAS GROUP (Anhy., Salt) ?—?—?	MINAS VIEJAS GROUP (Anhy., Salt)	LOUANN SALT / WERNER ANHY.
MIDDLE and LOWER JURASSIC / UPPER TRIASSIC		HUIZACHAL or NAZAS Fm. (RED BEDS)		HUIZACHAL	HUIZACHAL GROUP (RED BEDS)	EAGLE MILLS (RED BEDS)

Time scale markers (left margin): 100 MILLION YEARS, 141, 160, 195

Fig. 7. Stratigraphic correlation chart of Jurassic and lower Cretaceous in northeastern Mexico. From Wilson *et al.*, 1984 guidebook).

the Golden Lane or Tuxpan platform is not clear. It is built on a granite high with some schist. There may be Lower Cretaceous sediment only on its western margin. It is doubtful whether or not Lower Cretaceous shallow-water sediment exists either here or on the Cordova platform in the Vera Cruz area.

Middle Cretaceous

Continued subsidence caused a carbonate fringe around the whole Gulf of Mexico, and development of major platforms in Mexico occurred at this time (Fig. 8; Enos, 1983). Their facies pattern is well known from many studies of the Cuesta de El Abra near C. Valles, SLP (Fig. 12; Aguayo, 1978; Minero *et al.*, 1983).

There are differences between the Lower and Middle Cretaceous carbonate platforms: Middle Cretaceous platforms are more extensive and have higher relief; rudists were larger in Middle Cretaceous time and form more impressive buildups; coarser slope debris exists in Middle Cretaceous

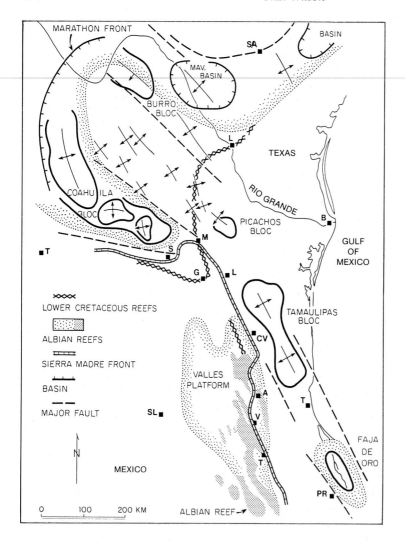

Fig. 8. Early Jurassic tectonic framework of northeastern Mexico — its control on distribution of Lower and Middle Cretaceous facies. From Wilson (1986, fig. 18).

strata; and Middle Cretaceous rocks show less lateral progradation and more aggradation.

The Middle Cretaceous Albian—Cenomanian trends of reefal carbonate with large abundant rudists outline major structural blocks. The belt of rudist facies borders the Coahuila block on the east and encircles the Sabinas Basin, trending down the southwest flank of the Burro—Salado high (Smith, 1981) and also encircles the intra-shelf Maverick Basin, which lies between the buried Quachita—Marathon front to the north and the Albian—Cenomanian Stuart City shelf margin to the south (Fig. 8).

The top of the Coahuila block (Acatita Formation) and the Llano uplift (Kirschberg Gypsum) both

developed sabkha and salinas with sulphate evaporites during Albian time. The East Texas embayment and the Maverick Basin also filled with gypsum although at slightly different times (Ferry Lake and McKnight Formations of Texas).

The Valles platform developed to its maximum extent in Albian—Cenomanian time. The platform grew so rapidly that its relief was 1000 m and its steep sides (up to 43°) furnished coarse debris (Xilitla area, Carrasco-V, 1977). The narrow southern extension of the Valles platform, termed the Actopan platform by Carrasco-V. (1971) was further narrowed by thrusting during Laramide time.

There is a disjuncture in the platform trend where

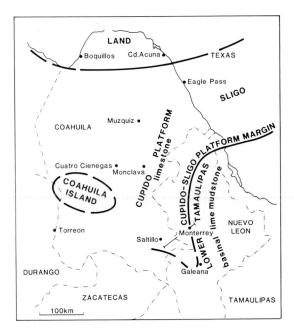

Fig. 9. Facies map of upper Cupido Formation (Lower Cretaceous-Lower Aptian) rimmed platform with trace of rudist reef margin. From Smith (1981, fig. 6). Shows trace of cross-section of Fig. 10.

height and furnished debris (Tamabra) into the east side of the Chicontepec–Mislantla trough just as the southern end of the Valles platform furnished debris into the trough from the west (Fig. 13; Carrasco-V, 1977).

The Valles platform, and perhaps the Golden Lane buildup, continued to develop into the late Cretaceous (Wilson, 1987). Elongate strips of shallow-water carbonate grew in certain places on top of the down-to-west tilted Valles platform. Unconformities developed both before and after Turonian time (Smith, 1987). Smith believed that late Cretaceous tectonism (i.e., faulting and block tilting) encouraged shallow-water carbonate development.

No particular basement lithology controls the distribution of the rudist reefs. Note that the Triassic–Liassic granite of Tamaulipas arch subsided and disappeared as a positive element, whereas the same granite beneath the Coahuila block and the Golden Lane–Tuxpan uplift supported positive areas which induced reef-rimmed Middle Cretaceous shallow platforms to form. This indicates that renewed Cretaceous subsidence and uplift of the early Mesozoic blocks were the most important controlling factors, not differential erosion and development of palaeotopography.

the Transverse Mexican Neovolcanic belt crosses. It separates the Valles platform from the Cordova platform. The author suggests that the presence of the Middle Cretaceous Actopan platform extension, the El Doctor, and the Toliman buildups, may be due to basement fragmentation caused by conjugate faulting off this E–W lineament. Likewise, the Tuxpan, or Golden Lane platform, grew to ≈ 1000 m

CONCLUSIONS

Islands in the late Jurassic transgressive sea, along the Tamaulipas arch, were surrounded by arkosic redbeds and Callovian–Kimmeridgian oolite shoals, forming local petroleum reservoirs. A similar wide shoal area existed between the northern end of the

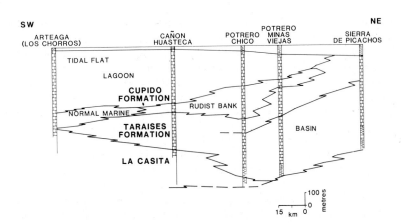

Fig. 10. Cross-section of Cupido Formation in northern Mexico from Saltillo to Monterrey (≈ 80 km). Trace of section shown on Fig. 9. From Wilson *et al.* (1984, fig. 24).

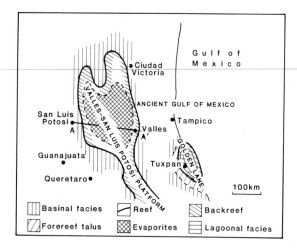

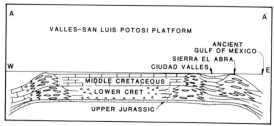

Fig. 11. Map and cross-section of Valles platform in early and middle Cretaceous. From Viniegra-O (1981) and Carrillo-Bravo (1971)

Valles–San Luis Potosi platform and the southern flank of the Coahuila block with much oolitic grainstone development.

Barremian–Aptian (Cupido–Sligo) reef-front loops around the Coahuila platform on its south and then east side, trends across the Sabinas Basin between the granite areas of La Mula and Monclova and the Lampazos high, crosses the La Babia fault and crosses the Rio Grande into Texas near Laredo.

Reef development of early Cretaceous age underlies the northern end of the Valles platform, and perhaps the Golden Lane, but was not developed over the Tamaulipas arch and Cordova platform. Pelagic carbonates developed over them because these areas subsided into the Gulf. This may also be the case across the Golden Lane area.

Albian through Cenomanian rudist reef development reached a maximum after deposition of the late Aptian transgressive shale (La Peña–Otates). The Sabinas Basin and Burro–Peyotes uplift and Coahuila block were bordered by rudist reefs and

the Valles–San Luis Potosi platform prograded from its northeastern corner to form an extensive rudist-rimmed platform whose southern border breaks up into smaller buildups on approaching the Transverse Mexican Neovolcanic belt. The Golden Lane atoll also developed in middle Cretaceous time over the Tuxpan granite uplift. Both Valles–San Luis Potosi and Golden Lane platforms have relief of ≈ 1000 m and have large caprinid rudists toward the shelf margins and extensive coarse debris downslope. These platforms continued building sporadically into the late Cretaceous but were affected by at least two unconformities owing to latest Cenomanian and Turonian sea-level drops.

Many factors control carbonate facies and build-ups, to wit: stages of organic evolution, oceanographic factors, timing and amount of eustatic sea-level fluctuation, and patterns of tectonic activity and subsidence rates. It is apparent that with such complex controls one factor alone may not be a dominant parameter. Nevertheless, it seems clear that palaeotopography resulting from: (1) exposure and weathering contrast between granite and schist in a semi-arid climate; (2) the presence of upfaulted–downfaulted Liassic blocks; and (3) continued gulfward subsidence, controlled both late Jurassic and Cretaceous carbonate facies development.

It seems obvious that since basement structure partially controls the distribution and trend of carbonate platforms, the reverse is also true. The distribution and orientation of carbonate platforms may be used to predict the presence, or absence, of basement blocks on which platforms must have formed.

ACKNOWLEDGEMENTS

The interpretations of basement lineaments, faults and horst blocks, and the sequence of geological events, are the authors opinions, but ideas have been stimulated, and the presentation refined, by discussions with, and critical reading of the manuscript by several helpful people. The author is especially grateful to Thomas H. Anderson, the University of Pittsburg; Zoltan deCserna, UNAM; Patricia W. Dickerson of Midland, Texas; James A. Peterson of USGS; and A.E. Weidie and W.C. Ward of the University of New Orleans who sharpened ideas about geological structure and notably improved the accuracy of the paper. The author is most grateful to Karen Gittins of the

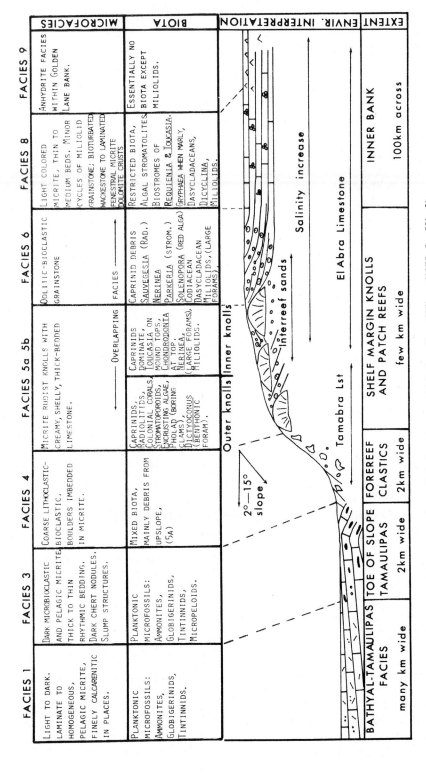

Fig. 12. Idealized middle Cretaceous facies across rimmed platform in central Mexico. From Wilson (1975, figs X1–3).

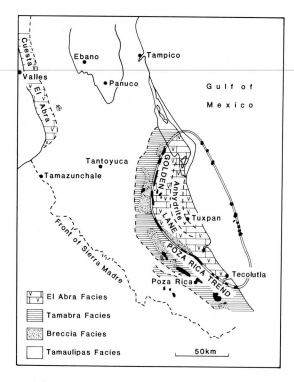

Fig. 13. Oil fields and carbonate facies of middle Cretaceous, eastern Mexico showing the Tuxpan—Golden Lane platform and eastern edge of Valles platform which is marked by the Cuesta de El Abra. From Wilson (1975, figs XI −7).

Department of Geological Sciences, University of Durham, for drafting Figs 4, 5, 6, 9, 10, 11 & 13.

REFERENCES

AGUAYO, J.E. (1978) Sedimentary environment and diagenesis of a Cretaceous reef complex, eastern Mexico: *An. Centro Ciencias del Mar y Limnol. Univ. Nac. Aut. Mex.* **5**, 83−140.

ANDERSON, T.H. & SCHMIDT, V.A. (1983) The Evolution of middle America and the Gulf of Mexico-Caribbean region during Mesozoic time. *Geol. Soc. Am. Bull.* **94**, 941−966.

ALFONSO-ZWANZIGER,J.W. (1978) Geologia regional del sistema sedimentario Cupido. *Bol. Assoc. Mex. Geol. Petroleros* **30** (1 & 2), 1−56.

BELCHER, R.C. (1979) *Depositional environments, palaeomagnetism and tectonic significance of Huizachal redbeds (lower Mesozoic) northeastern Mexico*. PhD Thesis, University of Texas, p. 276.

BLICKWEDE, J.F. (1981) *Stratigraphy and petrology of Triassic (?) "Nazas Formation", Sierra de San Julian, Zacatecas, Mexico*. Thesis, University of New Orleans, p. 93.

CARRASCO-V, B. (1971) Litofacies de la formacion El Abra en plataforma de Actopan, Hgo. *Inst. Mex. Petrolero Rev.* **3**, 5−26.

CARRASCO-V, B. (1977) Albian sedimentation of submarine autochthonous and allochthonous carbonates, east edge of the Valles-San Luis Potosi platform, Mexico. *SEPM Spec. Publ.* **25**, 263−272.

CARRILLO-BRAVO, J. (1971) La Plataforma Valles-San Luis Potosi. *Bol. Assoc. Mex. Geol. Petroleros* **XXIII**, 1−100, *Mex. Petrolero Rev.* **3**, 5−26.

CHARLESTON, S. (1981) A summary of the structural geology and tectonics of the state of Coahuila, Mexico, in Lower Cretaceous stratigraphy and structure, northern Mexico. *West Texas Geol. Soc. Publ.* **81−74**, 28−36.

CONKLIN, J. & MOORE, C. (1977) Environmental analysis of the Lower Cretaceous Cupido Formation, northeast Mexico. In: *Cretaceous Carbonates of Texas and Mexico*. (Eds Bebout, D. & Loucks, R.G.) Univ. Texas Bur. Econ Geol., Rept. Inv. 89, pp. 302−323.

DECSERNA, Z. (1970) Mesozoic sedimentation, magmatic activity, and deformation in northern Mexico. In: *Geologic Framework of the Chihuahua Tectonic Belt*. West Texas Geol. Soc., p. 99−117.

DENISON, R.E., KENNY, G.S., BURKE Jr., W.H. & HETHERINGTON JR., E.A. (1969) Isotopic ages of igneous and metamorphic boulders from the Haymond Formation (Pennsylvanian) Marathon Basin, Texas and their significance. *Geol. Soc. Am. Bull.* **80**, 245−256.

DENISON, R.E., BURKE, W.H., HETHERINGTON, E.A. & ORRO, J.B. (1970) Basement rock framework of parts of Texas, southern New Mexico and northern Mexico. In: *Geologic Framework of the Chihuahua Tectonic Belt*. West Texas Geol. Soc., pp. 3−14.

ENOS, P. (1983) Late Mesozoic paleogeography of Mexico. In: *Mesozoic Paleogeography of the West-Central United States*. Rocky Mountain Section SEPM, Denver, Colorado, Paleogeography Symp. 2, p. 133−158.

ERBEN, H.K. (1956) *El Jurassico Inferior de Mexico*, Contrib. al XX Congreso Geologico Internacional de la Inst. Geol. de UNAM, Mexico DF, p. 47.

GONZALES-GARCIA, R. (1969) *Area con posibilidades de produccion en sedimentos del Jurasico Superior (Caloviano−Titoniano)*. SEM. Explor. Petrolera Mesa Redonda 3 (2), Distrito Poza Rica: Inst. Mex. Petroleo, p. 1−19.

GONZALES-GARCIA, R. (1984) Petroleum exploration in the "Gulf of Sabinas", a new gas province in northern Mexico. In: *A Field Guide to Upper Jurassic and Lower Cretaceous Carbonate Platform and Basin Systems* (Eds Wilson, J.L. *et al.*) Gulf Coast Section SEPM, p. 64−76.

GURSKY, H.J. & RAMIREZ-R.C. (1986) *Notas preliminares sobre el descubrimiento de volcanitas acidas en el Cañon de Caballeros*. Actas Fac. Ciencias Tierra U.A. Nuevo Leon, Linares, Vol. 1, pp. 11−22.

HANDSCHY, J.W., KELLER, G.R. & SMITH, K.J. (1987) The Quachita system in northern Mexico. *Tectonics* **6**.

IMLAY, R.W. (1938) Studies of the Mexican geosyncline. *Bull. Geol. Soc. Am.***49**, 1683−1687.

IMLAY, R.W., CEPEDE, E., ALVAREZ, M. & DIAZ, T. (1948) Stratigraphic relations of certain Jurassic formations of eastern Mexico. *Am. Assoc. Petrol. Geol. Bull.* **32**, 1750–1761.

JONES, N.W., McKEE, J.W., MARQUEZ, D.B., TOVAR, J., LONG, L.E. & LANDON, T.S. (1984) The Mesozoic La Mula Island, Coahuila, Mexico. *Geol. Soc. Am. Bull.* **95**, 1226–1241.

LOPEZ-RAMOS, E. (1972) Estudio del basamento igneo y metamorfico de las Zonas Norte y Poza Rica (entre Nautla, Ver. y Jimenez, Tamps) *Bol. Assoc. Mex. Geol. Petroleros* **24**, 265–323.

LOPEZ-RAMOS, E. (1981) *Geologia de Mexico*: Tomo III (2nd edn.) p. 446.

LOPEZ-RAMOS, E. (1982) Geologia de Mexico: Tomo II (3rd edn) p. 454

McBRIDE, E.F., WEIDIE, A.E., WOLLEBEN, J.A., & LAUDON, R.C. (1974) Stratigraphy and structure of the Parras and La Popa basins, north-eastern Mexico. *Geol. Soc. Am. Bull.* **84**, 1603–1622.

McKEE, J.W., JONES, N.W. & LONG, L.E. (1984) History of recurrent activity along a major fault in northeastern Mexico. *Geology* **12**, 103–107.

McKEE, J.W., JONES, N.W. & ANDERSON, T.H. (1988) Las Delicias basin: a record of late Paleozoic arc volcanism in northeastern Mexico. *Geology* **16**, 37–40.

MEIBURG, P., CHAPA-GUERRERO, J.R., GROTEHUSMANN, I., KUSTUSCH, T., LENTZY, P., LEON-GOMEZ, H. & MANSILLA-TERAN, M.A. (1987) E1 basemento Precretacico de Aramberri-estructura clave para comprender el decollement de la cubierta jurasica/cretacica de la Sierra Madre Oriental, Mexico. *Actas Fac. Ciencias Tierra U.A. Nuevo Leon, Linares* **2**, 15–22.

MICHALZIK, D. (1986) Procedencia y parametros ambientales de los lechos rojos Huizachal en el area de Galeana, Nuevo Leon, Mexico: *Actas Fac. Ciencias Tierra U.A. Nueva Leon, Linares* **1**, 23–41.

MICHALZIK, D. (1988) *Trias bis tiefste Unter-Kreide der Nordostlichen Sierra Madre Oriental, Mexico—Facielle Entwicklung eines passiven Kontinentalrandes*. PhD Techn. Hochschule Darmstadt, DBR, p. 247.

MINERO, C.J., ENOS, P. & AGUAYO, J.E. (1983) *Sedimentation and Diagenesis of Mid-Cretaceous platform margin, East central Mexico*. Field Trip Guidebook, Dallas Geol. Soc., p. 168.

MIXON, R.B. (1963) Geology of the Huizachal redbeds, Ciudad Victoria area, southwestern Tamaulipas. In: *Perigrina Canyon and Sierra de El Abra*. Corpus Christi Geol. Soc. Ann. Field Trip, pp. 24–35.

OIVANKI, S.M. (1974) Paleodepositional environments in the upper Jurassic Zuloaga Formation (Smackover), northeastern Mexico. *Gulf Coast Assoc. Geol. Soc. Trans.* **24**, 258–278.

PADILLA Y SANCHEZ, R.J. (1982) *Geologic evolution of the Sierra Madre Oriental between Linares, Concepcion del Oro, Saltillo, and Monterrey, Mexico*. PhD, Univ. of Texas, p. 217.

PEDRAZINI, C. & BASANEZ, M.A. (1978) Sedimentacion del Jurassico medio-superior en el anticlinorio de Huyacocotla-Cuenca de Chicontepec, Estados de Hidalgo y Vera Cruz, Mexico. *Rev. Inst. Mex. Petroleo* **X**, 6–24.

SALVADOR, A. (1987) Late Triassic–Jurassic paleogeography and origin of Gulf of Mexico basin. *Am. Assoc. Petrol. Geol. Bull.* **71**, 419–451.

SANDSTROM, M. (1982) *Stratigraphy and environments of deposition of the Zuloaga Group, Victoria, Tamaulipas*. Gulf Coast Section, SEPM, 3rd Ann. Res. Conf., Program and Abstracts, p. 94–97.

SANSORES, M.E. & GIRARD, N.R. (1969) Bosqueo Geologico de la Zona Norte, *Inst. Mex. Petroleo, Seminario sobre Exploracion Petrolera. Mesa Redonda* **2**(1), 1–36.

SILVER, L.T. & ANDERSON, T.H. (1974) Possible left-lateral early to middle Mesozoic description of the southwestern North American craton margin (Abs.). *Geol. Soc. Am. Abs.* (with Programs) **6**, 955–956.

SMITH, C.I. (1981) Review of the geologic setting, stratigraphy and facies distribution of the Lower Cretaceous in northern Mexico. In: *Lower Cretaceous Stratigraphy and Structure, Northern Mexico*. West Texas Geol. Soc. Field Trip Guidebook, Publ. **81–74**, p. 1–27.

SMITH, B.A. (1987) *Upper Cretaceous stratigraphy and the Mid-Cenomanian unconformity of east central Mexico*. PhD Thesis, University of Texas, p. 200.

SUTER, M. (1984) Cordilleran deformation along the eastern edge of the Valles-San Luis Potosi carbonate platform, Sierra Madre Oriental fold-thrust belt, east-central Mexico. *Geol. Soc. Am. Bull.* **95**, 1387–1397.

SUTER, M. (1987) Structural traverse across the Sierra Madre Oriental fold-thrust belt in east-central Mexico. *Geol. Soc. Am. Bull.* **98**, 249–264.

TAVITAS-GALVAN, J.E. & SOLANO-MAYA, B.J. (1984) Estudio bioestratigrafico del subsuelo en el oriente de la plataforma de Valles-San Luis Potosi, estados de Tamaulipas y de San Luis Potosi. *Mem. III Congreso Latinoamericano de Paleontologia*, pp. 225–236.

VINIEGRA-O, F. (1981) Great carbonate bank of Yucatan, southern Mexico. *J. Petrol. Geol.* **3**, 247–278.

WILSON, H.H. (1987) The structural evolution of the Golden Lane, Tampico embayment. *J. Petrol. Geol.* **10**(1), 5–40.

WILSON, J.L. (1975) *Carbonate Facies in Geologic History*, Springer Verlag, New York, p. 471.

WILSON, J.L. (1986) Tectonic control of carbonate platforms. In: *Carbonate Depositional Environments, Pt. 2* (Eds Warme, J.E. & Shanley, K.W.) Colorado School Mines Quarterly, Vol 80, pp. 9–29.

WILSON, J.L. & PIALLI, G. (1977) A Lower Cretaceous shelf margin in northern Mexico. In: *Cretaceous Carbonates of Texas and Mexico* (Eds Bebout, D.G. & Loucks, R.) University Texas Bur. Econ. Geol. Rept. Inv. 89, pp. 286–294.

WILSON, J.L., WARD, W.C. & FINNERAN, J. (1984) *A field guide to Upper Jurassic and Lower Cretaceous carbonate platform and basin systems, Monterrey-Saltillo area, northeast Mexico*. Gulf Coast Section SEPM, p. 76.

Spec. Publs int. Ass. Sediment. (1990) **9**, 257–290

The Aptian–Albian carbonate episode of the Basque–Cantabrian Basin (northern Spain): general characteristics, controls and evolution

J. GARCÍA-MONDÉJAR

Departamento de Estratigrafía, Geodinámica y Paleontología, Universidad del País Vasco, 48080 Bilbao, Spain

ABSTRACT

The Mesozoic and Caenozoic sedimentary and tectonic history of the Basque–Cantabrian region of northern Spain was linked to the appearance and evolution of the Bay of Biscay. During the Aptian and Albian the Urgonian Complex of > 4000 m of rudistid limestones accumulated in places and shows rapid lateral facies changes into siliciclastic sediments. The influence of a warm and wet climate is indicated by the occurrence of a reefal fauna and local lignite deposits. Sedimentary environments were of the following types: fluvial, coastal plain, deltaic and siliciclastic shelf, carbonate platform, platform margin, carbonate talus and basin, and terrigenous talus. The Aptian section is characterized by only small bathymetric differences across the basin areas. The succeeding Albian carbonates are composed of thick platform and isolated bank limestones containing large reef mounds. Along some straight bank margins megabreccias were deposited as the result of tectonism and probable eustatic changes of sea-level. Carbonate deposition was interrupted by two periods of terrigenous deposition, which were widespread over the basin and occurred in the middle and the late Albian.

Four depositional sequences related to tectonic pulses and relative sea-level changes are identified. They are bounded by lowstands in the earliest Aptian, middle Aptian, early Albian, middle Albian and late Albian. Some of the lowstands coincide with terrigenous influx and the deposition of megabreccias and can be considered influenced by eustasy.

Tectonism was the most important factor controlling carbonate facies distribution. Halokinesis exerted a local control and block faulting caused differential subsidence. Limestone deposition was dominant over the highs and siliciclastic deposition occurred in the troughs. The NW–SE and SW–NE orientation of troughs and highs is interpreted to be the result of extensive movements related to the opening of the Bay of Biscay, with the major bounding faults probably coincident with basement faults.

INTRODUCTION

The Basque–Cantabrian region lies in the north of Spain between the Pyrenees and the Cantabrian Mountains (Fig. 1). It is made up of a thick sequence of Mesozoic–early Tertiary sedimentary rocks, formed as a consequence of the rifting and opening of the Bay of Biscay. More than 15 000 m of sediment were deposited from the beginning of the Triassic until the late Eocene, when the first compressional movements of the Pyrenean orogeny ended the subsidence that had prevailed during the Mesozoic. The Triassic is composed of thick units of red fluvial clastics overlain by claystones, evaporites and some limestones. The Lower–Middle Jurassic consists of platform carbonates. The Upper Jurassic and the Cretaceous are represented by a thick sequence of sediments, ranging from coarse alluvial clastics to reef carbonates and deep-sea turbidites. The pre-orogenic Tertiary consists mainly of deep-water turbidites and platform carbonates or siliciclastics.

Recent contributions that broadly cover the Basque–Cantabrian Basin are those of Rat (1959), Ciry *et al.* (1967), Brinkman & Lögters (1968), Feuillée & Rat (1971), Ramírez del Pozo (1971), Wiedmann (1979), Soler *et al.* (1981), Rat *et al.* (1983), Wiedmann *et al.* (1983) and García-Mondéjar *et al.* (1985a, b).

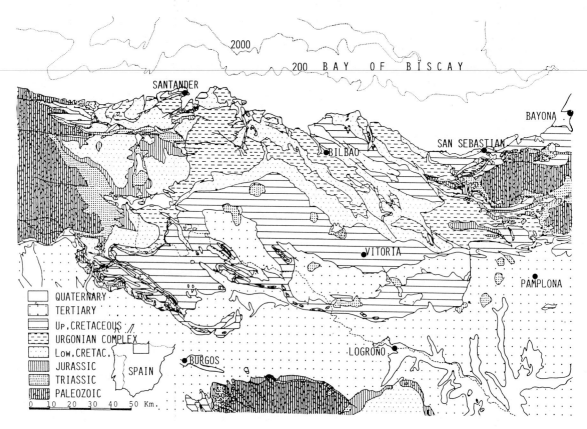

Fig. 1. Simplified geological map of the Basque−Cantabrian region, showing areas of outcrop of the Aptian−Albian, Urgonian Complex.

During the Mesozoic and early Tertiary the basin was subjected to a general extensional regime. Differential subsidence in fault-bounded areas took place during the early Triassic and during the late Jurassic−early Cretaceous (García-Mondéjar, 1979; García-Mondéjar *et al.*, 1985, 1986; Pascal *et al.*, 1976; Pujalte, 1981; Rat *et al.*, 1983; Wiedmann *et al.*, 1983). Three short-lived episodes of minor compression occurred at the end of the Triassic (García-Mondéjar *et al.*, 1986), in the Neocomian (Pujalte, 1981), and in the Albian (Rat, 1959; García-Mondéjar, 1979). The Albian movements formed a flysch trough in the northeast part of the basin, which deepened and widened during the late Cretaceous and persisted into the early Tertiary. Basaltic volcanism occurred during the late Cretaceous (Rossy *et al.*, 1979; Mathey, 1982).

The Upper Jurassic−Lower Cretaceous succession forms more than half of the total thickness of sediments in the basin. This interval has traditionally been subdivided into three main units or 'complexes' (Rat, 1959); the 'Wealden', 'Urgonian' and 'Supra-Urgonian' successions. The Urgonian is the subject of this paper and consists mainly of Aptian and Albian rudistid limestones. It followed a predominantly continental clastic phase of sedimentation (Wealden) and was succeeded by another clastic episode, ranging from fluvial to turbidite in origin (Supra-Urgonian). The Urgonian was deposited during the period when rifting gave way to spreading in the Bay of Biscay (Montadert *et al.*, 1979).

This paper deals firstly with the Urgonian sedimentary environments and facies, which include fluvial, coastal, marine platform (siliciclastic and carbonate), deltaic, reef, talus and intra-platform basin. Unconformities and their correlative conformities within the Urgonian define four major depositional sequences. A palaeogeographic map representative of each of them is included, showing the broad patterns of sedimentation that prevailed during the Aptian and Albian times. Finally, a discussion of the controls of the Urgonian facies sedimentation is presented. An appendix with fossils which characterize chronology of the sequences is included.

URGONIAN FACIES AND ENVIRONMENTS

The Urgonian Complex of the Basque—Cantabrian region was first defined by Rat (1959) as an 'extremely thick sedimentary complex of the Cretaceous including *Toucasia* limestones'. The discovery of several unconformities within the unit permitted the recognition of megacycles (García-Mondéjar, 1979). Recent major works specifically devoted to the carbonate episode are the dissertations of Rat (1959), García-Mondéjar (1979), Pascal (1984) and Fernández-Mendiola (1986). According to Rat (1959) the term Urgonian refers to a characteristic type of facies: 'massive limestones with different external aspect and microfacies, but with a common lack of terrigenous elements, important calcite cement and reef-building organisms, strong recrystallization, and rudists as the most characteristic biofacies, mainly of the *Toucasia* genus'.

The age of the Urgonian Complex was determined by Rat (1959) to be Aptian—Lower Albian, and later studies have extended it up to the Middle and Upper Albian (Duvernois *et al.*, 1972; García-Mondéjar & Pascal, 1978; Pascal, 1984; Fernández-Mendiola, 1986). As useful specimens of ammonites or planktonic foraminifera are scarce, dating of the Complex is based on several assemblage zones of different taxa and on range zones of different benthic foraminifera, mainly orbitolinids. From Middle Albian onwards, ammonites are more common so that a more precise biozonation of the uppermost Urgonian is then possible (Wiedmann, 1965).

The Urgonian Complex is > 4000 m thick and is made up of a succession of rudistid limestones, marls, mudstones, sandstones, breccias and conglomerates. Rudists are mostly requienid and caprotinid-types, and particularly from the genera *Toucasia* and *Polyconites*. Some other typical Urgonian fossils are corals, orbitolinids, ostreids, the oyster-like *Chondrodonta*, calcareous algae and monopleurid and radiolitid rudists.

The Urgonian crops out extensively in the northern part of the basin (Fig. 1), and consists of thick units of limestone with little lateral continuity, resting on palaeogeographic highs. The terrigenous sediments interbedded with the limestones were derived from Palaeozoic highlands on the margins of the basin.

The tectonic setting of the Urgonian Complex was an epicratonic basin bordering a proto-oceanic rift. In the basin, sedimentary environments ranged from continental to that of a relatively deep-marine intraplatform trough. Carbonate bioherms of the Urgonian Complex are mud- and reef-mounds like those occurring on margins of other Albian-aged carbonate platforms and banks. Megabreccias and other resedimented deposits are found adjacent to the bank margins; and they pass laterally into basinal black marls and mudstones with variable amounts of thin-bedded limestone or sandstone.

The descriptions that follow are mostly based on stratigraphic cross-sections from different parts of the region. The geographic locations of these sections are shown on the map in Fig. 2. A synthesis of the main Urgonian lithologies and sedimentary environments is shown in the schematic cross-section of Fig. 3. Reference to it will be made during the description of the particular Urgonian facies and environments.

Fluvial

Terrigenous sediments of fluvial origin are extensively represented in the southern outcrops of the region. Conglomerates of distal alluvial fan, channel sandstones of braided and meandering rivers, and mudstones of floodplain or abandoned channels have all been identified (García-Mondéjar, 1979). Conglomerates are siliceous (quartzites, quartz and chert) and have sandy matrix. They are clast- or matrix-supported, fill braided channels, and commonly show cross-stratified sets from migrating gravel bars (Fig. 4). Sandstones are predominantly quartz-arenites and subarkoses. They have abundant cross-stratified sets from migrating megaripples which are formed in braided channels. Less commonly, the sandstones contain lateral accretion surfaces from migrating point-bars in meanders. Mudstones occur between sandstone bodies, and are red or grey, without evidence of subaerial exposure. Transported wood fragments, which are common in the lowermost parts of the sandstone bodies, are the only fossils present. Thinning- and fining-upwards fluvial sequences are ubiquitous, particularly those of braided channel origin.

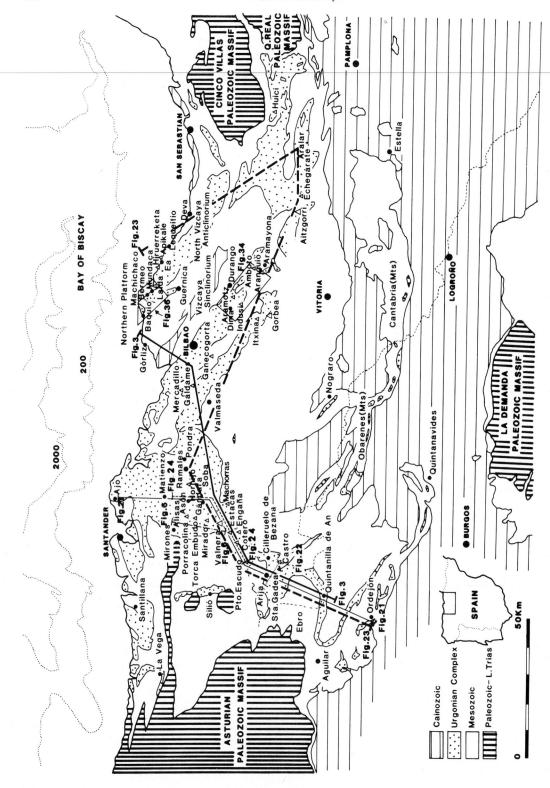

Fig. 2. Location map of cross-sections, block diagrams and points used in the text descriptions. The Aptian–Albian (Urgonian Complex) is differentiated within the Mesozoic.

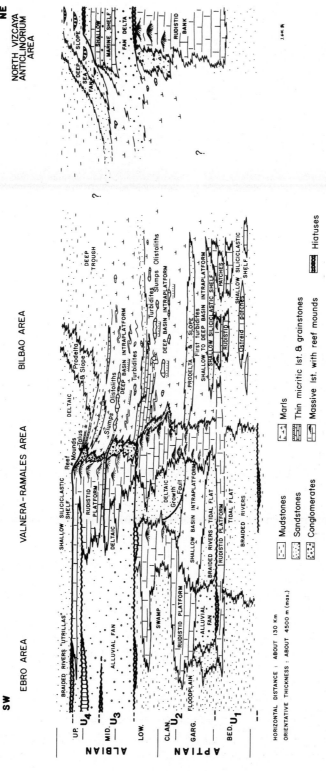

Fig. 3. Lithofacies assemblages and sedimentary environments in a transverse (SW–NE) and schematic cross-section. Adapted from García-Mondéjar (1985a, b).

Fig. 4. Quartz conglomerates and sandstones filling proximal braided fluvial channels. Quintanilla de An Formation (Ebro sedimentary sector). Hammer in the middle of the picture for scale.

Fig. 5. Coastal plain lignite-bearing mudstones and channel sandstones from the Arija area (Arija–Nograro sedimentary sector). Two fluvial cycles, both together ≈ 8 m thick, are shown in the picture.

Alluvial coastal plain

Close to the area of transition to platform limestones, the fluvial mudstones are very rich in remains of small plants preserved *in situ*. In some places, the original accumulation of vegetation has been converted into lignite (Fig. 5). Small sand-filled channels are interpreted to have meandered on a low-lying, coastal area, covered with a dense and herbaceous vegetation (marshes and swamps). In some areas of the coastal plain with a lower subsidence rate, channel deposits have a higher relative proportion, and flood basin deposits are much reduced in thickness.

Marine deposits appear at several intervals in the continental section. They consist of sandy marls and cross-stratified, bioclastic grainstones, which are attributed to subtidal and intertidal environments, respectively (García-Mondéjar, 1979).

Deltaic and siliciclastic shelf

The Urgonian sea received vast amounts of terrigenous materials through deltas of various types. In the Aptian, the delta-estuary type was more common, with well-sorted sands reworked in areas subjected to strong tidal currents. The best known example of this type is the Rio Miera delta (García-Mondéjar, 1979), which transported sand from the west to the centre of the basin. The resultant deposit, the Rio Miera Sandstone Formation, is lenticular in a transverse section and changes facies laterally into limestone (Fig. 6). It consists mainly of planar and trough cross-stratified sandstone, with evidence of

tidal influence including neap–spring cycles (bundle sequences), absence of fine sediment, cross-stratification with juxtaposed heterometric sets (backsets) (Fig. 7), and abundant reworked marine fossils (García-Mondéjar & Pujalte, 1981).

In the Albian, fan deltas were more common and coexisted with the delta-estuary type. They introduced conglomerates directly into the sea and were caused by high gradients related to highlands near the area of sedimentation. A large fan delta in the area of Valnera (Fig. 3) interdigitated with limestones and produced Gilbert-type lobes (García-Mondéjar, 1979). Another fan delta in the North Vizcaya Anticlinorium (Fig. 3) was subject to extensive reworking by waves (Robles *et al.*, 1988).

Wide terrigenous platforms occurred in the region only during the early Aptian and the early Upper Aptian. They were dominated by muds with a normal marine fauna (echinoids, orbitolinas, various bivalves, ammonites), and by sands sourced from stream mouth bars and reworked by tidal currents (García-Mondéjar & García-Pascual, 1982; García-Garmilla, 1987). Ostreid patches and bioclastic grainstones developed locally, and were followed by rudistid patches in a general process of replacement of the terrigenous platform by a carbonate one (Fig. 3).

Carbonate platform

Urgonian limestones are mostly micritic and contain a characteristic fossil association. Large biostromes

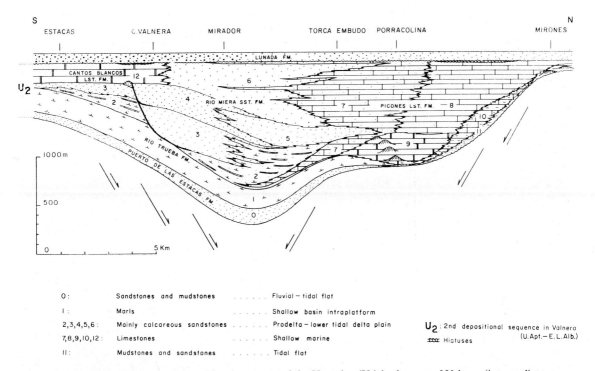

Fig. 6. Cross-section of the second depositional sequence of the Urgonian (U_2) in the area of Valnera (intermediate zone between the Arija—Nograro and Ramales sedimentary sectors). Deltaic—estuarine facies of the Rio Miere Sandstones Formation are shown in lateral transition to rudistid limestones. Numbers 0—12 represent the lithofacies identified. Modified from García-Mondéjar (1979).

Fig. 7. Detail of the Rio Miera deltaic—estuarine sandstones. Large-scale cross-stratification dipping to the right contain small-scale cross-strata dipping to the left (backsets). The reversal of flow direction indicated by these structures is interpreted as due to tidal action.

of rudists of the requienid type are the most common facies (Fig. 8) and represent the best environmental conditions under which carbonate platforms and isolated banks developed. Caprotinid rudists are abundant in places. Radiolitids, monopleurids and caprinids are much less common. Branching and massive corals occur with the rudists in platform facies, as well as various bivalves, gastropods, echinoids, the large foraminifera *Orbitolina* and several types of algae. The relative abundance of *Codiaceae*, *Dasycladaceae* and articulated coralline algae, gives a palaeobathymetry of <25 m for the Urgonian platforms, when compared with present day counterparts (Wray, 1977). Massive corals of laminar type are very abundant in the outer platforms (Fig. 9). They normally preceded the appearance of rudists during transgressions and formed wide bioherms. The oyster-like *Chondrodonta*, finally, was locally an important builder which constructed biostromes (Fig. 10) or bioherms near bank margins. Apart from the characteristic micritic facies (normally wackestones or floatstones), other common rock-types are packstones and grainstones. The Urgonian platforms and isolated banks must have

Fig. 8. Detail of a rudistid biotrome (Requienids, *Toucasia*) partially replaced by siderite, from the limestones of the Galdames sedimentary sector (late Lower Aptian).

been restricted to some degree during their formation, at least bathymetrically, as no fossils of open waters, like ammonites, are normally found within their facies. The carbonate platforms are never very wide, because of the deposition of terrigenous material from the south and the existence of many intra-platform troughs with basinal sediments (Fig. 3).

Platform margin

Urgonian limestone platforms change laterally to deeper facies at their outer margins. In some instances the change is gradual, with relatively gentle slopes, and in others it is more abrupt, with steep slopes (Fig. 3). In the first case, the slopes lack resedimented facies; limestone beds formed on them show a decrease in thickness as they became interbedded with deeper-water sediments. In examples with a high-energy deeper-water environment adjacent to the margin, the micritic beds change to packstones and grainstones and finally to sandstones (Fig. 11). In cases with a low-energy deeper-water environment nearby, the change is to thin-bedded impure limestones alternating with marls, with scarce metazoa (usually siliceous sponges). The transitional beds are then packstones—wackestones or coral-rich limestone tongues. Gently sloping margins were formed preferentially during the Aptian (Fig. 3).

Fig. 9. Upper Aptian coral facies from the Mundaca area northwest of Guernica sedimentary sector). Massive, planar-type corals dominate and a branching coral in living position is seen in the lower part of the photograph (top of the section to the left).

Platform margins with steep slopes were the sites where Urgonian reefs formed. These reefs are of mud mound and reef mound type (James, 1978) and show foreslopes with original dips >45° (García-Mondéjar, 1979). The origin and maintenance of the abrupt break at the platform edge can be related to synsedimentary tectonism, which created important differences in bathymetry probably because of differential subsidence in basement blocks (Fig. 12). The best exposed mud mound of the regin is at La Gándara (Fig. 13) (Rat, 1959; Pascal, 1974, 1984; García-Mondéjar, 1979, 1985a). It is lenticular in a section parallel to the platform margin and has a massive core, ≈ 70 m thick. Like other large carbonate mounds of the Basque—Cantabrian region, it is composed of micrite with scattered remains of metazoa such as rudists, corals or brachiopods, and

Fig. 10. Upper Aptian biostrome with the oyster-like bivalve *Chondrodonta* (top of the section to the left). Aralar sedimentary sector, Huici area.

abundant sponge spicules and *Bacinella*-type algal structures (Fig. 14). The flanking beds consist mainly of detrital fragments of rudists, massive and branching corals, echinoids, acanthochaetetids, brachiopods, pectinids and orbitolinas (Fig. 15). The actual mound structure of La Gándara suggests two different phases of construction: one interpreted to occur during a rapid relative rise in sea-level, consisting mostly of biologically-induced precipitation of micrite by algae below wave base and now forming the mound core. A second phase is interpreted to occur during a relative sea-level stillstand or fall and resulted in deposition of the flanking beds from detritus of the colonizing metazoa on the mound crest (García-Mondéjar, in press). The mound core and most of the flank beds, therefore, would not have been exactly synchronous, but representative of a pulse of rapid transgression followed by a slower regression. The relief of the clinoforms in a section parallel to this platform margin suggests the maximum palaeobathymetric difference between the mound crest and the base of the flank beds was ≈ 50 m (Fig. 13).

Platform margins with large reef mounds and steep foreslopes are characteristic of the Albian (Fig. 3). Some of them, affected by strong syn-sedimentary tectonism, became erosive margins, and created aprons of limestone blocks in the adjacent talus (Figs 16, 17 & 18). Alternatively, relative sea-level falls may have contributed in the same way to the formation of limestone megabreccias.

Fig. 11. Platform margin in the Asón area (Ramales sedimentary sector). A rapid lateral change from micritic, rudistid beds to packstones—grainstones and sandstones took place through a gentle slope. View from the east.

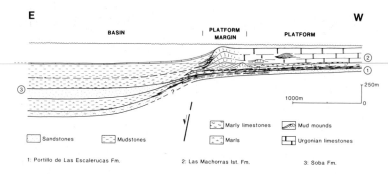

Fig. 12. Sedimentary model of platform-basin in the area between Valnera and Soba (Ramales sedimentary sector). The platform margin was typified by the upbuilding of mud mounds and was emplaced at the edge of a zone with a relative low rate of subsidence. Cyclic sedimentation suggests relative sea-level oscillations.

Fig. 13. 'La Gándara' mud/reef mound in the platform margin west of the Soba trough (Ramales sedimentary sector). The view is from the east and the mound is ≈ 70 m thick.

Carbonate talus and basin

Mass-transport deposits typify much of the talus adjacent to platforms where mud and reef mounds were thriving (Fig. 3). The autochthonous deposits formed off the margins of the carbonate platforms are characterized by a scarcity of metazoa, abundance of fine-terrigenous material and inclusion of finely-disseminated organic matter. Slumps, large olistoliths torn away from foreslopes, and several kinds of limestone breccia and megabreccia are the most common resedimented deposits in the talus. Megabreccias were formed where erosional processes were related to tectonism and possibly eustatic sea-level variations affected the platform margins. One of the best exposed megabreccias is in Baquio, at the coast in the north of the region. They exhibit a general thinning- and fining-upwards sequence (Figs 16 & 17) and show their largest clast size (> 50 m

across) near the lower erosional surface. Rockfall, modified grain flow, debris flow and concentrated turbidite current are the processes thought to have operated in a steep slope apron, at the foot of a carbonate margin which was being eroded (García-Mondéjar & Robador, 1984). An example of similar megabreccias longitudinally filling a trough occurs in Ea (North Vizcaya Anticlinorium) (Fig. 18; Agirrezabala & García-Mondéjar, in press).

Talus environments with a predominance of autochthonous deposits consist commonly of a rhythmic alternation of micritic limestone and marl (Fig. 19). These lack the typical Urgonian fossils but have in places a relative abundance of siliceous sponges. They normally appear thoroughly bioturbated and, in many places show slump beds (Fig. 19).

Basinal areas away from platform margins and

Fig. 14. Coral wackestones. Carbonate mound core ('La Gándara'). Micritic clotted matrix dominates. Some scattered floating metazoa (branching corals) lay parallel to the former seafloor. Irregular micritic lamination (algal?) on top of the lowermost specimen. Sponge spicules are common in the carbonate mud.

Fig. 15. Foraminiferal packstone. Carbonate mound flank ('La Gándara'). Densely packed fabric. Main bioclastic constituents are: orbitolinids (O), rudist fragments (R), and coral fragments (C). Bedding is shown by the horizontal alignment of major rudist bioclasts.

Fig. 16. Thick *ortho*-megabreccia with Urgonian limestone clasts up to 15 m from the Baquio area. Slope apron corresponding to the northern platform sedimentary sector. Height of the cliff: 50 m. People in the foreground give scale.

Fig. 17. *Ortho* and *para*-megabreccias with Urgonian limestone clasts and some calcareous turbidites alternating with black mudstones, from the Baquio area. They show no vertical organization and are attributed to a distal slope apron adjacent to the Northern Platform sector. Geologist at the bottom of the cliff for scale.

slope aprons were the preferential sites for marl accumulation (Figs 3 & 20). Mudstones and turbiditic sandstones or grainstones were intercalated at times within marls. All these sediments lack autochthonous shallow-water Urgonian reef-building organisms. The absence of light-dependent fossils suggests deposition below the local photic zone, which, according to bathymetric estimates made from the La Gándara mound (Fig. 12) as well as others, may have been in at least 100 m water depth.

The characteristic dark or black colours of most of the basinal sediments indicate a reducing environment on the seafloor. This may also have been the reason for the lack of benthic fossils that characterizes other intervals. Detailed palaeontological and geochemical studies of these facies have been done by Magniez & Rat (1972) and Pascal (1984). The presence of ammonites and other open-marine fossils suggests connection with adjacent oceanic areas.

Terrigenous talus

The upper part of the Urgonian Complex includes terrigenous units of talus slope deposition. They consist mainly of grey and black mudstones and calcareous or siliceous sandstones, but some are siliceous conglomerates. All units include mass-transport deposits, such as slumps, debrites or turbidites; all of them contain abundant plant debris and locally ammonites. Mudstones — and marls — are both bioturbated or interlaminated with sands; they usually contain ferruginous nodules and thin-bedded, sandy turbidites.

Sandstones appear within the mudstone units. They are lenticular bodies 100s to a few 1000s of metres wide and up to a few 10s of metres thick. Sharp bases and turbidite features are rare in these beds, and an origin of distal mouth bars of large deltaic distributaries has been proposed (Fig. 18; Agirrezabala & García-Mondéjar, in press). In other parts of the region (Soba Basin for instance, Fig. 12), there are amalgamated beds of fine-grained turbidites, filling channels 100 m wide and up to 15 m deep. The origin of the sandstones in this case is thought to be intermittent deposition from turbidity currents, within channels connected upslope to deltas.

Conglomerates fill channels excavated in mudstones and other talus deposits. They are attributed to either highly concentrated turbidity currents or debris flows, developed in talus slopes close to feeder fan-deltas (Fig. 18; Agirrezabala & García-Mondéjar, in press). The formation of the terrigenous talus units was contemporaneous with deposition of carbonates on Urgonian highs, so that some isolated limestone clasts or breccias from these highs are locally included within the terrigenous units (Fig. 18).

DEPOSITIONAL SEQUENCES

The Urgonian of the western part of the Basque–Cantabrian region has been shown to have a cyclic

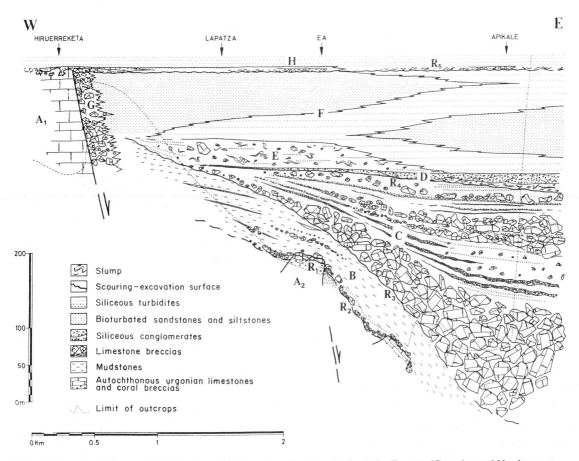

Fig. 18. Urgonian limestone megabreccias and terrigenous talus deposits from the Ea area (Guernica and Northern Platform sedimentary sectors). The slope apron megabreccias of unit C filled longitudinally (in N–S direction) the Ea trough. A–H: sedimentary units. R_1–R_5: local sedimentary breaks. After Agirrezabala & García-Mondéjar (in press).

Fig. 19. Talus deposits consisting predominantly of autochthonous micritic limestones alternating with marls. A slump bed is seen in the centre of the picture. Somorrostro area, to the northwest of Bilbao.

Fig. 20. Monotonous black marls with a slump fold 5 m thick. Basinal sediments of the Soba area (west of Bilbao sedimentary sector).

organization, based on unconformity surfaces and alternations of terrigenous and carbonate formations (García-Mondéjar & Pujalte, 1977; García-Mondéjar, 1979) (Fig. 21). The cyclicity has been attributed to repeated tectonic pulses in the basin and referred to as depositional sequences in the sense of Mitchum *et al.* (1977), (García-Mondéjar & Pujalte, 1981). An example of some of the unconformities present in the southwestern part of the region, used as limits of sequences, is shown in Fig. 22.

Recent studies have traced the unconformities and their correlative conformities over much of the Basque−Cantabrian region, and four main cycles or depositional sequences have been identified (U_1 to U_4, Fig. 23). Each sequence is normally made up of a lower terrigenous unit, in places conglomeratic and continental, and an upper fine-grained terrigenous or marine carbonate unit. The complete cycle suggests a transgressive episode followed by a regressive one. The top regressive parts of the cycles are locally marked by significant erosional surfaces.

U_1 sequence

The first Urgonian sequence, U_1, is of early Aptian age. It is best established in the western part of the region because of the presence of two boundary unconformities, locally angular. The lower limit has been identified as an erosional surface, locally an angular unconformity, between two distinct fluvial units in the area of Valnera (García-Mondéjar & Pujalte, 1975). In the rest of the region the correlative conformity is considered to be the base of the Lower Aptian transgressive deposits. The upper limit is an angular unconformity in the southwestern part of the region, either between fluvial and fluvial sediments or between marine and fluvial sediments

(top of U_1 in Fig. 22). It is also a sharp boundary between shallow platform limestones and fluvial sandstones in the southwestern part of the region (García-Mondéjar, 1979) (Fig. 24 top of U_1). It is a sharp boundary in the rest of the region, locally erosional with palaeochannel morphologies, separating shallow-water platform limestones from marine sandstones, dark mudstones or marls. A field sketch of one of these palaeochannel morphologies, taken from the central part of the region close to Bilbao, is shown in Fig. 25 (García-Mondéjar & García-Pascual, 1982). Draping marls resting on the erosional surface are considered to have been deposited in deeper water than the underlying Urgonian limestones, because of the open-sea conditions suggested by their constituent ammonites, belemnites, echinoids and ostreids, and because of the presence of resedimented deposits (talus limestone breccias) in a nearby outcrop.

U_2 sequence

The second sequence, U_2, of the Upper Aptian−Lowermost Albian, is defined by two quite noticeable breaks in the sedimentation. The lower sequence boundary is the top of sequence U_1. The upper sequence boundary, which can be dated approximately at the base of the foraminifera zone *Simplorbitolina conulus* Schroed., is represented in different places by an angular unconformity or a disconformity showing palaeokarst features and dolomitization. Gorliz, in the northwestern part of the North Vizcaya Anticlinorium, is one of the places where the paleokarst is better exposed. Metre-scale cavities of dissolution in limestones, localized in joints, are filled with sandstones of shallow-marine origin (Fig. 26). A minimum of 50 m has been calculated for the relative sea-level fall in

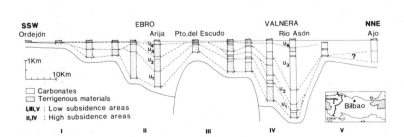

Fig. 21. General cross-section of the Urgonian Complex in the western part of the region, showing sequence organization (U_1-U_4) and the broad constituent facies. The Ebro trough in the south was predominantly filled with continental deposits, while the Valnera trough in the north was filled mostly with marine deposits. I−V: areas of differential subsidence. After García-Mondéjar (1979).

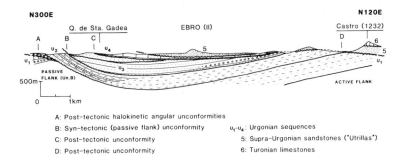

Fig. 22. NW—SE geological section of the Ebro sedimentary sector in the area of Santa Gadea, showing several unconformities (A—D), the four Urgonian depositional sequences (U_1—U_4), and the passive and active flanks of a tectonically mobile zone. Modified from García-Mondéjar (1979).

A: Post-tectonic halokinetic angular unconformities
B: Syn-tectonic (passive flank) unconformity
C: Post-tectonic unconformity
D: Post-tectonic unconformity

u_1-u_4: Urgonian sequences
5: Supra-Urgonian sandstones ("Utrillas")
6: Turonian limestones

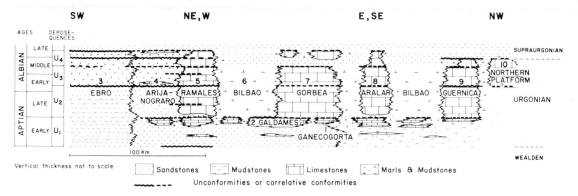

Fig. 23. Schematic and composite cross-section of the Basque—Cantabrian Urgonian Complex, showing the general organization in platforms and basins, the depositional sequences distinguished (U_1—U_4) and the main sedimentary sectors referred to in the text (1—10).

that area (García-Mondéjar & Pujalte, 1983). Extensive dolomitization in the northwestern part of the region, in association with local erosional surfaces, can be attributed, at least partially, to the same break in sedimentation. An example of an irregular replacement of great masses of Urgonian limestone by dolomite is shown in Fig. 27, taken from Matienzo. The whole U_2 sequence records the history of at least one transgression followed by one regression. Another regressive period, locally characterized by an erosional surface, appears in the sedimentary sector of Ramales (Figs 23 & 24) and is used to differentiate the sub-sequences U_2^1 and U_2^2.

U_3 sequence

The third sequence, U_3, of early—middle Albian age, is limited at its base by the unconformity on top of U_2. Its upper boundary is in some places like Gorbea, a sharp contact between Urgonian carbonates below and regressive sandstones above. In the

southwestern part of the region, the upper boundary appears locally as a subtle angular unconformity separating fluvial from fluvial sediments (García-Mondéjar, 1979). In the western central part of the region, it appears as a sharp contact between Urgonian limestones and regressive sandstones (Fig. 24, Portillo de las Escalerucas Formation on top of Picón del Fraile Limestones Formation). The boundary in this platform area can be followed into the adjacent basin, where autochthonous marls of the sequence U_3 are replaced by prodelta sands and mudstones of the U_4, laterally associated with limestone megabreccias (García-Mondéjar, 1985a, fig. 2.23). In the northeast part of the region, an erosional contact in the area of Ea between limestone megabreccias and siliceous conglomerates of talus environment, is attributed to the same upper boundary of U_3 (Fig. 18).

At the beginning of the formation of sequence U_3, when sedimentation was re-established in the areas that previously had been emergent, coarse-

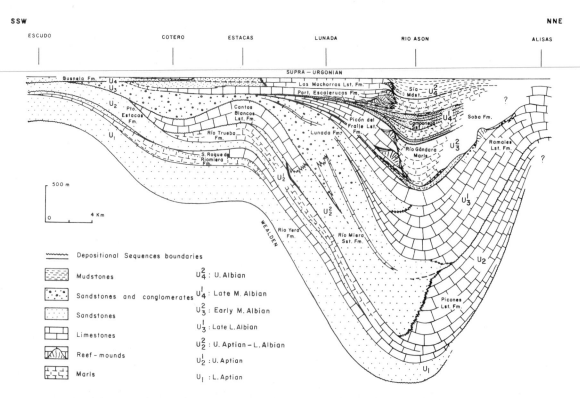

Fig. 24. Stratigraphic cross-section of the Urgonian in the Valnera area (west of the region) restored to the base of the Supra-Urgonian. From south to north fluvial, mixed platform, deltaic, reef and basin major environments are represented. U_1-U_4 refer to depositional sequences and $U_2^1-U_4^2$ to depositional sub-sequences. Modified from García-Mondéjar (1979).

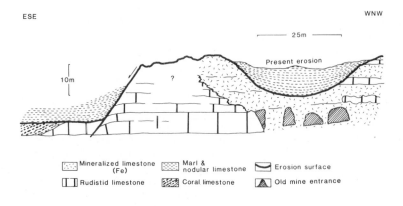

Fig. 25. Erosional surface with palaeochannel morphology on top of platform limestones of the Galdames Formation (upper part of the first Urgonian sequence, U_1, in an outcrop near Bilbao). Draping marls above the limestones correspond to sequence U_2. After García-Mondéjar & García-Pascual (1982).

grained sediments associated with fan deltas covered wide areas. The history of the sequence was thereafter the history of a transgression. This was interrupted by at least one regressive pulse, which produced a local canyon-like surface in platform margin limestones of the Ramales area (Fig. 23), later filled with sandstones. This regressive pulse also resulted in a correlative sandstone and megabreccia interval in the basinal area of Soba (western Bilbao sedimentary sector, Fig. 23). This pulse permits the division of U_3 into the sub-sequences U_3^1 and U_3^2.

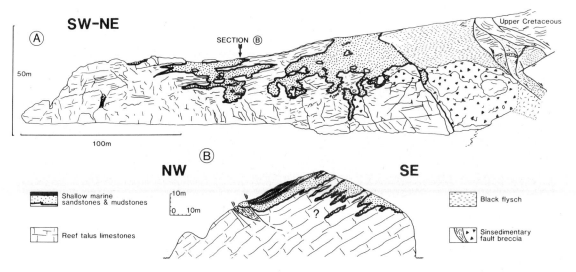

Fig. 26. Paleokarst structures on top of limestones of the sequence U_2 in the area of Gorliz (north of Bilbao). Marine sandstones from the sequence U_3 filled the dissolution cavities. After García-Mondéjar & Pujalte (1983).

Fig. 27. Extensive dolomitization (dark areas) in limestones from the second Urgonian sequence U_2. Ramales sedimentary sector in the area of Matienzo. The thickness of the section shown in the picture is $\approx$ 400 m.

U_4 sequence

The last sequence, U_4, is of middle–late Albian age. Its base is the top of the previous sequence U_3, except in some localized areas, where the pre-U_4 sequence erosion was very intense and left behind no deposits from the sequence U_3. An example of this is in the Escudo area (Fig. 24), where an angular unconformity separating two fluvial units is clearly evident (García-Mondéjar, 1979). This unconformity disappears very quickly towards the north as the thickness increases in that direction (Fig. 28, corresponding to the Cotero section in Fig. 24). The upper limit of the sequence U_4 in the southwestern part of the region is an angular unconformity, which separates the marine carbonates of the top of the sequence from fluvial sandstones of the overlying Supra–Urgonian Complex (Fig. 22; García-Mondéjar, 1979). A regression from marine carbonates to fluvio-deltaic terrigenous sediment in the central western part of the region has been correlated with this boundary (García-Mondéjar, 1979). In basinal areas, like those separating the sedimentary sectors of Gorbea from Aralar, and Guernica from Northern Platform (Fig. 23), the base of an interval of resedimented deposits which includes some limestone megabreccias, may represent the upper limit of the sequence U_4. The sharp lithological change at the top of some of the carbonate remnants of the Urgonian Complex (for instance in the Gorbea sedimentary sector, Fig. 23), is attributed to the same upper limit.

An important break in sedimentation occurred during the formation of sequence U_4 in the central, western and northeastern parts of the region (Fig. 28). In some areas of the Ramales sedimentary sector, fluvial or fluvio-deltaic terrigenous materials succeeded the marine carbonates. In the northeast

Fig. 28. Angular unconformity separating materials from the sub-sequences U_4^1 and U_4^2 in the area of Cotero (west of the region). The 20 m thick limestone interval of U_4^2 (Upper Albian) represents the last Urgonian transgression in the area. After García-Mondéjar (1979).

of the region (Guernica sector), some palaeokarst features on top of limestone highs correspond to the same break (Fig. 18). The break was followed in that area by burial of the last limestone remnants and flysch sedimentation. Future identification of this important boundary in other sectors will permit the subdivision of the sequence U_4 into two new sequences valid for all the region. The limestones of sequence U_4 were localized with respect to previous episodes of limestone formation (Feuillée, 1971) (Fig. 23), and reflect the progressive disappearance of the Urgonian carbonate episode in the region, in favour of the succeeding Supra—Urgonian siliciclastic stage.

The Urgonian Complex of the Basque—Cantabrian region is then considered to be made up of four main cycles, with characteristics of depositional sequences. Two of the cycles roughly belong to the Aptian and the other two to the Albian (Fig. 29). The sedimentological study of the constituent facies of the cycles has shown that repeated transgressions and regressions determined their characteristics to a great extent. Transgressions reflect, at least in part, relative sea-level rises. Regressions are more difficult to explain in terms of relative sea-level variations, as changes in the rate of terrigenous influx from land areas can cause regressions during continuous relative sea-level rises. Nevertheless, the erosion surfaces which accompany the regressions can help to elucidate the respective origins of these.

Sequence U_1 has erosive limits locally, the lower one separating fluvial from fluvial sediments and the upper one fluvial from fluvial, shallow marine from fluvial or shallow marine from deeper-marine sediments. As tectonic movements are implied here because of the presence of local angular unconformities coincident with the limits, subaerial erosion caused by relative sea-level falls may have occurred prior to and at the end of the formation of the sequence. The upper boundary of sequence U_2 has erosive characteristics and separates fluvial from fluvial, shallow marine from deltaic, and shallow marine from deeper-marine sediments. Paleokarst features are locally evident, and tectonism also played an important role as demonstrated by a related angular unconformity. A relative sea-level fall of at least 50 m has been calculated in some areas for the upper limit of this cycle (Fig. 26).

Sequence U_3 also has an upper boundary characterized by erosion within fluvial sediments, and is associated with a local, minor angular unconformity. In the platforms, large quantities of siliciclastics are deposited on the carbonates, and in the adjacent basins limestone megabreccias and prodeltaic siliciclastics replace vertically monotonous marls. A relative sea-level fall is also considered here.

Lastly, sequence U_4 has an upper boundary which is in places an angular unconformity separating marine from fluvial sediments. It is also a contact between platform carbonates and fluvio-deltaic siliciclastics, and in troughs it is the base of resedimented deposits including limestone megabreccias. Tectonism is also implied and a relative sea-level fall is proposed. A curve summarizing the most important relative oscillations of the sea-level during the

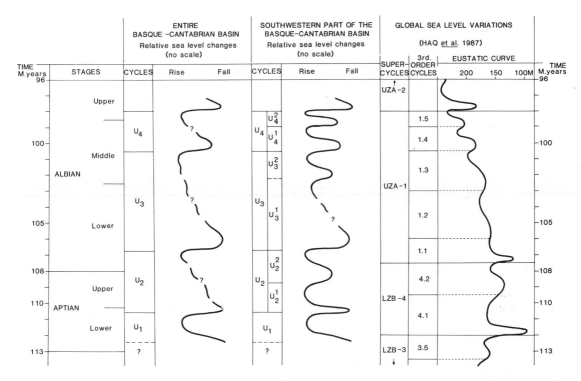

Fig. 29. Urgonian depositional sequences (U_1–U_4) valid for most parts of the region and depositional sub-sequences (U_2^1–U_4^2) valid for the southwestern part of it. The corresponding relative sea-level changes are also shown as well as the eustatic curve of Haq *et al.* (1987), in order to make comparisons.

Urgonian time is shown in Fig. 29.

In the west of the region (centre and south), a better expressed cyclicity in terms of depositional sequences is developed because the area was less affected by the alpine tectonism and shows a greater thickness of sediment and excellent outcrops. Each of the units U_2, U_3 and U_4 has been subdivided into two sub-units, according to the presence of local unconformities or their correlative conformities. The identification of the resultant seven sequences in other parts of the region is being presently investigated, so that, for the moment, they appear as the best representative testimony of the major phenomena that took place in the basin during the Aptian—late Albian interval. These sequences and their relative sea-level variations are shown in Fig. 29.

PALAEOGEOGRAPHY

The palaeogeography of the Basque—Cantabrian Basin during the Aptian and Albian was strongly conditioned by tectonism and, most probably, by eustacy. Climate and sedimentary processes were other influential factors.

Four palaeogeographic maps (Figs 30, 32, 34 & 38) representing the Urgonian sequences have been constructed. Each map shows a synthesis of the main lithological types, sedimentary environments, clastic dispersal systems, axis of subsidence and inferred source areas of one particular sequence. In order to reconstruct more accurately the palaeogeography, a palinspastic restoration has been made, based on published geological maps and sections (Rat, 1959; Lotze, 1973; Soler *et al.*, 1981), and the author's field data. The position of the Arija—Nograro and Ramales sedimentary sectors (Fig. 23) has been maintained at the present distance from the coastline, as they constitute a central, slightly deformed belt flanked by two thrusted areas. SW—NE shortening of $\approx$ 30 km is calculated after restoration.

At the beginning of the first Urgonian sequence

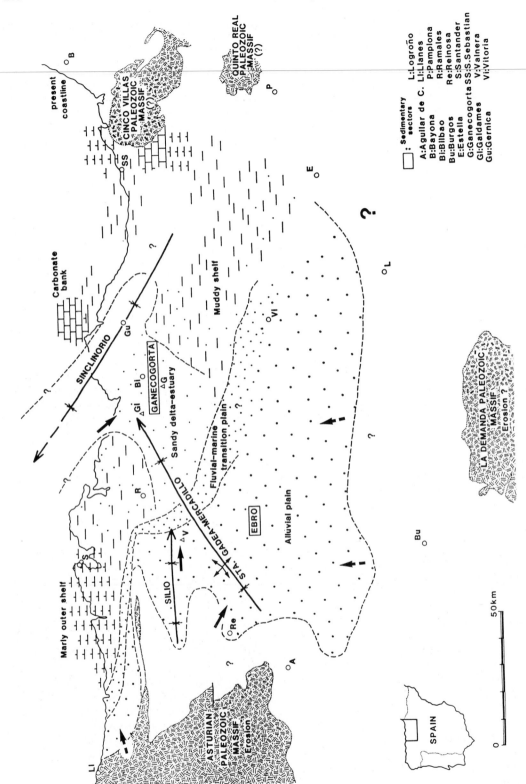

Fig. 30. Palaeogeographic synthesis of the lower part of the sequence U₁ (early Lower Aptian), in the Basque—Cantabrian Basin, with palinspastic restoration from surface data. The Ebro and Ganecogorta sedimentary sectors are represented. Arrows refer to clastic dispersal systems.

(U_1, Lower Aptian), the sedimentary basin was to a large extent the site of mainly terrigenous deposition (Ebro and Ganecogorta sedimentary sectors, Fig. 23, Fig. 30). A wide belt of fluvial sediments in the south and west (Ebro) was feeding into a northerly shallow-marine shelf (Ganecogorta). The high proportion of mudstones with respect to sandstones in the southern part of the fluvial belt suggests low gradients for that subaerial part of the basin (García-Mondéjar, 1979). In the west, braided river facies filled the Silió trough (Pujalte, 1974; García-Mondéjar & Pujalte, 1975), and are indicative of higher gradients. A possible source area in the south might have been the La Demanda Palaeozoic massif (Fig. 30), among other possible emergent Palaeozoic areas now buried. In the west, the Asturian Palaeozoic massif was probably being eroded at that time.

Outcrops of tidal facies (tidal-channel sandstones, mudflats, skeletal grainstones) constitute a belt of transitional deposits which originally separated fluvial from marine domains. Farther north was the normal marine realm. A large sandy delta-estuary was formed in the centre of the basin, approximately at the confluence of the Santa Gadea—Mercadillo and Synclinorium axis of subsidence (Fig. 30). It received open-marine influence from the northwest. On both sides of the estuary, in areas with lower rates of subsidence, were deposited shelf mudstones and sandstones with local distributary mouth bars and open-marine fauna (ammonites). In the northwestern part of the basin (Fig. 30), relatively deep-water marls with abundant ammonites (Collignon *et al.*, 1979) are laterally equivalent to the estuarine sandstones. In the southeastern part, mudstones dominated the seafloor and sandstones of stream mouth bar were also present, the latter indicating a fluvial influence from the south. Finally, in the northeastern part, where the rate of subsidence was very low, marls and nodular limestones with orbitolinas, ostreids and some corals were deposited, and a carbonate bank was developed to the north of Guernica.

The Santa Gadea—Mercadillo and Synclinorium axes of subsidence are the main tectonic features of the basin during the formation of the lower part of sequence U_1 (Fig. 30). They suggest extension in NW—SE and SW—NE directions, respectively; the former one slopes towards the northeast, and the second one presumably towards the northwest. Another though smaller axis is in the central western part of the region (Silió) and shows a clear polarity towards the east. It disappears at the confluence with the Santa Gadea—Mercadillo axis and suggests extensive movements in a N—S direction.

The lower, terrigenous part of sequence U_1 is nearly 1000 m thick in the centre of the basin (Ortega, 1983), < 100 m in the carbonate northern areas and ≈ 200 m in the eastern part. The upper carbonate part of the sequence (Fig. 31) (Galdames sedimentary sector in Fig. 23, not represented in the map of Fig. 30), shows a maximum thickness of ≈ 200 m in the centre of the basin; it is the more significant testimony of the long transgression that characterized the sequence U_1.

The second Urgonian sequence U_2 (Upper Aptian—lowermost Albian) is represented mainly by carbonate sediments (Fig. 32). At the beginning of its formation, sandstones, black shales and sandy marls were deposited in many areas, following the erosive processes that have been attributed in a previous section to a relative sea-level fall. Limestone sedimentation was never interrupted in low subsidence zones, so that it expanded from them into some of the adjacent areas during the time span of U_2.

Low-energy fluvial deposition continued in the southern part of the basin (Ebro sector), with deposition of red, flood-basin mudstones and light-coloured channel sandstones. No high gradients are deduced from sedimentological analysis, so that rivers that flowed through straight to meandering channels, were close to the continental base level.

The transition zone between the continental and

Fig. 31. Field aspect of the Lower Aptian rudistid limestones in Galdames (upper part of sequence U_1). Tabular micritic biostromes of rudists are the main constituent facies.

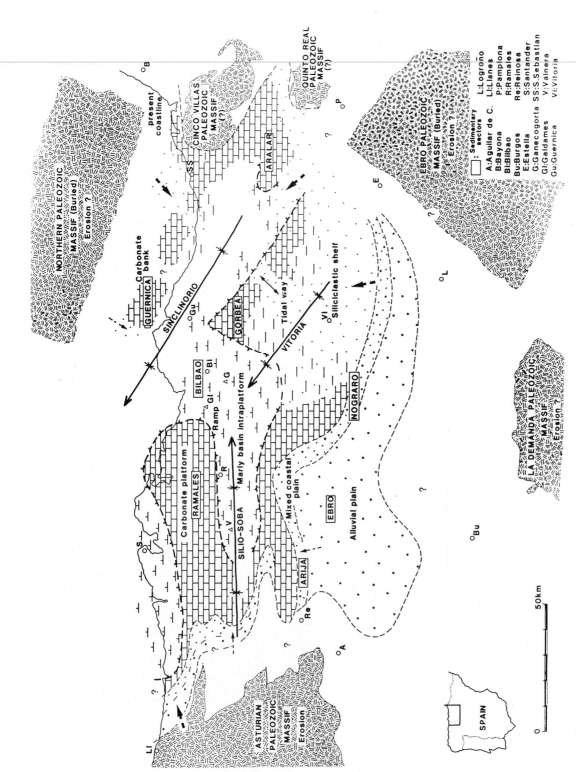

Fig. 32. Palaeogeographic synthesis of the lower part of U$_2$ sequence (early Upper Aptian) in the Basque–Cantabrian Basin, with palinspastic restoration from surface data. The Ebro, Arija–Nograro, Ramales, Bilbao, Gorbea, Aralar and Guernica sedimentary sectors are represented. Arrows refer to clastic dispersal systems.

marine realms (Arija−Nograro sector) was characterized by the important development of marshes in the low-lying, fluvial coastal areas. High-and low-energy tidal deposits alternated with the fluvial materials as the shoreline fluctuated with time.

Carbonate platforms grew on relative low subsidence zones (Ramales, Gorbea, Aralar and Guernica sedimentary sectors, Fig. 32), while shallow intra-platform basins dominated by marly sedimentation occurred between platform areas (Bilbao sedimentary sector). The transition areas between platforms and basins were characterized by gentle slopes, with few, if any, signs of mass-transport deposition. Incipient platform margins with small reef mounds typified those transition areas (e.g. facies number 9 in Fig. 6). In some places like Hornijo (Ramales sedimentary sector, Fig. 32) gently to steeply inclined clinoforms (10°−40°) characterized the platform margin, just prior to the formation of the thick mud mounds which typified the sequence U_3 (Fig. 33) (García-Mondéjar, 1985b).

The main geomorphologic features of the seafloor during the sedimentation of the U_2 sequence were the Silió−Soba, Vitoria and Synclinorium troughs (Fig. 32). The confluence of the three in the area of Bilbao was probably determined by the intersection of several deep-seated, normal faults, with NE−SW and NW−SE orientations. The Silió−Soba trough was filled with marls during the early Upper Aptian time (U_2^1 time); later on it received a thick succession of deltaic sandstones of western provenance, which is considered to belong to another sub-sequence (U_2^2, uppermost Aptian−lowermost Albian). The Vitoria shallow trough was open to the northwest (Fig. 32); it represented a back-reef area (lagoon) where much of the siliciclastic material of southern provenance was trapped (García-Rodrigo & Fernández-Alvárez, 1973). The Synclinorium trough was predominantly filled with marls. It is thought to have been the main connection through to the opening Bay of Biscay trough, situated to the northwest of the region.

Marls with ammonites in the northwestern part of the region (Rat, 1959) (Santander area, Fig. 32) suggest the presence of a shallow intra-platform basin at the beginning of the U_2 time. Nevertheless, soon afterwards shallow-water limestones, very reduced in thickness, dominated in that area. A similar situation of little thickness and rudistid limestone predominance is found in the northeastern outcrops, so that another low subsiding zone, probably cor-

Fig. 33. Platform−basin transition in the Ramales sedimentary sector. Limestones of the 'Ramales barrier' in the Hornijo area (Rat, 1959) face to the right (south) the basinal sediments of the Soba trough. The section of the cliff in the middle of the picture is ≈ 250 m thick. It shows a lower unit of prograding limestones (U_2) separated from an upper tabular unit with mud mounds (U_3) by an angular unconformity. From García-Mondéjar (1985).

responding to an individualized basement block, is deduced for that area. Fine-grained sandstones and mudstones in the areas of Guernica and San Sebastián (Gu and SS, Fig. 32), are inferred to correspond to dispersal systems of northern provenance. Close to San Sebastián they contain carbonaceous deposits and can be considered of transitional continental−marine environment. The emergent land of Voort (1964) to the north of the present outcrops, might thus have been present as early as the late Aptian.

Apart from this hypothetical northern source area for the terrigenous materials, Palaeozoic uplifted blocks must have existed to the south and southeast of the basin (probably La Demanda and Ebro, Fig. 32) to give an apron of fluvial sediments. In the west the Asturian Palaeozoic Massif continued to be emergent, as testified by the deltaic deposits of Rio Miera (Fig. 6), derived from the west. The maximum thickness of the U_2 sequence, ≈ 1500 m, has been measured in the Rio Miera area.

The palaeogeography of the basin at the beginning of the sequence U_3 (Lower−early Middle Albian) was characterized by a rather reduced area of marine sedimentation and a relatively wide surface subjected to erosion, karstification and dolomitization (Figs 22, 26 & 27, respectively). Coastal onlap occurred afterwards and was characterized by a general transgression, punctuated by a phase of regression towards the early–middle Albian transition, at least in the west of the basin.

During the deposition of the sequence U_3 the Urgonian reefs acquired their most characteristic configuration. A palaeogeographic synthesis of this interval is presented in Fig. 34, showing the eight main sedimentary sectors distinguished. Because of the strong differential subsidence of the basin in U_3 times, the incipient intra-platform troughs that had been formed in the late Aptian deepened considerably; as a consequence, thick carbonate build ups, composed of stacked mud mounds–reef mounds, appeared on their margins (Fig. 33). The most important of these troughs was the Synclinorium–Huici (Fig. 34), which had several troughs of smaller size converging into it, and was probably linked with the Bay of Biscay proto-oceanic rift to the northwest of the region.

In the western and southern areas, coarse terrigenous materials of sand and gravel sizes were transported by strong fluvial currents along high-gradient slopes (Ebro sector). Braided rivers flowed directly into the sea forming fan deltas, many of them of the Gilbert type (García-Mondéjar, 1979). Oscillations of the shoreline gave rise with time to a mixed clastic and carbonate coastal plain belt (Arija–Nograro sector), which separated fluvial from carbonate platform sedimentary environments (Fig. 34).

The Ramales platform (Fig. 34) was the major 'carbonate factory' of the region. It was transitional with the terrigenous deposits of the Santillana, Silió and Santa Gadea troughs in the west, and with the basinal Soba–Bilbao deposits in the east. About 2000 m of thickness, which includes the best developed Urgonian reefs of the region, are recorded there (Fig. 24). Limestone megabreccias were deposited intermittently as slope aprons in front of the reef belt, and a variety of terrigenous materials were spread over both the platform and the basin areas.

The Gorbea sedimentary sector was also limited by troughs (Fig. 34). Its linear northeastern and northwestern platform margins suggest basement faulting, with a higher rate of subsidence in the troughs. A palaeogeographical reconstruction of the northern 'corner' of this carbonate buildup is shown in Fig. 35 (Duranguesado area; Fernández-Mendiola & García-Mondéjar, 1983). Slope apron megabreccias, adjacent to the margins, indicate the formation of high-gradient foreslopes and the partial destruction of these margins. An aspect of the western platform margin with development of clinoforms

is shown in Fig. 36, corresponding to the Itxina limestones (Fernández-Mendiola, 1986). These clinoforms show clear downlap relationships with respect to the underlying sediments and prograde towards the Bilbao Basin (Fig. 34). The present height of the escarpment in Fig. 36 is 328 m and calculations relative to the height of the primitive foreslope give a minimum relief of $\approx$ 150 m. If, on the other hand, a minimum slope of 1° is accepted for the adjacent basin talus containing gravitational deposited sediment derived from the platform margin, 100 m of supplementary relief for the seafloor could be added. That gives a minimum palaeobathymetry of 250 m for the basinal sediments in the Bilbao area. This agrees well with results obtained from the opposite margin of the same local basin (Soba trough), since a relief of more than 100 m for the La Gándara original foreslope (Fig. 13) has been calculated.

In the Aralar and Guernica sedimentary sectors (Fig. 34), carbonate platform conditions prevailed similar to the ones described above. They were limited by several troughs, the Synclinorium–Huici being one of the largest. An example of a local trough with important development of slope apron megabreccias is in Bermeo, in the western part of the Guernica sector (Fig. 34). It was limited by a halokinetic synsedimentary dome in the southeast and by a carbonate escarpment margin in the north (Fig. 37) (García-Mondéjar & Robador, 1989).

Basement structural trends controlled the shape and dimensions of both the troughs and the highs, indicating extensional patterns in the NW–SE and SW–NE directions, respectively. In relation to source areas, most of the terrigenous materials are interpreted to come from uplifted blocks to the southeast of the region, perhaps from La Demanda and Ebro Palaeozoic Massifs (Fig. 34). In the west, the Asturian Massif continued as an emergent area subjected to erosion, and in the north, coarse fan-delta clastics of northern provenance indicate the presence of an uplifted northern Palaeozoic massif.

The sequence U_4 (late Middle–early Upper Albian) was built through successive pulses, which recorded the complete replacement of the Urgonian carbonate facies by the terrigenous Supra-Urgonian facies (Fig. 23). A palaeogeographic synthesis of this interval is shown in Fig. 38. The most important feature of it is the Black Flysch trough, which appeared with the late Albian. At the beginning of the formation of the sequence, megabreccias and a clastic wedge of prodeltaic to proximal turbidite channel

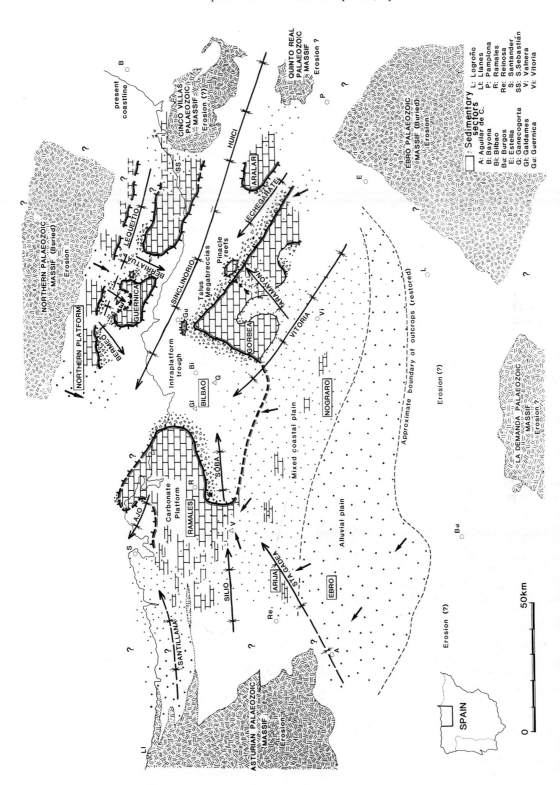

Fig. 34. Palaeogeographic synthesis of U₃ sequence (early−middle Albian) in the Basque−Cantabrian Basin, with palinspastic restoration from surface data. The Ebro, Arija−Nograro, Ramales, Bilbao, Gorbea, Aralar, Guernica and Northern Platform sedimentary sectors are represented. Arrows refer to clastic dispersal systems.

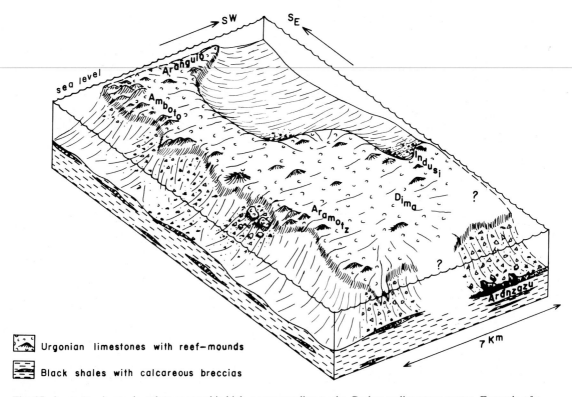

Urgonian limestones with reef—mounds

Black shales with calcareous breccias

Fig. 35. Aramotz−Aranguio palaeogeographic high, corresponding to the Gorbea sedimentary sector. Example of a carbonate bank from the sequence U₃ with rectilinear, high-gradient outer margins and mud mounds. From Fernández-Mendiola & García-Mondéjar (1983).

Fig. 36. Western platform margin of the Gorbea sedimentary sector in Itxina (sequence U₃). Foreslope clinoforms are seen in the foreground downlapping on to a limestone tongue. Horizontal beds in the background correspond to the reef crest. The height of the cliff is 328 m and the estimated original relief of the clinoforms is ≈ 150 m.

were formed in the west of the region (Ramales and Bilabo sectors) (Figs 12 & 24). They are interpreted as lowstand deposits since a contemporaneous local angular unconformity has been identified in the Ebro sector (Fig. 22). In the centre of the basin, the big 'Valmaseda Delta' (Rat, 1959; Pujalte & Monge, 1985) prograded towards the northwest and partially filled the Bilbao Basin. In the eastern part of the basin, carbonate platform and basin deposition practically disappeared in the Gorbea sector (Fernández-Mendiola, 1986) and was significantly reduced in the Aralar (Floquet & Rat, 1975) and Guernica (Agirrezabala *et al.*, in press.) sectors.

Subsequently, platform carbonate deposition recovered in localized places and followed the succeeding relative sea-level rise. In the western part of the region, transgressive marls were deposited on both basinal and platform areas, following the previous clastic wedge stage (Fig. 12). Prograding limestones rebuilt the platform to its former limit and grew vertically (Figs 12 & 24). These processes

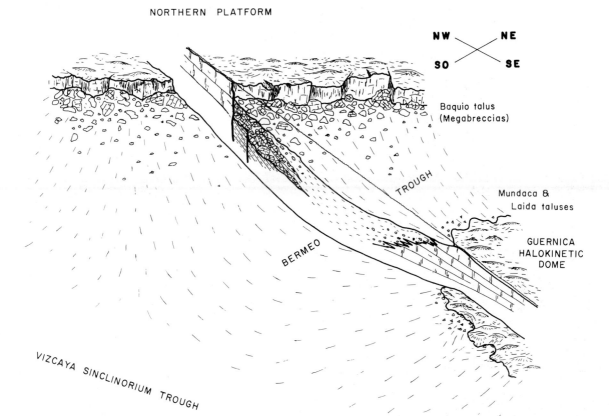

Fig. 37. Palaeogeographic reconstruction of the Bermeo trough, placed between the Guernica and Northern Platform sedimentary sectors (see Fig. 34). A slope apron adjacent to a margin of escarpment-type in the north and a depositional talus with some megabreccias in the southeast face each other. After García-Mondéjar & Robador (1986−87).

can be considered representative of a highstand stage. In the eastern part of the region some carbonate banks and the Aralar platform limestones were the result of the same transgression (Fig. 38).

The relative sea-level fall that occurred towards the middle−late Albian transition brought about erosion in the continental environment (Figs 22 & 24) and karstification on some carbonate highs (Fig. 18). A subsequent transgression deposited the last Urgonian limestones in the south of the basin (Fig. 22). That event was coincident with the appearance of the Black Flysch trough in the northeast area (Fig. 38; Rat, 1959; Feuillée, 1967; García-Mondéjar, 1979; Wiedmann *et al.*, 1983; Pascal, 1984; García-Mondéjar *et al.*, 1987; Badillo *et al.*, 1988).

Limestones are best represented in the northwestern part of the basin (Ramales platform). In the centre of the basin the former Bilbao trough was practically filled with tidally influenced, deltaic sediments (Pujalte & Monge, 1985). And, in the east, a sandy and muddy storm-influenced shelf developed, and probably transgressed on to the Ebro massif (Fig. 38).

The Black Flysch trough was filled mainly laterally. The presence of coarse-grained proximal turbidites and distal deltaic deposits on its adjacent northern margin suggests the existence of a marine shelf and a belt of alluvial fans, both probably very narrow, and an emergent land to the north of the present outcrops (García-Mondéjar *et al.*, 1987; Badillo *et al.*, 1988).

The maximum thickness of the U_4 sequence was close to 1000 m in the western and central areas (Fig. 24), where, because of the NW−SE extension, the Soba and Bilbao troughs were still two axes of

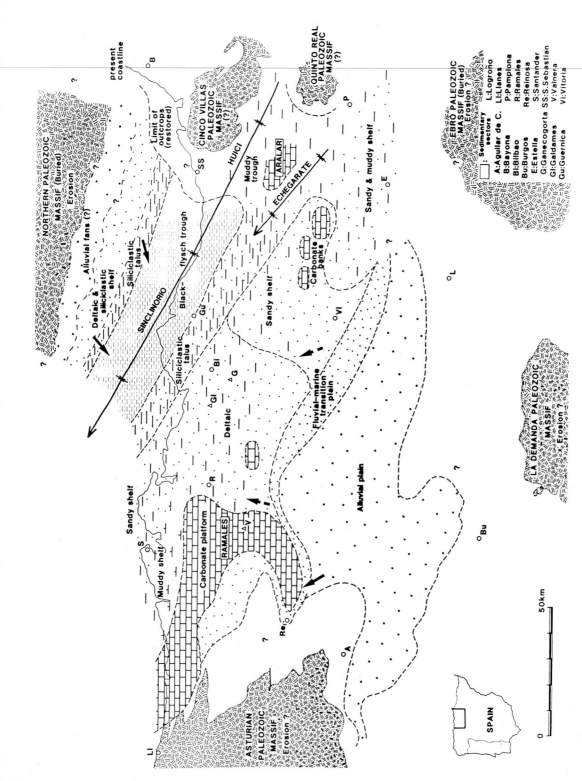

Fig. 38. Palaeogeography of the Basque–Cantabrian Basin at the beginning of the late Albian (upper part of sequence U4). The arrows refer to the main clastic dispersal systems.

subsidence. Nevertheless, the main syntectonic feature was the flysch trough, which resulted from enlargement of the former Synclinorium—Huici trough (Fig. 38) and suggests extension in a NE—SW direction. The source areas during this interval must have been situated to the southeast and to the north of the basin (Ebro? and Northern Massifs, respectively). The La Demanda, Asturias, Cinco Villas and Quinto Real Massifs (Fig. 38) could have had some influence, but they were more probably predominantly passive.

The last areas of Urgonian carbonate sedimentation were definitely abandoned when a new late Albian tectonic pulse created an angular unconformity in the southwest (Fig. 22; Pascal *et al.*, 1976; García-Mondéjar, 1979). Subsequently, great amounts of terrigenous materials inundated the area of sedimentation ('Utrillas' fluvial sandstones).

CONTROLS ON SEDIMENTATION

Four factors were probably most influential in determining the nature of the Urgonian deposits: climate, sedimentary processes, tectonism and eustacy.

Climatic conditions during the Urgonian must have been warm and wet. Urgonian limestones contain reef-building organisms, which indicate warm, shallow and well-lit marine waters. The humid character of the climate is deduced from the abundant plant remains present in the coastal plain facies, and from the widespread fluvial sediments which lack evidence of arid conditions, such as evaporites or calcretes.

Sedimentary processes controlled the type and distribution of facies. Intermittent terrigenous influx from emergent areas, caused by synsedimentary tectonic pulses and probably eustatic sea-level falls, prevented carbonates from being deposited over wide areas. Tides redistributed sands over platforms and troughs, while platform-margin buildups restricted the circulation of water during periods of lowered sea-level.

Tectonism was probably the most important control on sedimentation, determining basin configuration, thickness and geometry of the units, angular unconformities and palaeobathymetry. The Urgonian was coincident with the maximum development of differential subsidence in the basin. This is demonstrated by the abundance and size of mass-gravity deposits associated with foreslope environments and by the outstanding differences in thickness

over short distances (e.g. Fig. 24).

Differential subsidence was already important early in the Aptian and followed block-faulting in the basement that probably exploited pre-existing major faults of Hercynian age. During the Albian the entire basin was subdivided into well-defined highs and troughs (Fig. 34), displaying a major NW—SE structural trend and a subordinate SW—NE trend. A northward displacement of depocentres of some of the troughs is discernible during the Aptian and Albian (e.g. Figs 21 & 24) (García-Mondéjar, 1979), which culminated with the formation of the Black Flysch trough in the northeast of the region in the late Albian. As this migration continued northwards in the late Cretaceous and Tertiary (Soler *et al.*, 1981), it is clear that it was a major feature of basin evolution in the region. It probably resulted from the progressive eastward extension of the main rifting axis of the Bay of Biscay.

The most important tectonic movements that affected the basin during Urgonian sedimentation occurred: (1) at the beginning of the Aptian (García-Mondéjar & Pujalte, 1975); (2) in the early—late Aptian transition; (3) in the early Lower Albian; (4) in the middle Albian; (5) in the middle—late Albian transition; and (6) in the late Albian (García-Mondéjar, 1979). All of them have been inferred from local erosion surfaces which, at least in one place, correspond to slight angular unconformities (e.g. Fig. 22).

The early Albian movements were perhaps the most important of all, as they created local disconformity, angular unconformity, new troughs, new source areas or rejuvenation of the existing ones, and high fluvial gradients. Slight N—S compressional effects in the west of the region have been described in association with this episode (García-Mondéjar, 1979).

The movements at the beginning of the late Albian were also important, as they created local unconformities (Fig. 28; García-Mondéjar, 1979; García-Mondéjar *et al.*, 1987, fig. 36), and brought about the emplacement of the flysch series. Subsequent movements in the late Albian, at the top of the Urgonian, can be considered the final pulse of a series of tectonic movements which created a mobile zone in the southwest of the region (Fig. 22).

Halokinesis of Upper Triassic clays and evaporites was a second type of tectonic control on sedimentation. According to other studies it became very intense in the basin since the Upper Cretaceous (e.g. Brinkman & Lögters, 1968). Nevertheless, its synsedimentary influence has been detected in deposits of the Aptian and Albian (García-Mondéjar, 1979; Badillo, 1982; García-Mondéjar, 1982; Vadala et al., 1981; Reitner, 1982; Antigüedad et al., 1983; Hines, 1985). Synsedimentary halokinesis created local relief on the seafloor, over which reef-building organisms were concentrated. As a result, thick carbonate caps with mud mounds, local unconformities (Fig. 39), radial slopes and other related features were formed on those salt-induced highs. In continental Urgonian environments, several angular unconformities testify the development of diapirism during sedimentation (Fig. 40).

The succession of local unconformities in the Urgonian of the Basque–Cantabrian Basin is primarily attributed to tectonic pulses, a feature also recognized in many sequences of similar age from different basins. Nevertheless, the unconformities corresponding to the Aptian and the Middle Albian and, in general, all the Aptian and Albian unconformities distinguished, have little evidence of tectonic action in many parts of the basin. Therefore they could also imply other mechanisms besides tectonism. In relation to this, Kent (1976) considered that the Aptian–Albian interval was an important phase of global change in the rheology of the materials of the crust, with a move from taphrogenic to thermo-tectonic subsidence and an increase in the rate of sea floor spreading. Cooper (1977) reported that many of the probably eustatic Cretaceous falls in sea-level were associated with unconformities, and Vail et al. (1984) and Haq et al. (1987) have made a genetic connection between unconformities and eustatic sea-level falls. Haq et al. (1987) identified two supercycles and nine third-order cycles in the Aptian

and Albian, each cycle consisting of a succession of both regressive and transgressive eustatic episodes.

As each of the proposed tectonic pulses for the Basque–Cantabrian Basin brought about a relative sea-level fall, and was followed by a relative sea-level rise, the resultant regressive–transgressive cycles could also be considered to be local reflections of global eustatic cycles, provided a correlation with them can be established.

A comparison between curves of relative sea-level changes in the Basque–Cantabrian Basin and the eustatic curve of Haq et al. (1987) is shown in Fig. 29. The relatively small differences observed in the timing of the late Aptian and Albian sea-level falls and rises could be due to imprecise biostratigraphy, and there is good agreement in general terms between the eustatic curve and the curve for the western part of the basin. The early Aptian timing has more discrepancy, so that the sea-level fall reported in the Basque–Cantabrian region close to the early–late Aptian transition, has no equivalent in the eustatic curve. Reciprocally, the sea-level fall of the eustatic curve at 112 Ma would not be represented in the example described. For these early Aptian cases, the Basque–Cantabrian cycles would correspond more exclusively to the local tectonism, perhaps not yet fully associated to the major phenomena that were taking place in the Bay of Biscay at this time. Nonetheless, the comparison made demonstrates that most of the Urgonian relative changes in sea-level in the Basque–Cantabrian area, were certainly influenced by the eustatic changes established in the Aptian and Albian (e.g. Hallam, 1984; Kauffman, 1986; Haq et al., 1987).

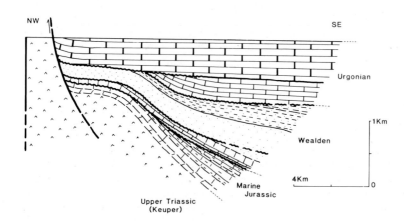

Fig. 39. Urgonian angular unconformities associated with the Pondra diapir (Ramales sedimentary sector). The lower Urgonian unit corresponds to the sequence U_2 (at least in part) and the upper one is attributed to the sequence U_3. After Badillo (1982), and García-Mondéjar (1982).

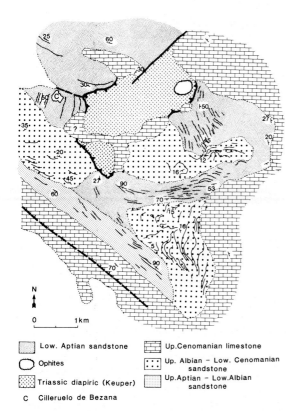

Low. Aptian sandstone

Ophites

Triassic diapiric (Keuper)

C   Cilleruelo de Bezana

Up.Cenomanian limestone

Up. Albian – Low. Cenomanian sandstone

Up.Aptian – Low.Albian sandstone

Fig. 40. Map of the Cilleruelo de Bezana diapir, in the Arija–Nograro sedimentary sector. Several Urgonian angular unconformities are associated with the diapir, and Cenomanian limestones rest on top of any of the older units, including the diapiric Keuper. The discontinuous lines in the sandstone units are key mapping beds which reveal the structure. After García-Mondéjar (1979).

ACKNOWLEDGEMENTS

The final version of this paper was completed during a short visit to the Open and Oxford Universities, UK. Gratitude is expressed to staff of the Earth Sciences Department of the Open University for all the facilities provided, and especially to Dr Chris Wilson for revising the English version of the text and comments on it, and for continuous help during my stay in Milton Keynes. John Taylor, also of the Open University, assisted in the drafting of some of the figures.

Drs W.J. Kennedy and H.G. Reading of Oxford University, allowed the author to work in the University Museum and Earth Sciences Department and helped with the classification of ammonites and the English revision of the first part of the paper, respectively. However, all opinions expressed are the responsibility of the author.

The manuscript was improved by critical reviews from M. Esteban, R.J. Ross Jr., J.F. Sarg and J.L. Wilson. Begoña Bernedo processed the manuscript, and María Jesús Sevilla and Luis Miguel Agirrezabala helped in the final composition. The paper has been written under research project 310.08–11/86 from the Universidad del País Vasco.

APPENDIX

Chronostratigraphy of the depositional sequences
(Author's and co-workers' data)

Sequence U_1 (Lower Aptian)

Ammonites: *Deshayesites* sp., *Cheloniceras (Cheloniceras) meyendorffi* (d'Orbigny). Foraminifera: *Chofatella decipiens* (Schlum.), *Palorbitolina lenticularis* (Bluemenb.), *Praeorbitolina cormyi* (Schroed.), *Mesorbitolina lotzei* (Schroed.), *Iraqia simplex* (Henson). Collignon *et al.* (1979) reported ammonites species of the earliest Aptian in the northwest part of the region.

Sequence U_2 (Upper Aptian–early Lower Albian)

Ammonites: *Epicheloniceras tschernyschewi* (Sinzow), *Cheloniceras* sp. Foraminifera: *Mesorbitolina parva* (Dougl.), *Mesorbitolina minuta* (Dougl.), *Mesorbitolina texana* (Roemer), *Hensonina lenticularis* (Henson), *Simplorbitolina manasi* (Ciry and Rat), *Simplorbitolina conulus* (Schroed.).

Sequence U_3 (Lower Albian–early Middle Albian)

Ammonites: *Hypacanthoplites* sp.?, *Douvilleiceras* sp. (gr. *inaequinodum*, Quendst). Foraminifera: *M. minuta*, *M. texana*, *H. lenticularis*, *S. manasi*, *S. conulus*.

Sequence U_4 (late Middle Albian–early Upper Albian)

Ammonites: *Kosmatella demolyi* Brest., *Eogaudryceras shimizui gaonail*? Wiedmann, *Tegoceras camatteanum*? Orb., *Lyelliceras lyelli*? Orb., *Puzosia* sp., *Anapuzosia* sp., *Brancoceras* sp. Foraminifera: *Mesorbitolina aperta* (Erman),

Neorbitolinopsis conulus (H. Douvil.), *Orbitolina (Orbitolina)* (d'Orb.). Rudists: *Caprina choffati* Douv.

REFERENCES

AGIRREZABALA, L.M. & GARCÍA-MONDÉJAR, J. (in press) La serie de talud urgoniano de Ea (Bizkaia): caracteres sedimentológicos e implicaciones paleogeográphicas. *Libro Homenaje a Rafael Soler*.

ANTIGÜEDAD, I., CRUZ-SANJULIAN, J., FERNANDEZ-MENDIOLA, P.A. & GARCIA-MONDEJAR, J. (1983) Argumentos sedimentológicos e hidrogeoquímicos sobre la existencia de un diapirismo de materiales triásicos en el área de Dima (Vizcaya). *Bol. Geol. Min.* **XCIV−VI**, 489−495.

BADILLO, T. (1982) Estudio geológico del sector de Ramales de la Victoria (Prov. de Santander). *Kobie* **12**, 139−171.

BADILLO, T., AGIRREZABALA, L.M. & GARCIA-MONDEJAR, J. (1988) Caracteres generales de la sucesión Albiense superior del Flysch Negro entre Elantxobe y Deba (Bizkaia y Gipuzkoa). *Congr. Geol. España, comun.* **1**, 35−38.

BRINKMANN, R. & LÖGTERS, H. (1968) Diapirs in western Pyrenees and foreland, Spain. In: *Diapirism and Diapirs* (Eds Braunstein, J. & O'Brien, G.D.) Am. Assoc. Petrol. Geol. Bull. pp. 275−292.

CIRY, R., RAT, P., MANGIN, J. Ph., FEUILLÉE, P., AMIOT, M., COLCHEN, M. & DELANCE, J.H. (1967) Compte-rendu de la réunion extraordinaire de la Société Géologique de France en Espagne: des Pyrénées aux Asturies. *Compt. Rend. Soc. Géol. France* **9**, 389−444.

COLLIGNON, M., PASCAL, A., PEYBERNES, B. & REY, J. (1979) Faunes d'ammonites de l'Aptien de la région de Santander (Espagne). *Ann. Paléontol. (Invertébrés)* **65**, 139−156.

COOPER, M.R. (1977) Eustacy during the Cretaceous: its implications and importance. *Palaeogeog. Palaeoclimatol. Palaeoecol.* **22**(1), 1−60.

DUVERNOIS, CH., FLOQUET, M. & HUMBEL, B. (1972) *La Sierra de Aralar, Pyrénées basques espagnoles. Stratigraphie-structure.* Université de Dijon, Institut des Sciences de la Terre, p. 277.

FERNANDEZ-MENDIOLA, P.A. (1986) *El Complejo Urgoniano en el sector oriental del Anticlinorio de Bilbao.* Tesis Doctoral, Universidad del País Vasco, p. 421.

FERNANDEZ-MENDIOLA, P.A. & GARCÍA-MONDÉJAR, J. (1983) Estudio geológico del Anticlinorio de Bilbao en el sector del Duranguesado. *Kobie (Bilbao)* **XIII**, 299−324.

FEUILLÉE, P. (1967) Le Cénomanien des Pyrénées basques aux Asturies. Essai d'analyse stratigraphique. *Mém. Soc. Géol. France* **46**, 343.

FEUILLÉE, P. (1971) Les calcaires biogéniques de l'Albien et du Cénomanien Pyrénéo-Cantabrique: problèmes d'environment sedimentaire. *Palaeogeog. Palaeoclimatol. Palaeoecol.* **9**, 277−311.

FEUILLÉE, P. & RAT, P. (1971) Structures et paléo-geographies Pyrénéo-Cantabriques. In: *Histoire Structurale du Golfe de Gascogne*. Publications de l'Institute Français du Pétrole, Collection Colloques et Séminaires, Technip, Paris, Vol. 22, Bd2, p.v.1−1−v.1−48.

FLOQUET, M. & RAT, P. (1975) Un exemple d'interrelation entre socle, paléogéographie et structure dans l'arc pyrénéen basque. La Sierra d'Aralar. *Rév. Géog. Phys. Géol. Dynam.* **XVII**, 497−512.

GARCÍA-GARMILLA, F. (1987) *Las formaciones terrígenas del "Wealdense" y del Aptiense inferior en los anticlinorios de Bilbao y Ventoso (Vizcaya, Cantabria): Estratigrafía y Sedimentación.* Tesis Doctoral, Universidad del País Vasco, p. 340.

GARCÍA-MONDÉJAR, J. (1979) *El Complejo Urgoniano del sur de Santander.* Tesis Doctoral, Universidad de Bilbao. Ann Arbor University Michigan, Microfilms International, 1980, p. 673.

GARCÍA-MONDÉJAR, J. (1982) Tectónica sinsedimentaria en el Aptiense y Albiense de la región Vasco-Cantábrica occidental. *Cuad. Geol. Ibérica* **8**, 23−36.

GARCÍA-MONDÉJAR, J. (1985a) Aptian and Albian reefs (Urgonian) in the Asón-Soba area. In: Sedimentation and Tectonics in the Western Basque−Cantabrian area (northern Spain) during Cretaceous and Tertiary Times (Eds Milá, M.D. & Rosell, J.) *Field-guide excursion 9. 6th European Regional Meeting of Sedimentology*, Int. Assoc. Sedimentol., Lleida, pp. 329−352.

GARCÍA-MONDÉJAR, J. (1985b) Carbonate platform-basin transitions in the Soba reef area (Aptian−Albian of western Basque−Cantabrian region northern Spain). In: *Sedimentation and Tectonics in the Western Basque−Cantabrian area (northern Spain) during Cretaceous and Tertiary Times* (Eds Mila, M.D. & Rosell, J.) *6th European Regional Meeting of Sedimentology*, Abstracts book, Int. Assoc. of Sedimentol., pp. 172−175.

GARCÍA-MONDÉJAR, J. & GARCÍA-PASCUAL, I. (1982) Estudio geológico del Cretácico inferior del anticlinorio de Bilbao, entre los ríos Nervión y Cadagua: *Kobie* **XII−1982**, 101−137.

GARCÍA-MONDÉJAR, J., HINES, F.M., PUJALTE, V. & READING, H.G. (1985) Excursion No. 9: Sedimentation and tectonics in the western Basque−Cantabrian area (northern Spain) during Cretaceous and Tertiary times. *6th European Regional Meeting of Sedimentology, Excursion Guidebook* (Eds Milá, M.D. & Rosell, J.), Lleida. Int. Assoc. Sedimentol., pp. 307−392.

GARCÍA-MONDÉJAR, J. & PASCAL, A. (1978) Précisions stratigraphiques et sédimentologiques sur les terminaisons calcaires sud-occidentales du système urgonien basco-cantabrique (Espagne du Nord). *Bull. Soc. Géol. France* **20**, 179−183.

GARCÍA-MONDÉJAR, J. & PUJALTE, V. (1975) Contemporaneous tectonics in the Early Cretaceous of central Santander province, north Spain. *IX Int. Congress on Sedimentology*, Nice. Tectonics and Sedimentation IV, pp. 131−137.

GARCÍA-MONDÉJAR, J. & PUJALTE, V. (1977) Ciclos sedimentarios mayores del Jurásico superior-Cretácico inferior de Santander. *VIII Congreso Nacional de Sedimentología (Abstracts)*, Oviedo-León. VI.

GARCÍA-MONDÉJAR, J. & PUJALTE, V. (1981) El Jurásico superior y Cretácico inferior de la región vasco-cantábrica (parte occidental). *Libro Guía Jornadas de Campo*. Grupo Español del Mesozoico, p. 134.

GARCÍA-MONDÉJAR, J. & PUJALTE, V. (1983) Origen, karstificación y enterramiento de unos materiales carbonatados albienses (Punta del Castillo, Górliz, Vizcaya). *X Congreso Nacional de Sedimentología, Libro de Abstracts*, pp. 3.9—3.12.

GARCÍA-MONDÉJAR, J., PUJALTE, V. & ROBLES, S. (1986) Características sedimentológicas, secuenciales y tecto-estratigráficas del Triásico de Cantabria y norte de Palencia. *Cuad. Geol. Ibérica* **10**, 151—172.

GARCÍA-MONDÉJAR, J., PUJALTE, V. & ROBLES, S. (1987) Evolución de los sistemas sedimentarios del margen continental cantábrico durante el Albiense y Cenomaniense, en la transversal del litoral vizcaíno. *Congreso de Geología, II Congreso Mundial Vasco, Libro Guía de las Excursiones Científicas*, Bilbao, p. 73.

GARCÍA-MONDÉJAR, J. & ROBADOR, A. (1984) Carbonate slope deposits in the Urgonian of Baquio (Vizcaya, northern Spain). *5th European Regional Meeting of Sedimentology* (Abstracts), Marseille, Int. Assoc. Sedimentol., pp. 187—188.

García-Mondéjar, J. & Robador, A. (1986—87) Sedimentacion y paleogeografía del Complejo Urgoniano (Aptiense—Albiense) en el área de Bermeo (region Vasco-Cantábrica Septentrional. *Acta Geol. Hispanica* **21—22**, 411—418.

GARCÍA-RODRIGO, B. & FERNÁNDEZ-ALVAREZ, J.M. (1973) Estudio geológico de la provincia de Alava. *Mem. Inst. Geol. Min. España, Madrid* **83**, 198.

HALLAM, A. (1984) Pre-Quaternary sea-level changes. *Ann. Rev. Earth Planet. Sci.* **12**, 205—244.

HAQ, B.U., HARDENBOL, J. & VAIL, P.R. (1987) Chronology of fluctuating sea-levels since the Triassic. *Science* **235**, 1156—1167.

HINES, F.M. (1985) Sedimentation and tectonics in north-west Santander. *6th European Regional Meeting, Excursion Guidebook* (Eds Milá, M.D. & Rosell, J.) Int. Assoc. Sedimentol., pp. 371—389.

JAMES, N.P. (1978) Facies models 10. Reefs. *Geosci. Can.* **5**(1), 16—26.

KAUFFMAN, E.G. (1986) High-resolution event stratigraphy: regional and global Cretaceous bio-events. In: *Lecture Notes in Earth Sciences*, Vol. 8, *Global Bio-Events* (Ed. Walliser, O.) Springer Verlag, Berlin, pp. 279—335.

KENT, P.E. (1976) Major synchronous events in continental shelves. *Tectonophy.* **36** (1—3), 87—91.

LOTZE, F. (1973) *Geologische Karte des Pyrenäisch—Kantabrischen Grenzgebietes. 1:200000.* Akademie der Wissenschaften und der Literatur, Mainz.

MAGNIEZ, F. & RAT, P. (1972) Les foraminifères des formations à spongiaires et Tritaxia dans l'Aptien-Albien Cantabrique (Espagne). *Rev. Española Micropaleont.*, No. extraordinario, pp. 159—178.

MATHEY, B. (1982) El Cretácico superior del Arco Vasco. In: *El Cretácico de España*. Universidad Complutense, Madrid, pp. 111—135.

MITCHUM, R.M. Jr., VAIL, P.R. & THOMPSON, S. III, (1977) The depositional sequence as a basic unit for stratigraphic analysis. In: *Seismic Stratigraphy-Applications to Hydrocarbon Exploration* (Ed. Payton, Ch.E.) Am. Assoc. Petrol. Geol., Tulsa, Oklahoma, Mem. 26, pp. 53—62.

MONTADERT, L., ROBERTS, D.G., de CHARPAL, O. &

GUENNOC, P. (1979) Rifting and subsidence of the northern continental margin in the Bay of Biscay. *Initial Reports of the Deep Sea Drilling Project*, Washington (US Government Printing Office), Vol. XLVIII, pp. 1025—1059.

ORTEGA, R. (1983) *El Cretácico inferior de la región minera de Bilbao entre los ríos Cadagua y Mercadillo*. Tesis de Licenciatura, Universidad del País Vasco, p. 239.

PASCAL, A. (1974) Un faciès type de l'Urgonien cantabrique (Espagne): les micrites à Rudistes. *Compt. Rend. Acad. Sci. Paris* **279**, 37—40.

PASCAL, A. (1984) *Les systèmes biosédimentaires urgoniens (Aptien—Albien) sur la marge nord-ibérique*. Thèse, Université de Dijon, p. 561.

PASCAL, A., RAT, P. & SALOMON, J. (1976) Sédimentation, stratigraphie et dynamique dans le complexe continental et marin basco—cantabrique (Jurassique terminal—Albien). *4ème Réunion Ann. Sci. Terre*, Paris, p. 320.

PUJALTE, V. (1974) Litoestratigrafía de la facies Weald (Valanginiense superior-Barremiense), en la provincia de Santander (Norte de España). *Bol. Geol. Min.* **85—1**, 10—21.

PUJALTE, V. (1981) Sedimentary succession and palaeo-environments within a fault-controlled basin: the "Wealden" of the Santander area, northern Spain. *Sed. Geol.* **28**, 293—325.

PUJALTE, V. & MONGE, C. (1985) A tide dominated delta system in rapidly subsiding basin: the Middle Albian—Lower Cenomanian Valmaseda Formation of the Basque—Cantabrian region, northern Spain. *6th European Regional Meeting of Sedimentology* (Abstracts), Lleida, Int. Assoc. Sedimentol., pp. 381—384.

RAMÍREZ DEL POZO, J. (1971) Bioestratigrafía y microfacies del Jurásico y Cretácico del Norte de España (región Cantábrica). *Mem. Inst. Geol. Min. España* **78**, 357.

RAT, P., (1959) *Les pays crétacés basco—cantabriques (Espagne)*. Thèse Université de Dijon, Vol XVIII, p. 525.

RAT, P., AMIOT, M., FEUILLÉE, P., FLOQUET, M., MATHEY, B., PASCAL, A. & SALOMON, J. (1983) Vue sur le Crétacé Basco—Cantabrique et Nord-Ibérique. Une marge et son arrière-pays, ses environments sédimentaires. *Mem. Geol. Univ. Dijon* **9**, 191.

REITNER, J. (1982) Die Entwicklung von Inselplattformen und Diapir-atollen in Alb des Basko-kantabrikums (Nordspanien). *N. Jb. Geol. Paläont. Abh.* **165**, 87—101.

ROBLES, R., GARCÍA-MONDÉJAR, J. & PUJALTE, V. (1988) A retreating fan-delta system in the Albian of Biscay, northern Spain: facies analysis and palaeotectonic implications. In: *Fan-Deltas: Sedimentology and Tectonic Settings* (Eds Nemec, W. & Steel, R.J.) Blackie, Glasgow, pp. 197—211.

ROSSY, M., MATHEY, B. & SIGAL, J. (1979) Précisions sur l'âge du magmatisme crétacé supérieur du synclinorium de Biscaye (Espagne). *7ème Réunion Ann. Sci. Terre*, Lyon, p. 411.

SOLER, R., LOPEZ-VILCHEZ, J. & RIAZA, C. (1981) Petroleum geology of the Bay of Biscay. In: *Petroleum Geology of the Continental Shelf of North-West Europe*. Institute of Petroleum, London, pp. 474—482.

VADALA, P., TOURAY, J.C., GARCÍA-IGLESIAS, J. & RUIZ, F.

(1981) Nouvelles données sur le gisement de Reocín (Santander, Espagne). *Chron. Recherch. Min.* **462**, 43–59.

VAIL, P.R., HARDENBOL, J. & TODD, R.G. (1984) Jurassic unconformities, chronostratigraphy, and sea-level changes from seismic stratigraphy and biostratigraphy. In: *Interregional Unconformities and Hydrocarbon Accumulation* (Ed. Schlee, J.S.) Am. Assoc. Petrol. Geol. Mem. 36, pp. 129–144.

VOORT, H.B. (1964) Zum flysch problem in den Westpyrenäen. *Geol. Runds.* **53**, 220–233.

WIEDMANN, J. (1965) Sur la posibilité d'une subdivision et des corrélations du Crétacé inférieur ibérique. Colloque sur le Crétacé inférieur, Paris. *Bull. Recherch. Géol. Min.* 819–823.

WIEDMANN, J. (1979) Itinéraire géologique à travers le Crétacé moyen des Chaînes Vascogotiques et Celtibériques (Espagne du Nord). *Cuad. Geol. Ibérica* **5**, 127–214.

WIEDMANN, J., REITNER, J., ENGESER, T. & SCHWENTKE, W. (1983) Plattenktonik, fazies und subsidenzgeschichte des basko–kantabrischen kontinentalrandes während Kreide und Alttertïar. *Zitteliana* **10**, 207–244.

WRAY, J.L. (1977) *Calcareous Algae*. Elsevier, Amsterdam, p. 185.

Spec. Publs int. Ass. Sediments. (1990) **9**, 291–323

Evolution of the Arabian carbonate platform margin slope and its response to orogenic closing of a Cretaceous ocean basin, Oman

K. F. WATTS* *and* C. D. BLOME

*Department of Geology and Geophysics, Geophysical Institute, University of Alaska, Fairbanks 99775,
USA Geological Survey, MS 919 Denver, Colorado 80225, USA*

ABSTRACT

The Jurassic to Cretaceous Mayhah Formation formed along the passive continental margin slope between the extensive shallow-marine Arabian carbonate platform and the deep-oceanic Hawasina Basin (South Tethys Sea). Near Wadi Qumayrah, it is overlain by pre- and syn-orogenic sediments of the Upper Cretaceous Qumayrah Formation. These strata record the initial response of the platform margin to a late Cretaceous orogeny and impending ophiolite emplacement resulting from closing of an ocean basin. Tectonism along the platform margin, supply of platform carbonate sediment, and palaeoceanographic effects all influenced slope sedimentation.

During the passive margin phase, redeposited oolitic calcarenite, megabreccia and calcirudite of Jurassic age formed a base-of-slope apron along an east-facing block-faulted platform margin. Redeposited limestone and slump deposits become uncommon up-section suggesting a transition into basinal sedimentation. With tectonic quiescence this basin developed ramp-like margins along which mass movements were unlikely. It may have been a relatively shallow embayment or marginal basin adjacent to the deeper Hawasina Basin.

A well-documented drowning of the platform during late Jurassic–early Cretaceous led to formation of bedded chert in this marginal basin. The chert is primarily silicified limestone indicating a shallower-depositional setting (above the carbonate compensation depth, CCD) or greater supply of carbonate sediment than for equivalent radiolarian chert at the type locality, Jebel Sumeini. Basinal carbonate sedimentation became re-established in post-Valanginian time. In the uppermost Mayhah Formation, slump deposits and intra-formational breccias indicate steepening of the slope, possibly resulting from tectonic downbowing associated with basin closing.

Cenomanian to Coniacian chert and siliceous mudstone of the Qumayrah Formation formed below the CCD. Siliceous sedimentation resulted from tectonically-induced deepening of the ocean basin, upwelling in a relatively narrow closing ocean basin and/or global sea-level highstand and shallowing of the CCD. Beds of calcirudite and calciturbidite in the unit were derived from an uplifted shelf edge and local rudist banks. A thick lenticular synorogenic megabreccia may have been derived from erosion of adjacent anticlinal highs.

The Arabian carbonate platform margin was influenced by repeated intervals of tectonism along the continental margin in Oman. Block-faulting associated with Triassic and Jurassic rifting events led to the development of steep escarpment slopes and base-of-slope debris aprons. Tectonic quiescence led to reduced platform margin slopes, so the marginal basin at Wadi Qumayrah developed ramp margins. With changing plate motions, the outer part of the platform was drowned due to a major rise in sea-level, in part resulting from rapid tectonic subsidence in the latest Jurassic and earliest Cretaceous. Later, a carbonate ramp was able to prograde over the drowned platform. In late Cretaceous time, steepening and deepening of the slope, accompanied by faulting on the Arabian platform, preceded oceanic nappe emplacement as the Oman continental margin approached a northward-dipping subduction zone.

* Research conducted at Earth Science Board, University of California, Santa Cruz 95064, USA.

INTRODUCTION

A passive continental margin developed along the northeastern margin of the Arabian carbonate platform in Permian and Triassic times and existed in this area until the Oman Mountains orogeny in the late Cretaceous (Glennie *et al.*, 1973; Glennie, 1974; Searle *et al.*, 1983; Watts & Garrison, 1986). Rocks which formed along this passive continental margin were tectonically thrust over the Arabian carbonate platform as a series of nappes during the Oman Mountains orogeny (Fig. 1; Glennie *et al.*, 1973; Glennie, 1974; Searle *et al.*, 1983). With the development of the Oman Mountains fold and thrust belt, the Aruma foreland basin formed over the outer part of the Arabian carbonate platform. An understanding of the evolution of the Arabian carbonate platform and slope provides valuable insights into the relationships between carbonate sedimentation, tectonic processes, sea-level fluctuations and palaeoceanography.

Carbonate slope deposits of the Jurassic to Lower Cretaceous Mayhah Formation (Fm) and the siliceous sediments of the superjacent Upper Cretaceous Qumayrah Formation record the response of the Arabian carbonate platform margin slope and Oman continental margin to Jurassic rifting and Cretaceous convergent tectonics. Much of the sediment in the slope sequence was derived from the coeval Arabian carbonate platform and shelf edge. Studies of these platform-derived slope sediments provide

information on the Arabian platform margin which is otherwise hidden beneath structurally higher thrust sheets across most of the Oman Mountains. The carbonate slope was particularly sensitive to tectonically-induced steepening of the slope which led to the deposition of megabreccias and other coarse calciclastic sediments. Major platform drowning events are related to tectonic subsidence of the outer Arabian carbonate platform in addition to global sea-level highstands (Murris, 1980; Harris *et al.*, 1984). During these drowning events, widespread intervals of radiolarian chert in the slope and basinal sequences formed in response to reduced supply to periplatform carbonate sediment (Glennie, 1974; Watts & Garrison, 1986). Here, the nature of the slope sequence at Wadi Qumayrah is documented, and its significance interpreted in terms of the tectonic and palaeoceanographic evolution of the Arabian carbonate platform, Oman continental margin and Tethys Sea.

In this paper, the Qumayrah Facies of the Muti Formation of Glennie (1974) is renamed the Qumayrah Formation and included in the Aruma Group along with the Muti Fm and Fiqa Fm. The type locality (M-4) is located just north of the village of Qumayrah (Fig. 1) at UTM 47.4N 17.5E. It is described later in text and shown in Figs 2, 3, 6, 7, 15, 18, 19 & 20.

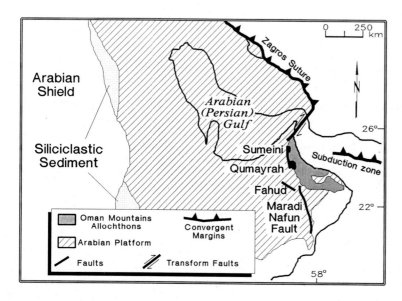

Fig. 1. Map of the Arabian platform showing the major tectonic elements in the Oman region.

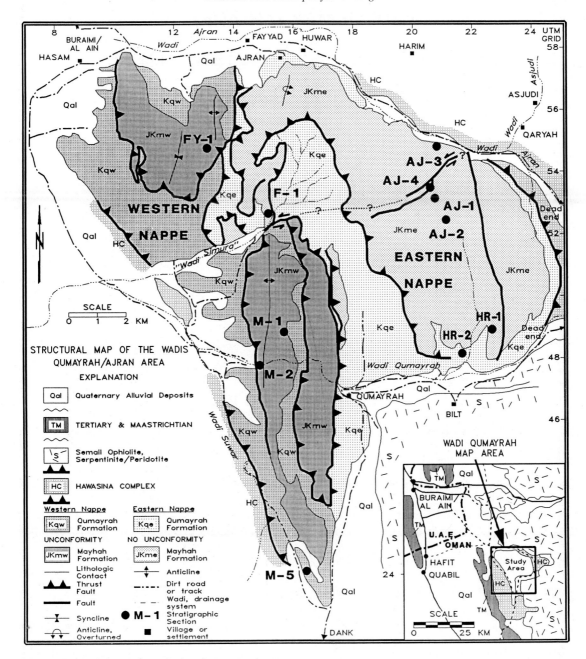

Fig. 2. Map showing the major structural elements of the Wadi Qumayrah study area and the location of measured stratigraphic sections.

REGIONAL STRUCTURAL GEOLOGY AND TECTONIC SETTING

The Oman Mountains are a major collisional mountain belt which formed in late Cretaceous time in response to convergent tectonics along the South Tethys Sea (Glennie, 1974; Gealey, 1977; Coleman, 1981; Searle *et al.*, 1983). A number of allochthonous thrust sheets (nappes) comprised of rocks of the Semail Ophiolite, the Hawasina Complex and the

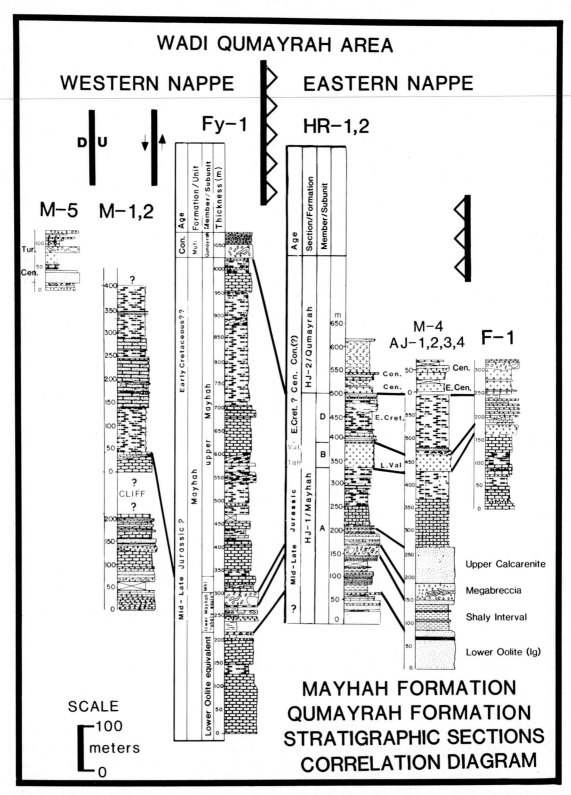

Fig. 3. Correlation diagram for stratigraphic sections in the Wadi Qumayrah study area. See Fig. 2 for locations of measured sections and Fig. 4 for key to symbols used. More detailed stratigraphic columns are shown in Figs 8 and 15.

Sumeini Group were progressively emplaced over the Arabian carbonate platform (Fig. 1). These nappes apparently represent different parts of a telescoped passive continental margin which have been tectonically stacked and shortened. The most distal nappe, Semail Ophiolite (ocean crust?) was thrust-faulted over basinal and continental rise sediments of the allochthonous Hawasina Complex and slope sediments of the Sumeini Group. Younger out-of-sequence thrust-faulting has further complicated the structure of these deformed rocks.

Structure of the Wadis Qumayrah area

The Mayhah Fm and associated rocks near Wadi Qumayrah form a major structural high or culmination (Fig. 2). This structure is part of a large tectonic window beneath the Semail Ophiolite, the As Judi culmination of Searle (1984). In the study area, two major thrust sheets, termed the *eastern and western nappes*, are each comprised of Mayhah Fm and Qumayrah Fm rocks. Neither the nature of the basement upon which it formed nor the amount of tectonic displacement is known, because of structural detachment at the base of the Jurassic—Cretaceous slope sequence. The front of each nappe is marked by a major thrust fault, and the plunging flanks indicate lateral ramp margins of the thrust sheets (hanging wall cut-offs). Out-of-sequence thrust faults locally place the Mayhah Fm over the Hawasina Complex (and the Semail Ophiolite at one locality; Watts, 1985).

STRATIGRAPHY AND FACIES ANALYSIS

Several lithologic units within the Mayhah Fm and Qumayrah Fm were mapped and described in the Wadi Qumayrah study area (Watts, 1985). The lower part of the sequence, the Middle to Upper Jurassic A Member of the Mayhah Fm, begins with coarse redeposited limestone which can be divided into four mappable sub-units and passes upward into fine-grained calcilutite and marlstone (Figs 3 & 5). Locally, an interval of bedded chert, the Tithonian to Valanginian B Member of the Mayhah Fm, and thin beds of chert—clast breccia of the C Member, are key marker horizons. Thin-bedded calcilutite of the Valanginian to Cenomanian D Member of the Mayhah Fm marks the return to carbonate sedimentation whereas slumps and calcirudite in the upper

part indicate rejuvenation/steepening of the slope. Siliceous sediments of the Upper Cretaceous Qumayrah Fm indicate major changes in patterns of sedimentation at the platform margin and are overlain by synorogenic megabreccias.

Major differences in stratigraphy documented between the eastern and western nappes indicate different depositional settings and possible tectonic shortening. In the Jurassic part, several stratigraphic differences suggest that the western nappe represents relatively proximal upper slope deposits (Fig. 7) whereas the eastern nappe represents relatively distal base-of-slope apron deposits (Fig. 6). Megabreccias and other distinct markers allow correlation between nappes. The western nappe lacks distinct uppermost Jurassic—Cretaceous bedded cherts of the B Member that characterize the eastern nappe and occur at Jebel Sumeini, the type locality (Watts & Garrison, 1986). Without the intervening B Member chert, the thin-bedded calcilutites of the upper A and D Members cannot be distinguished, so the upper part of the succession in the western nappe is simply referred to as the upper Mayhah Fm (Figs 3 & 7). Aspects of each lithologic unit figure importantly in interpreting the tectonic history, palaeogeography and palaeoceanography of this part of the Oman continental margin.

Lower A Member, Mayhah Formation — calcarenite, calcirudite and megabreccia

Eastern nappe

Thick intervals of redeposited calciclastic (coarse carbonate) sediment form a stratigraphic succession of four mappable sub-units The A Member in the eastern nappe (Fig. 8) consists of the lower oolite calcarenite, 'shaly interval', megabreccia and upper calcarenite sub-units.

At the base, the thick-bedded lower oolite sub-unit is primarily composed of ooids and has graded bedding and interbeds of orange calcilutite in the upper part. Numerous beds of calcirudite, oolitic and intraclastic calcarenite, calcareous shale, and marlstone comprise the '*shaly interval*', the middle sub-unit of the lower A Member. Beds of calcirudite commonly contain slope-derived blocks and lesser clasts of fossiliferous shallow-marine limestone including coralline limestone. The unit is no older than middle—late Jurassic based upon the presence of the foraminifera *Nautiloculina circularis* in a

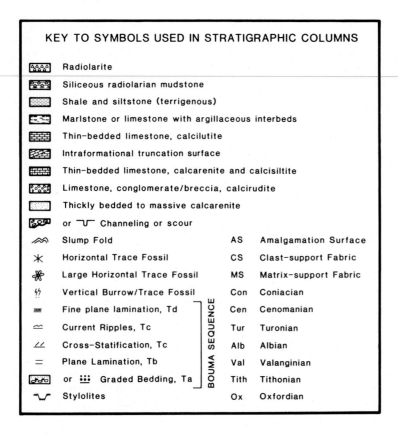

KEY TO SYMBOLS USED IN STRATIGRAPHIC COLUMNS

- Radiolarite
- Siliceous radiolarian mudstone
- Shale and siltstone (terrigenous)
- Marlstone or limestone with argillaceous interbeds
- Thin-bedded limestone, calcilutite
- Intraformational truncation surface
- Thin-bedded limestone, calcarenite and calcisiltite
- Limestone, conglomerate/breccia, calcirudite
- Thickly bedded to massive calcarenite
- or ⎍ Channeling or scour
- Slump Fold
- Horizontal Trace Fossil
- Large Horizontal Trace Fossil
- Vertical Burrow/Trace Fossil
- Fine plane lamination, Td
- Current Ripples, Tc
- Cross-Statification, Tc
- Plane Lamination, Tb
- or ⋮⋮⋮ Graded Bedding, Ta
- Stylolites

BOUMA SEQUENCE

AS	Amalgamation Surface
CS	Clast-support Fabric
MS	Matrix-support Fabric
Con	Coniacian
Cen	Cenomanian
Tur	Turonian
Alb	Albian
Val	Valanginian
Tith	Tithonian
Ox	Oxfordian

Fig. 4. Key to symbols used in stratigraphic columns in Figs 3, 8 and 15.

redeposited calcarenite in section HR-1 (Robert Scott, Amoco written comm., 1983).

A megabreccia forms a thick (23–37 m) and widespread marker bed which can be correlated between the eastern and the western nappes (Figs 8 & 9). It contains huge blocks of thin-bedded slope limestone (up to 1 km across) and rounded boulders of coralline limestone. The upper calcarenite sub-unit consists of a thick sequence of lenticularly bedded, channellized, redeposited oolitic calcarenite and intraformational calcirudite (Fig. 8). In the upper part, calcirudite is interbedded with thin-bedded calcilutite and passes gradationally into thin-bedded calcilutite of the upper A Member.

Fig. 5. [*Opposite*] Outcrop photographs of sections HR-1 and Fy-1, the stratotypes for the eastern nappe and western nappe.

(A) In the eastern nappe, section HR-1 (dashed line) shows a complete but overturned sequence of the Mayhah Formation and its conformable contact with the stratigraphically higher Qumayrah facies. The A Member of the Mayhah Formation is further subdivided into the lower oolite (1g), the shaly interval with megabreccias (sh), upper calcarenite (ug), and the upper A Member (a). The A Member lies stratigraphically below the distinctive dark-coloured chert (b) which is followed by light-coloured limestone the D Member (d) of Mayhah Formation. Chert of the Qumayrah Formation (q) forms the rubble-covered slope along the axis of the south-plunging syncline.

(B) Section Fy-1 is the most complete sequence of Mayhah Formation limestones in the western nappe. The base of the section is cut by an out-of-sequence thrust fault which truncates synclinally folded lower Mayhah Formation which is thrust over the Qumayrah Formation (q). The Mayhah Formation consists of thin-bedded lime mudstone of the lower oolite equivalent ('lg'), the shaly interval with megabreccias (sh), the lower Mayhah Formation (Mhl), and the upper Mayhah Formation (Mhu). The upper part of the section is shown in Fig. 5C (below); note the lack of the B Member chert interval in this succession.

(C) The upper part of section Fy-1 is exposed along the axis of the major syncline at Jebel Fayad. Thin-bedded limestones of the upper Mayhah Formation (Mhu) are unconformably overlain by the Qumayrah Formation (q) which here consists of megabreccia, other calciclastic sediments and minor radiolarian chert.

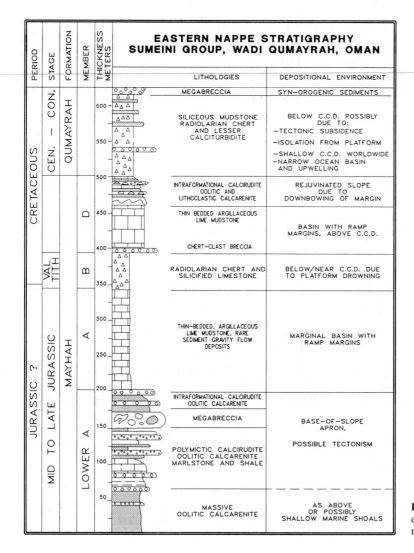

Fig. 6. Stratigraphy and environmental interpretations for rocks of the eastern nappe.

Western nappe

In the western nappe, beds of redeposited calcirudite and calcarenite in the Lower Mayhah Fm are generally much more localized and discontinuous than those in the eastern nappe. However, the widespread megabreccia provides a means of correlating the two areas. No lower oolitic sub-unit exists in the western nappe. In a similar stratigraphic position in section Fy-1, a thick interval of thin-bedded calcilutite is similar to that in the upper Mayhah Fm (Fig. 8). This interval could represent a thrust repetition of the upper Mayhah Fm which was later folded along with the overlying units to form the eastern flank of a major syncline (Figs 2 & 5B).

An interval of shale, marlstone and calcirudite, the '*shaly interval equivalent*', is only locally exposed in the western nappe near section Fy-1 (Fig. 8). The lithologies are similar to that of the 'shaly interval' of the eastern nappe. Two megabreccias occur within the lower part of the Mayhah Fm in the western nappe. The upper megabreccia is thick (32 m), laterally extensive and probably correlates with the megabreccia of the eastern nappe. It contains huge blocks of contorted slope-derived limestone (up to

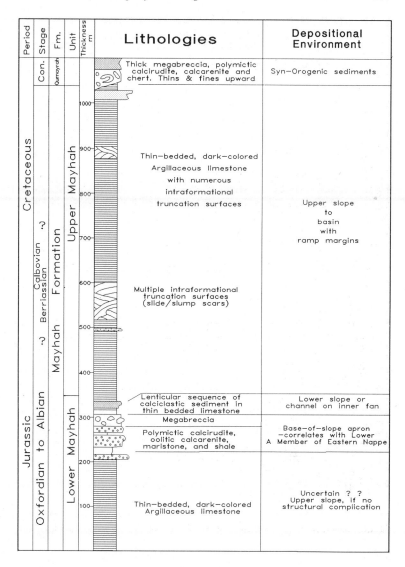

Fig. 7. Stratigraphy and environmental interpretations for rocks of the western nappe.

15 m across) and rare boulders of coralline limestone (Fig. 9).

Cliff-forming limestone of the 'lower Mayhah Fm' sub-unit overlies the 'shale equivalent' (near section Fy-1) and is more widely exposed elsewhere in the western nappe (Figs 2, 5 & 8). Abrupt lateral facies changes occur with the primary lithofacies, thin-bedded calcilutite, passing laterally into localized, lenticular sequences of thick-bedded calcirudite and oolitic calcarenite. The calcilutite is dark coloured and contains a sparse fauna of radiolarians, sponge spicules and minor pelagic bivalves. Lenticular beds of calcarenite and calcirudite are commonly channellized. Typically, genetically related calciclastic beds are organized into discrete sequences that form lenticular bodies of coarse sediment. The calciclastic beds in section Fy-1 are different from those of section Fy-2 (only 13 km to the south) and are separated by intervening thin-bedded calcilutite (Fig. 8). Along Wadi Qumayrah, large rotated slump-blocks and intraformational truncation surfaces, draped by thin-bedded calcilutite, occur above bedded sequences of redeposited limestone in the lower Mayhah Fm (Fig. 10). The lower Mayhah Fm

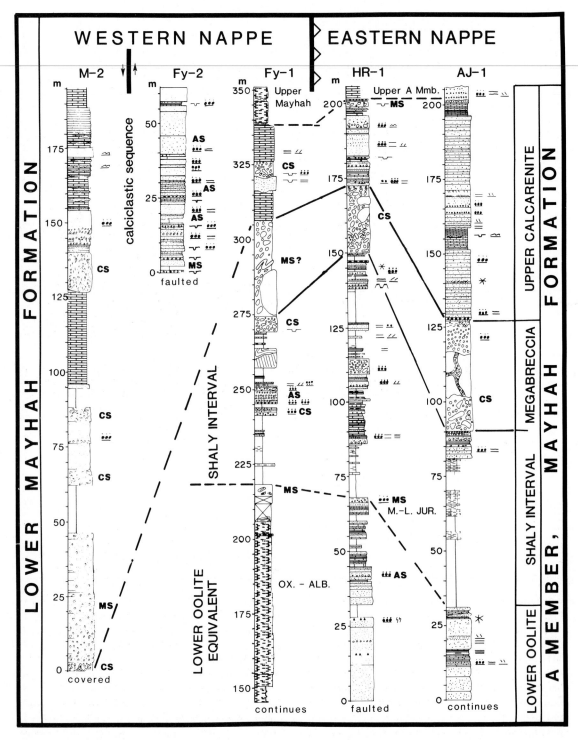

Fig. 8. Detailed stratigraphy of the lower part of the A Member of the Mayhah Formation. See location map (Fig. 2); less-detailed stratigraphic columns (Fig. 3) and key (Fig. 4).

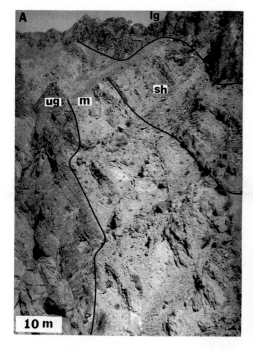
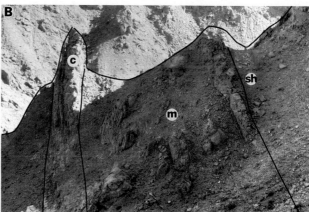

Fig. 9. Outcrop photographs of the megabreccia in the lower part of the Mayhah Formation.

(A) Megabreccia (m) in overturned sequence at section HR-1 contains numerous blocks of slope-derived limestones and rare boulders of reefal limestone. The shaly interval (sh) contains lenticular beds of calcirudite and calcarenite. The shaly interval is underlain by the massive, cliff-forming, lower oolite sub-unit (lg) and is overlain by cliff-forming, calciclastic sediments of the upper calcarenite sub-unit (ug). View north from section HR-1.

(B) Vertically bedded sequence (top-to-left) at section Fy-1 in the western nappe. Megabreccia (m) contains large blocks of slope-derived limestones and rounded boulders of reefal limestone. Note lenticular bed of calcirudite beneath the megabreccia in the shaly interval (sh). A lenticular sequence of bedded calcirudite and calcarenite (c) overlies the megabreccia and passes upsection into thin-bedded calcilutite.

passes gradationally upward into argillaceous calcilutite of the upper Mayhah Fm.

Interpretation — base-of-slope channels and debris apron

Thick intervals of redeposited calciclastic sediment in the lower A Member represent sediment gravity flows deposited at the base of a carbonate slope (Fig. 11). In the eastern nappe, the depocentre for calciclastic sediments migrated through time, but numerous beds of redeposited limestone coalesced to form a laterally continuous base-of-slope apron (*cf.* Read, 1982, 1985; Mullins & Cook, 1986). Lenticular, channelized calciclastic sequences in the western nappe represent conduits that supplied much of the calciclastic sediment to the more widespread

apron. Rotated limestone slump blocks and slide scars in associated thin-bedded calcilutites (Cook & Taylor, 1977; Davies, 1977) also indicate a carbonate slope environment. Ooids in calcarenite and calcirudite were derived from coeval ooid shoals at the platform margin. Coralline limestone suggests that reefs also existed at the shelf edge.

A correlatable megabreccia occurs in both the eastern and western nappes indicating that the redeposited sediments in these areas were genetically linked. Huge blocks of thin-bedded slope sediments slid downslope and were incorporated into the megabreccias; numerous intra-formational truncation surfaces in the western nappe could mark these slide scars. Rounded boulders of neritic and reefal limestone were eroded from coeval or older shallow-marine deposits. Although coarse calciclastic

Fig. 10. Outcrop photographs of the lower part of the Mayhah Formation in the western nappe along Wadi Qumayrah.

(A) Multiple intra-formational truncation surfaces (arrow) in thin-bedded lime mudstones of the lower Mayhah Formation are visible on the cliff-face. Location is along south wall of Wadi Qumayrah above section M-2.

(B) Numerous rotated slump blocks (b) are draped by thin-bedded lime mudstones of the lower Mayhah Formation. Location is the north side of Wadi Qumayrah opposite section M-2 and Fig. 10A.

sediments can be generated by a variety of slope processes (e.g. over-steepening, seismicity, tectonic uplift, sea-level falls), the megabreccias may indicate that this abrupt Jurassic platform margin and steep slope formed along fault scarps resulting from rifting along the Oman margin (see below). Elsewhere in Tethys, similar thick and widespread megabreccias formed due to cataclysmic collapse of carbonate platform margins and are thought to have been generated by seismicity (Johns *et al.*, 1981; Mutti *et al.*, 1984; Watts & Garrison, 1986).

Upper A Member, Mayhah Formation — thin-bedded limestone

Coarse redeposited calciclastic sediments of the lower Mayhah Fm pass gradationally upward into thin-bedded limestone of the upper Mayhah Fm. In the eastern nappe, a thick interval of such thin-bedded limestone comprises the upper A Member and is sharply overlain by ribbon chert of the Upper Jurassic/Lower Cretaceous B Member of the Mayhah Fm (Figs 3 & 6). In the western nappe, a similar but thicker interval of thin-bedded limestone

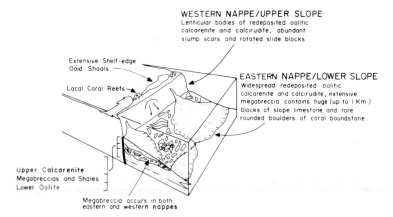

Fig. 11. Facies model for the middle Jurassic lower part of the Mayhah Formation, a base-of-slope debris apron.

which forms the upper Mayhah Fm lacks the B Member chert interval, and is unconformably overlain by the Upper Cretaceous Qumayrah Fm (Figs 3, 6 & 7).

The upper A Member is a thick interval, (170 m), mostly consisting of thin- and evenly-bedded, dark-coloured limestone containing calcified sponge spicules, radiolarians, peloids and micrite. In the lower part (140 m), the dominant lithology is fine-grained calcarenite and calcisiltite composed of similar grains. In the upper part, beds of argillaceous calcilutite alternate with thin partings and interbeds of marlstone and shale.

Sedimentary structures in the calcarenite include grading, parallel lamination, cross-lamination and ripples (Ta-e to Tc-e calciturbidites). Beds of calcilutite are typically bioturbated with locally preserved calcisiltite laminae. Minor thick beds of oolitic calciturbidite occur in the lower part. Such redeposited sediment becomes rarer upsection, but thin beds and laminae of calcarenite containing ooids persist throughout the unit. Slumps and slides are uncommon; however, a local debris sheet containing folded intra-formational clasts occurs near the upper contact in the northern part of the area (UTM 55.0 °N 18.4 °E).

Interpretation — thin-bedded basinal limestone

The thin-bedded argillaceous limestone lacks fossils except for calcified sponge spicules and radiolarians suggesting it probably formed in a deep-water, peri-platform basinal setting (Fig. 12; Scholle *et al.*, 1983). The general lack of redeposited calciclastic

sediment and slumps in the unit favour the interpretation of retrogradation of basinal thin-bedded limestones over a base-of-slope apron. The rarity of calciclastic sediment indicates that basin slopes were gentle, possibly a carbonate ramp (Fig. 12). Local intra-formational calcirudites were derived from rare slides and debris flows. Ooids in thin beds and laminae of calcarenite were derived from coeval ooid shoals possibly on the shallow part of the ramp. Considering palinspastic reconstructions and interpretations of the underlying units, the presence of bioturbation suggests deposition in a basin below the oxygen minimum zone. In contrast, correlative dark-coloured and unbioturbated rocks in the more proximal western nappe (see below) formed within the oxygen minimum zone higher on the ramp. The presence of oxygenated bottom waters beneath the oxygen minimum zone would imply that oceanic currents circulated through this marginal basin (Demaison & Moore, 1980). The abundance of reworked pelagic fossils and peloids in the calcarenite and calcisiltite also indicates winnowing by bottom currents.

B Member chert, Mayhah Formation — eastern nappe

The B Member is a widespread interval of chert which is a key marker horizon for geological mapping and correlation. It correlates with coeval radiolarian chert in the Hawasina Complex and a platform drowning event on the Arabian platform (Glennie *et al.*, 1973; Glennie, 1974; Connally & Scott, 1985;

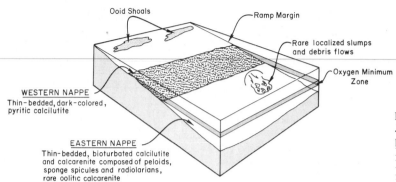

Fig. 12. Facies model for the upper Jurassic upper A Member, lower part of the Lower Cretaceous D Member and the upper Mayhah Formation, marginal basin with ramp margins.

Watts & Garrison, 1986). In the Wadi Qumayrah area, this chert occurs in the eastern nappe but is absent in the western nappe (Figs 3, 6 & 7). Within the eastern nappe, the B Member thins westward (toward the platform), from 34 to 15 m over a distance of ≈ 10 km (Fig. 2). The contacts with thin-bedded limestones above and below are sharp and breccia beds in the C Member contain tabular clasts of chert apparently derived from the B Member (Fig. 13, see below).

Chert of the B Member near Wadi Qumayrah differs from red radiolarite at the type section, Jebel Sumeini. Much of the chert in the Wadi Qumayrah area is green in colour, vitreous, lacks significant pelagic red clay and contains radiolarians which are extensively recrystallized. Radiolarian faunas extracted from these cherts yielded ages ranging from Tithonian to late Valanginian–Hauterivian. The Valanginian fauna is younger than the previously inferred age range of the B Member and is similar to faunas of the Hawasina Complex (Wahrah Formation near Ibri; Blome & Garrison, 1982).

Interpretation—chert formed below or near CCD/platform drowning event

Much of the outer Arabian carbonate platform was drowned during Tithonian (or Berriasian) to late Valanginian time (Fig. 14; Glennie, 1974; Hassan et al., 1975; Searle et al., 1983; Connally & Scott, 1985; cf. Schlager, 1981; Kendall & Schlager, 1981). Conally & Scott (1985) stated that deep-water limestone of the Salil Fm formed on this drowned platform as late as the late Valanginian and that shallow-marine carbonates of the Habshan Fm prograded over this drowned platform by latest Valanginian to Hauterivian time. The chert of the B Member may have formed in response to platform

submergence and the concomitant reduction in the supply of shelf-derived carbonate.

The B Member chert at Wadi Qumayrah represents silicified radiolarian lime mud rather than primary highly siliceous radiolarian oozes as indicated by its locally calcareous nature and because it is typically vitreous, lacking significant terrigenous clay (cf. Nisbet & Price, 1974; Scholle et al., 1983). Because clasts of lithified chert occur in the overlying breccia (see below), the silicification was early diagenetic and could be related to interaction with silica-rich ocean bottom waters that were undersaturated with respect to carbonate. Either the supply of periplatform carbonate to the Qumayrah area was greater than elsewhere or this calcareous chert formed in a shallow basin above but near to the CCD rather than in a deep basin as inferred for the chert at Jebel Sumeini and in the Hawasina Complex (Watts & Garrison, 1986). Considering the stratigraphic relationships between the eastern and western nappes, and other Sumeini Group exposures, the depositional setting may have been a relatively shallow-marginal basin or embayment into the Arabian carbonate platform (Fig. 14).

Reasons for the lack of the B Member chert in the western block are not fully understood, but the lack could indicate deposition above the CCD. Alternatively, the chert interval in the western nappe could have been removed by erosion either within the sequence or beneath the unconformity at the base of the Qumayrah Fm. Intra-formational truncation surfaces within the upper Mayhah Fm may represent slump scars from which the chert–clast breccias of the C Member were eroded (see below).

C Member, Mayhah Formation—thin breccias beds

The C Member at the type locality, Jebel Sumeini, is

Fig. 13. Outcrop photographs of the B Member–Mayhah Formation, eastern nappe.

(A) Light-coloured limestones of the D member (d) overlie dark-coloured cherts of the B Member (b) of the Mayhah Formation. In the lower part of the D Member, perhaps representing the C Member (c), a number of thin widespread beds of breccia contain clasts of chert. Near section AJ-3, 45 m (480 m on composite section in Fig. 3).

(B) Closer view of a breccia bed shows the clast-supported fabric developed between sub-horizontal, tabular clasts of chert in a granular calcarenite matrix. Location is near Wadi Ajran at UTM 54.9°N 20.0°E.

composed of very thick megabreccias (up to 190 m thick; Watts & Garrison, 1986). Near Wadi Qumayrah, such thick megabreccias do not occur, and, relatively thin beds of chert–clast breccia overlie the B Member (Figs 13 & 15). These sheet-like breccias occur throughout the eastern nappe and although they are too thin to map individually, they form a distinct marker at the top of the B Member which is useful in mapping areas of structural complexity.

The breccias contain clasts of chert apparently derived from the underlying B Member and the matrix consists of granular calcarenite. Clasts range in size from cobbles to slabs up to 0.2–0.3 m in length. Breccia beds are up to 5.5 m thick in the west (section F-1) and typically have a clast-supported fabric with clasts oriented sub-parallel to bedding (Figs 13 & 16). The breccias are interbedded with limestones of the D Member and underlain by a thin interval of thin-bedded calcilutite immediately over the B Member. Chert–clast breccias occur as much as 10 m above the B Member.

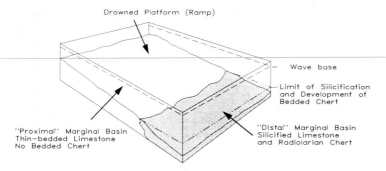

Fig. 14. Facies model for the Tithonian to Valanginian late Jurassic to early Cretaceous B Member, Mayhah Formation, platform drowning event.

Interpretation – locally derived debris flows

To the north, thick megabreccias of the C Member represent extensive submarine debris avalanche deposits (Watts & Garrison, 1986; Searle *et al.*, 1983). Such large mass flows may have originated due to tectonic steepening of the slope and possibly could have been triggered by seismicity (*cf.* Mutti *et al.*, 1984). The Wadi Qumayrah areas was little affected by this platform margin collapse (90 km to the north). The beds of chert–clast breccia, however, may be a distal equivalent of the megabreccias generated by these far-removed but cataclysmic events. However, the age of the megabreccia at Jebel Sumeini is poorly known (pre-Cenomanian?) so it might instead correlate with calcirudite in the D Member (see below). Alternatively, the underlying B Member chert could have formed an inherently unstable substrate leading to its repeated failure and downslope movement as debris flows without being triggered by seismicity or tectonic steepening.

D Member, Mayhah Formation – eastern nappe

The D Member* and its transition into siliceous sediments of the Qumayrah Fm is highly significant because it records the initial response of the Oman continental margin to the closing of the Hawasina Basin (South Tethys Sea). In the lower part, the D Member comprises a thick interval of thin-bedded argillaceous calcilutite similar to the upper A Member

---

* Glennie (1974) referred to the upper part of the Mayhah Formation at Wadi Qumayrah as the C Member. Because its lithology is more similar to thin-bedded limestone of the D Member than to C Member megabreccia, these rocks are referred to as the D Member here.

(Fig. 15). In the upper part, numerous sediment gravity flows, slumps and slides indicate steepening of the slope prior to basin closing. The unit ranges in age from Valanginian to Cenomanian as bracketed by radiolarian dating of the underlying B Member and overlying Qumayrah Fm. Thin limestones of the D Member interfinger with chert–clast breccias of the C member (?) and sharply overlie the B Member chert.

Intrastratal deformation in the upper part of the D Member is the first indication of steepening of the slope. Numerous, thin to thick, lenticular beds of redeposited oolitic and intraclastic calcarenite are common in the upper part of the unit (Fig. 16). Intra-formational calcirudite occurs in beds up to 4 m thick, traceable for several kilometres. The bases of these beds are irregular and typically contain folded or ripped up slabs of thin-bedded calcilutite (Fig. 16B). Relatively fine-grained calciturbidites commonly form the tops of these sediment gravity flow deposits (*cf.* Krause & Oldershaw, 1979).

The coarse-grained deposits are interbedded with moderately bioturbated, thin-bedded calcilutite which locally is interlaminated with calcarenite. Calcarenite laminae are composed of abundant radiolarians, pelagic foraminifera, sponge spicules, peloids and rare ooids or intraclasts. Shale and marlstone interbeds are locally common. Chert is minor but becomes more common up-section from the intrastratal deformation. The chert occurs in very thin beds rich in radiolarians and as replacement nodules and stringers along burrows.

The upper contact with siliceous radiolarian-bearing mudstone of the Qumayrah Fm is sharp and apparently conformable. On closer examination, the contact is somewhat gradational; chert occurs in the upper D Member and the lower Qumayrah is locally calcareous.

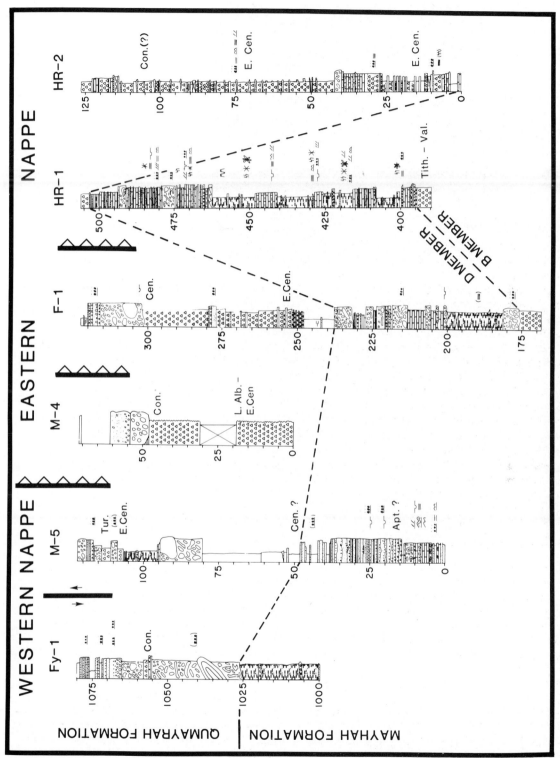

Fig. 15. Detailed stratigraphic columns showing the transition between the calcareous Mayhah Formation and the siliceous Qumayrah Formation. In the eastern nappe, sections HR-1 and F-1 show the entire D Member and uppermost part of B Member. See location map (Fig. 2), less-detailed stratigraphic columns (Fig. 3) and key to symbols used (Fig. 4).

Fig. 16. Outcrop photographs of the D Member of the Mayhah Formation, eastern nappe.

(A) Thick, lenticular beds of calcarenite (g) are interbedded with radiolarian calcilutite in the upper part of the D Member. Toward the top (left), the breccia bed (b), pictured in Fig. 16B has an irregular erosional base. Location is near section HR-1, 475 m.

(B) Breccia bed (debris sheet) in the upper part of the D Member has a clast-supported fabric and contains abundant folded slabs of thin-bedded intra-formational calcilutite which suggest that this deposit originated as a slump or slide before evolving into a debris flow. The base of the deposit is erosional and locally shows folded rip-up clasts. Section HR-1, 477 m.

Interpretation — restoration of carbonate sedimentation followed by steepening of slope

Carbonate sedimentation resumed with deposition of thin-bedded calcilutite of the D Member along the Oman continental margin, following a major platform drowning event and formation of the B Member chert (Glennie, 1974; Watts & Garrison, 1986). The underlying B Member chert apparently formed an unstable substrate which periodically failed moving downslope as debris flows that were deposited as thin chert–clast breccia beds of the C Member. Following these early slope readjustments, evenly bedded calcilutite and calcarenite largely composed of pelagic microfossils and carbonate mud marked the return to monotonous basinal carbonate sedimentation (Fig. 12).

Later in the early Cretaceous (near Cenomanian), the sudden onset of slumps and sediment gravity flows could signify steepening of the slope resulting from tectonic downbowing of the continental margin (Fig. 17, see below). Thick lenticular beds of oolitic calcarenite were derived from shallow-water ooid

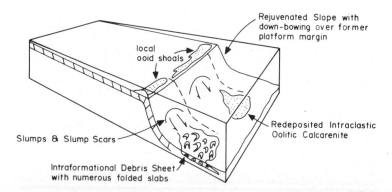

Fig. 17. Facies model for the pre-Cenomanian upper part of the Mayhah Formation, rejuvenation-tectonic steepening of slope.

shoals at the shelf-edge. Widespread debris sheets containing clasts of folded slope limestone originated as mass movements (slumps or slides) and evolved into debris flows (*cf.* Cook & Mullins, 1983). The sudden onset of such gravity-induced mass-flows suggest rapid (i.e. tectonic) steepening of the slope rather than gradual progradation and progressive steepening of the slope. The recurrence of calciclastic sediments in the same area as Jurassic megabreccias may indicate reactivation of basement structures, along the same trends. Further downbowing of the slope below the CCD eventually may have led to siliceous sedimentation of the Qumayrah Fm (see below).

Upper Mayhah Formation — western nappe

In the western nappe, the upper part of the Mayhah Fm consists of a thick, monotonous sequence of thin-bedded limestone which notably lacks the B Member chert interval (Figs 3, 5 & 7). The upper Mayhah Fm may correlate with the Jurassic Member A and/or the Cretaceous D Member of the eastern nappe, but stratigraphic relationships between the areas are incompletely understood. However, the lower Mayhah Fm of both nappes is similar in many respects, suggesting that these thrust sheets were genetically linked.

The upper Mayhah Fm of the western nappe consists of thin-bedded, dark-coloured, finely-laminated, argillaceous limestone with no mappable marker beds. It generally weathers to form steep slopes, but locally forms cliffs. Multiple intra-formational truncation surfaces are common in the lower and middle parts of the unit (Fig. 18). Thin beds and laminae of calcarenite are rare and locally show planar and convolute lamination.

In the upper part, the upper Mayhah Fm shows

evidence of soft sediment deformation and locally includes intra-formational calcirudite. In the south (section M-5, Fig. 15), these are overlain by a sequence of beds of coarse redeposited calcarenite and calcirudite with graded bedding and numerous amalgamation surfaces. These deposits contain abundant *Orbitulina* foraminifera and fragments of rudists. The contact between the Mayhah Fm and the overlying Qumayrah Fm in the western nappe is unconformable (Figs 3, 7 & 15).

Interpretation — dark-coloured organic-rich upper slope deposits

Because the western nappe is farther west and tectonically lower than the eastern nappe, it probably formed closer to the Arabian carbonate platform, possibly on the upper slope (Figs 12 & 17). The lack of B Member chert in the western nappe could signify a different palaeogeographic setting (upper vs. lower slope) and associated palaeoceanographic effects, or may have resulted from post-depositional erosion of the chert interval (see above). Locally abundant intra-formational truncation surfaces formed as slide scars on a carbonate slope. If the B Member chert originally existed in the area, it could have been eroded from these slide scars or from beneath the pre-Qumayrah unconformity.

As suggested previously, the dark-coloured, organic-rich, laminated calcilutite which comprises most of the unit formed within oxygen minimum zone conditions on the upper slope (Fig. 12). The general lack of coarse-grained sediments indicates that the slope was a gentle ramp and that rare sediment gravity flows either did not occur or by-passed this slope to be deposited in the deeper-marginal basin represented by the A and D Member of the eastern nappe. Where the uppermost Mayhah

10 m

Fig. 18. Outcrop photographs of the upper Mayhah Formation, western nappe. Multiple intra-formation truncation surfaces in the upper Mayhah Formation. Section Fy-1, ≈ 500 m.

Fm is not eroded beneath the pre-Qumayrah unconformity (section M-5), soft sediment deformation and sediment gravity flow deposits indicate steepening of the slope (as did the upper D Member, above). The localized sequence of redeposited bioclastic calcarenite and calcirudite in the south (section M-5) formed as a debris apron derived from adjacent rudist banks at the platform margin. The angular unconformity at the top of the unit indicates synorogenic deformation and erosion associated with the Qumayrah Fm.

Qumayrah Formation — chert and calcirudite

The Qumayrah Fm is an extremely varied lithological unit composed of chert, siliceous mudstone and redeposited calcirudite and calcarenite. In the eastern nappe, it conformably overlies the Mayhah Fm and consists of a thick interval of chert with minor beds of calcarenite overlain by thick beds of calcirudite (including megabreccia) (Figs 3 & 6). In the western nappe, it unconformably overlies deformed limestone of the Mayhah Fm and consists of chert and redeposited calcarenite and calcirudite. Radiolarian faunas extracted from cherts range in age from Cenomanian to Coniacian. This timespan is longer and younger than previously inferred by Glennie (1974).

It is believed that the high degree of variability within the unit relates to complex interactions of tectonics and sedimentation. Unfortunately, these rocks are relatively poorly exposed beneath talus cover and are commonly highly deformed along the footwalls of major thrust faults. The nature of lateral facies transitions and the inter-relationships of the different measured sections are poorly understood. Thus, the following discussion is only a small part of the complex history of the response of the Arabian platform margin to the closing of the adjacent Hawasina ocean basin (South Tethys Sea; Robertson, 1987; Patton & O'Connor, 1988).

Eastern nappe

Cenomanian purple siliceous mudstone and radiolarian chert abruptly overlie thin-bedded limestone of the Mayhah Fm (Figs 15 & 19). Thin beds of calcarenite are uncommon but contain a variety of platform-derived material including *Orbitulina* and bioclastic rudist bivalve fragments and slope-derived intraclasts of radiolarian calcilutite. The matrix contains abundant pelagic radiolarians which contrast with the redeposited neritic fossils. Graded bedding is common and flute casts occur locally, indicating that these represent calciturbidites.

Near Qumayrah village (section M-4), two thick channellized beds of calcirudite overlie ≈ 50 m of Cenomanian to Coniacian age radiolarian chert and minor thin beds of calcarenite (Fig. 15). Coniacian chert in the upper part contains radiolarians similar to those in coeval sediments associated with the Semail Ophiolite. The calcirudite contains clasts of

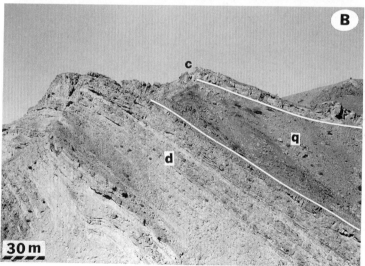

Fig. 19. Outcrop photographs of the Qumayrah Formation-eastern nappe.

(A) Thin-bedded calcarenite (calciturbidite) interbedded with dark-coloured siliceous mudstones of the Qumayrah Formation. Section HR-2, 50 m.

(B) Light-coloured thin-bedded calcilutite and lenticular beds of calcirudite of the D member–Mayhah Formation (d) are sharply overlain by dark-coloured siliceous mudstone and chert of the Qumayrah Formation (q). This Cenomanian to Coniacian radiolarite is overlain by a sequence of bedded calcirudite and calcarenite (c). Section F-1, 200–320 m.

slope-derived calcilutite, bioclastic calcarenite and fossil fragments. At section F-1, similar thick channellized beds of calcirudite, containing clasts as large as 10 m across, form the basal part of a thinning- and fining-upward sequence of redeposited limestones (Figs 15 & 19B).

Western nappe

Along Wadi Qumayrah in the western nappe, a thick interval of radiolarian chert overlies the Mayhah Fm along an angular unconformity (Fig. 20A). Farther up-section, a thick channellized calcirudite cuts into the radiolarite. Farther to the south (section M-5), the rudist-bearing calciclastic sequence in the upper Mayhah Fm is conformably(?) overlain by poorly exposed marlstone and a lenticular sequence of calcirudite and calcarenite of the Qumayrah Fm (Fig. 15). This is overlain by Cenomanian to Turonian radiolarian chert and another lenticular sequence of beds of graded calcirudite with boulders of neritic rudist-bearing limestone (Fig. 20B).

At Jebel Fayad (section Fy-1), a thick (32 m),

Fig. 20. Outcrop photographs of the Qumayrah Formation—western nappe.

(A) Angular unconformity between thin-bedded limestones of the upper Mayhah Formation (m) and poorly exposed radiolarian cherts (r) of the Qumayrah Formation. The chert is overlain by thick calcirudites/megabreccias (c) of the Qumayrah Formation. Location is south of Wadi Qumayrah at UTM 474°N 160°E.

(B) Thick sequence of bedded calcirudite and calcarenite (above person) overlies a 3 m thick section of radiolarian chert. Farther to the right beneath these cherts, another 2.3 m thick bed of clast-supported calcirudite has an outsized boulder (b) at top. Section M-5, 114–122 m. Photograph courtesy of Mary Anne McKittrick.

lenticular megabreccia unconformably overlies deformed thin-bedded limestone of the Mayhah Fm (Figs 7 & 15). It contains large folded slabs derived from the underlying Mayhah Fm as well as blocks of rudist-bearing neritic limestone. The unconformity has significant erosional relief, and the Mayhah Fm is much thinner at the crest of an adjacent anticline to the east (Fig. 2). Local erosional remnants of Coniacian radiolarian chert occur above the megabreccia and beneath a thinning- and fining-upward sequence of bedded calcirudite and calcarenite.

Interpretation—synorogenic debris flows over deep-marine chert

The Qumayrah Fm represents synorogenic deposits which formed in response to closing of the Hawasina Basin (South Tethys Sea; Robertson, 1987; Patton & O'Connor, 1988). These sediments accumulated immediately before the emplacement of the Semail Ophiolite over the Oman continental margin. Thus, they record the initial response of the Arabian platform margin to impending collisional tectonics.

The predominance of siliceous mudstone and

radiolarian chert in the Qumayrah Fm of the eastern nappe and locally in the western nappe suggests deposition below the CCD; (Fig. 21; Garrison & Fischer, 1969). Red to purple siliceous mudstone with abundant radiolarians in the matrix accumulated slowly in an oxidizing environment (below the oxygen minimum zone). Unlike the well-sorted radiolarian chert, the siliceous mudstone was not deposited by currents, but instead represents pelagic sediment that settled through the water column (*cf.* Nisbet & Price, 1974; Scholle *et al.*, 1983). Rare thin-bedded calcarenites were deposited rapidly as calciturbidites and were therefore less affected by dissolution below the CCD (Scholle, 1971). The abrupt transition from limestone of the Mayhah Fm into siliceous sediments of the Qumayrah Fm could be due to rapid rising of the CCD and/or tectonic subsidence of the continental margin slope below the CCD (Fig. 21). A shallower CCD during Cenomanian to Coniacian time developed adjacent to the Arabian platform margin because:

1 the global CCD was at an unusually shallow depth at this time (Fig. 26; Van Andel, 1975; see below);

2 the global Cretaceous highstand of sea-level was accompanied by attendant rise in the CCD (Fig. 26; Haq *et al.*, 1987; see below);

3 rates of carbonate sediment supply were reduced. Regional evidence indicates that the Oman margin was uplifted as a peripheral bulge as it approached a northward dipping subduction zone (Glennie, 1974; Robertson, 1987; Patton & O'Connor, 1988; see under tectonics below). An emergent platform margin and/or the ancestral Aruma foreland basin would have served as an effective barrier preventing periplatform ooze from reaching the continental margin slope (Fig. 21); and

4 closing of the Hawasina Basin would have intensified oceanic circulation, promoting upwelling and resultant siliceous sedimentation (*cf.* Jenkyns & Winterer, 1982).

Intraclasts in calcarenite and neritic limestone boulders in calcirudite were derived in part from erosion of the emergent shelf-edge, possibly along a tectonically-induced flexure at the continental margin which formed as a peripheral bulge or swell (*cf.* Robertson, 1987; Patton & O'Connor, 1988). A sea-level lowstand in Coniacian time might have also promoted the erosion and redeposition of calciclastic sediments in the uppermost Qumayrah Fm (Fig. 26). The calcirudite is thicker and more common in the relatively proximal western nappe *versus* the distal eastern nappe. Abundant *Orbitulina*, encrusting calcareous algae and fragments of rudist and *Inoceramus* bivalves were apparently derived from coeval or older rudist banks at the platform margin—peripheral bulge (Fig. 21). Well-rounded skeletal fragments were probably abraded in high-energy, wave-agitated environments (Watts, 1985).

Thick beds of calcirudite and/or megabreccia overlying the Coniacian chert are synorogenic and derived from collapse of the platform margin and slope. Slope instabilities could have been due to tectonic steepening resulting from downflexure of the continental margin during the final stages of an approach of the Oman margin toward a northward dipping subduction zone. The very thick, lenticular megabreccia at Jebel Fayad (section Fy-1) was derived, in part, from erosion of adjacent anticlinal

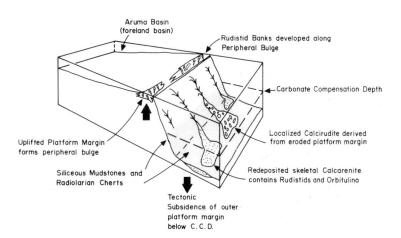

Fig. 21. Facies model for the Cenomanian lower part of the Qumayrah Formation, downbowing of slope beneath CCD.

highs which indicate that synorogenic deformation was affecting the slope sequence (Fig. 22). Folded slabs of Mayhah Fm indicate that these thin-bedded limestones were only semi-lithified when they were incorporated into a large debris flow that moved westward (?) toward the Aruma foreland basin and were deposited in the adjacent synclinal depression. Submarine mass movements and debris flows moved off over-steepened palaeotopographic/ structural highs and were deposited in relatively deep water as indicated by a thin localized interval of Coniacian radiolarian chert interbedded with these calcirudites.

INTERPRETIVE SYNTHESES

Evolution of slope and basinal facies through time

The slope bounding the Arabian platform margin in the Wadi Qumayrah area had a long complex history that ranged in age from middle late Jurassic to late Cretaceous. Patterns of sedimentation were effected by tectonism along the Oman continental margin, palaeoceanography and inter-relationships between carbonate platform, slope and basinal sedimentation.

In middle to late Jurassic time, a thick apron of redeposited calcirudite and oolitic calcarenite formed along the base of an escarpment slope (Fig. 23A). Ooid shoals, which formed at the shelf-edge, shed large quantities of carbonate sand into deeper-water environments. Channellized beds of calcirudite containing a variety of neritic and slope-derived clasts filled gullies on the slope which spread out and coalesced to form a base-of-slope apron. Widespread megabreccia and other calcirudites

formed due to collapse of the platform margin generated by seismicity, tectonic steepening and/or possible faulting along the platform margin.

Later in the Jurassic, coarse calciclastic sedimentation nearly ceased and basinal limestones formed along a gently sloping ramp-type basin margin (Fig. 23B). Cessation of faulting along the platform margin would have allowed the previous escarpment slope to become stabilized and a wedge of sediment accumulated to form a more gently sloping ramp. Being more calcareous, the basin here was either not as deep as the Hawasina Basin proper, or received more periplatform carbonate sediment. The depositional setting may have been some sort of embayment or marginal basin along the outer Arabian platform.

In the latest Jurassic to early Cretaceous (Tithonian to Valanginian), the outer Arabian carbonate platform was drowned (Figs 23C and 24; Glennie 1974; Connally & Scott, 1985). As a result, radiolarian chert formed across much of the Hawasina Basin. In the Wadi Qumayrah area, silicified radiolarian lime mudstone formed near the CCD either in a relatively shallow marginal basin or in an area with continued supply of peri-platform carbonate ooze. No chert exists in the more proximal western nappe; either this area was farther above the CCD or the chert interval was removed by erosion.

With progradation of shallow-marine carbonates across the previously drowned ramp in early Cretaceous (Valanginian or younger) time, carbonate sedimentation resumed over the slope and basin (Fig. 23D). Monotonous, thin-bedded calcilutite formed along the ramp-like slopes of the Qumayrah marginal basin. Later, steepening of the slope led to mass movements and sediment gravity

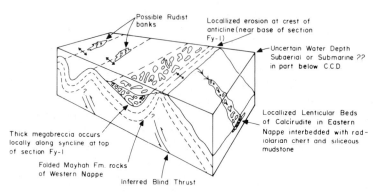

Fig. 22. Facies model for the Coniacian upper portion of the synorogenic Qumayrah Formation. In the western nappe, slope sediments of the Mayhah Formation were deformed and eroded unconformably beneath megabreccias of the Qumayrah Formation.

flows. Ooids were derived from coeval shelf-edge shoals and intraclasts were derived from submarine erosion of slope sediments (Fig. 23E).

In Cenomanian (mid-Cretaceous) time, carbonate sedimentation ceased abruptly and siliceous mudstone and radiolarian chert of the Qumayrah Fm formed below the CCD (Fig. 23F). The global level of the CCD was relatively shallow at this time (Fig. 26; Van Andel, 1975) and eustatic sea-level was near its Cretaceous maximum (Fig. 26; Haq *et al.*, 1987). In addition, continued steeping and downbowing of the slope could have caused the area to subside below the CCD (Fig. 25). The Oman margin was uplifted and eroded at this time (Fig. 24; Glennie 1974; Robertson, 1987; Patton & O'Connor, 1988) and would have formed a barrier to peri-platform ooze produced on the Arabian platform. Normal faulting at Fahud field indicates that tectonic readjustments were affecting the outer Arabian carbonate platform (Figs 23F & 25; Tschopp, 1967). The Aruma foreland(?) basin may have developed at this time, preventing carbonate sediment of the Arabian carbonate platform from reaching the area, but it was later obliterated by erosion over the westward-migrating peripheral bulge or swell (Patton & O'Connor, 1988). Closure of the Hawasina Basin would have caused the formation of relatively narrow basins with vigorous circulation and upwelling of nutrient-rich waters triggering siliceous plankton blooms and resultant siliceous sedimentation (*cf.* the Gulf of California).

In Coniacian (late Cretaceous) time, a megabreccia and other calcirudites were generated due to uplift and erosion of the platform margin and deformation and steepening of the slope (Figs 23G & 25). Limestone clasts in the megabreccia were derived from erosion of both shallow-marine and slope sediments indicating collapse of the entire platform margin. A notable sea-level lowstand at this time could have helped to trigger mass movements that carried coarse calciclastic sediments derived from the platform margin downslope (Fig. 26). Deformation and synorogenic sedimentation of the Qumayrah and Mayhah Fms could be related to either initial folding and thrust-faulting of the slope sequence (Fig. 23, G1) or to large gravity slides which moved down the oversteepened slope (Fig. 23, G2). The deformation of the Mayhah Fm and synorogenic sedimentation of the Qumayrah Fm preceded emplacement of the higher nappes and are precursors of the major collisional event (Fig. 23G–J).

Between Coniacian and Maastrichtian (late Cretaceous, Campanian?) time, the Hawasina Complex and Semail Ophiolite were progressively emplaced over the Arabian carbonate platform margin along with the Wadi Qumayrah slope deposits (Fig. 23H). Later out-of-sequence thrust-faulting placed the western nappe over the Hawasina Complex and locally over the Semail Ophiolite (Fig. 23I). Thus, the Oman Mountains orogeny began with flexural tectonism and steepening of the slope in Cenomanian to Coniacian time and was followed by nappe emplacement and deformation that was complete before Maastrichtian time (Fig. 25; Robertson, 1987). Younger relatively undeformed Maastrichtian and Tertiary post-orogenic deposits unconformably overlie the ancestral Oman Mountains (Fig. 24).

Arabian carbonate platform, sea-level fluctuations and slope sedimentation

The Arabian carbonate platform served as the source of much of the sediment which was redeposited on to the carbonate slope (Fig. 24). The dramatic effect that sea-level fluctuations have upon carbonate platform sedimentation is also reflected in carbonate slope sediments. In particular, platform drowning events are particularly well recorded as condensed sequences of radiolarian chert (Fig. 24). Although some of the drowning events for the Arabian platform margin appear to be tectonically induced (Murris, 1980; Harris *et al.*, 1984), relative changes in sea-level play a major role in both platform and slope sedimentation (Fig. 26).

The Jurassic Sahtan Formation contains abundant oolitic calcarenite, typical of Jurassic platform margins (Fig. 24; Glennie, 1974). Redeposited oolitic calcarenite is also common in Jurassic slope and basinal deposits, particularly in the lower part of Wadi Qumayrah sequence (Glennie *et al.*, 1973; Glennie, 1974; Watts & Garrison, 1986). Ooid shoals might have been favoured along steep platform margins such as that inferred for the Wadi Qumayrah area (Fig. 23A). With the later development of a marginal basin with gently sloping ramp margins, supply of oolitic calcarenite was greatly reduced suggesting that the ooid shoals either became inactive or migrated away from the area.

In the latest Jurassic (Tithonian) to early Cretaceous (Valanginian), the outer part of the Arabian carbonate platform was drowned as recorded by the lower part of the Thamama Group (Glennie 1974; Murris, 1980; Harris *et al.*, 1984; Connally & Scott,

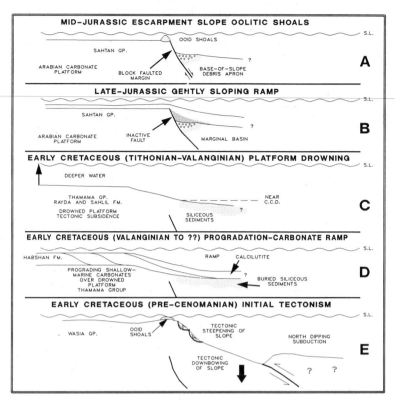

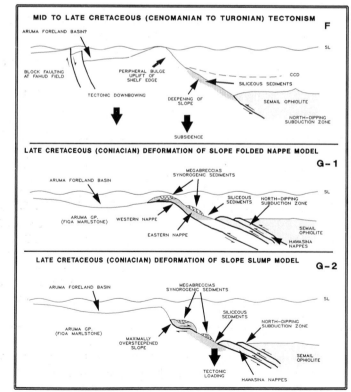

Fig. 23. Schematic interpretive cross-sections illustrate the tectonic evolution of the Oman continental margin.

(A) Block-faulting and base-of-slope debris apron along the Jurassic passive continental margin.

(B) Late Jurassic development of a gently sloping ramp along a marginal basin which formed due to tectonic quiescence.

(C) Latest Jurassic early Cretaceous drowning of the outer Arabian platform and resultant siliceous sedimentation (chert) in the slope and basinal sequences.

(D) Valanginian restoration of shallow-marine carbonate sedimentation due to progradation of the Arabian carbonate platform.

(E) Pre-Cenomanian tectonic downbowing of slope led to steepening of slope and generation of sediment gravity flows.

(F) Cenomanian to Turonian development of the Aruma foreland(?) basin, peripheral bulge at shelf-edge, siliceous sedimentation with lesser calciturbidites on the slope and block-faulting at Fahud oilfield.

(G) Coniacian megabreccias and deformation of the slope sequence before oceanic nappe emplacement. Model G1 suggests that the deformation was due to early phases of compressional tectonics. Alternatively, Model G2 shows deformation due to sliding of the western nappe down a tectonically steepened slope.

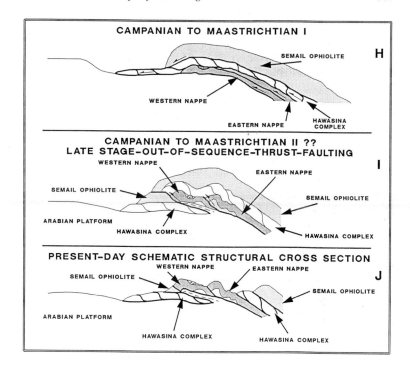

Fig. 23. Cont.
(H) Campanian to Maastrichtian I successive emplacement of Hawasina Complex and Semail Ophiolite nappes.
(I) Campanian to Maastrichtian II late-stage out-of-sequence thrust-faulting of the Mayhah and Qumayrah Formations over the previously emplaced nappes.
(J) Present-day schematic structural cross-section across the Wadi Qumayrah study area.

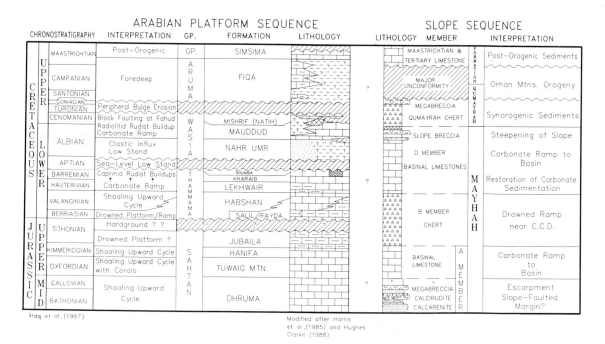

Fig. 24. Relationship between stratigraphy of the Arabian platform sequence and the Mayhah Formation and Qumayrah Formation slope sequence.

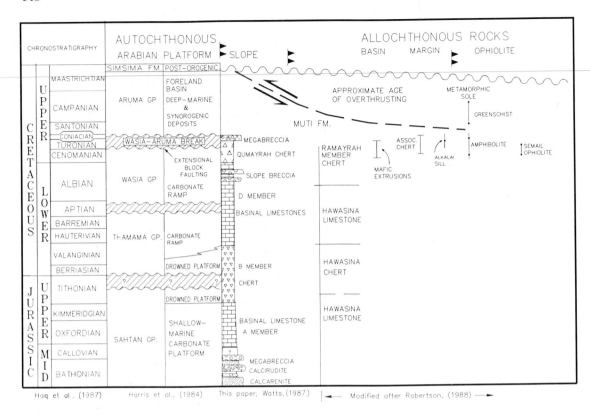

Fig. 25. Correlation chart for the Arabian platform sequence, Mayhah and Qumayrah Formations slope deposits, Hawasina Basin sediments, and Semail Ophiolite illustrates the relationship between these tectonically emplaced rocks. Note the widespread inception of tectonism during Cenomanian time.

1985; Watts & Garrison, 1986). The drowning event was preceded by a major unconformity (hardground?) at the end of the Jurassic (Fig. 24; Hughes-Clarke, 1988). Deep-water limestones, radiolarian calcilutite of the Rayda Fm and argillaceous calcilutite of the Salil Fm, formed due to tectonic downwarp and deepening of the Oman carbonate ramp margin (Figs 23 & 25; Harris *et al.*, 1984). In the Wadi Qumayrah slope sequence, radiolarian chert and calcareous radiolarian chert of the B Member of the Mayhah Fm formed in response to reduced supply of carbonate sediment. At Jebel Sumeini and in the Hawasina Complex, Tithonian to early Cretaceous red radiolarian cherts are widespread indicators of this platform drowning event in the deeper-water successions (Fig. 24; Watts & Garrison, 1986). The upper part of the Habshan Fm (Yamama Fm) reflects the progradation of shoal-water conditions over this previously drowned ramp during Valanginian time (Figs 23D & 24). Chert from the

uppermost B Member of the Mayhah is also of Valanginian age and is overlain by thin-bedded limestone, indicating restoration of carbonate supply due to the propgradation of shallow-marine carbonates over the outer Arabian carbonate platform.

The upper part of the Thamama Group (Hauterivian to lower Aptian Lekhwair Fm, Kharaib Fm and Shuaiba Fm) represents another shallowing-upward sequence culminating in caprinid rudist buildups of the Shuaiba Fm (Fig. 24; Harris *et al.*, 1984). Coarse-grained carbonate apron deposits in the southern part of the Wadi Qumayrah area contain abundant fragments of rudists and *Orbitulina* foraminifera and may have been derived from local buildups (section M-5; Figs 2 & 15).

A major unconformity exists between the Thamama Group and Wasia Group, but it has not been recognized within monotonous thin-bedded calcilutite of the D Member of the Mayhah Fm. (There is no age control between the Valanginian

B Member and Cenomanian Qumayrah Fm.) The Upper Aptian to Cenomanian Wasia Group is another major depositional sequence. In the lower part, the Nahr Umr Fm contains abundant siliciclastic sediment which apparently formed on a shallow-marine shelf during a long interval of relatively low sea-level and reduced sedimentation rates (Harris *et al.*, 1984). A late Albian transgression led to carbonate deposition of Mauddud Fm followed by progressive shallowing-upward into local radiolitid rudist-reef buildups (Harris *et al.*, 1984). Locally, porous calcarenites of the Cenomanian Mishrif Fm (and Natih Fm) formed upon tilted upthrown fault blocks such as at the Fahud oil fields (Figs 23F & 24; Tschopp, 1967a; Harris & Frost, 1984). Thus, major changes in depositional patterns in the Arabian carbonate platform succession occurred in Cenomanian time in response to tectonism along the Oman margin and the early development of the Aruma foreland (?) basin. Major changes are also recorded in the Cenomanian slope sequence, where the transition from carbonate to siliceous facies might be related to tectonism and/or reduced sedimentation of periplatform carbonate along the Oman continental margin. This transition was preceded by steepening of the slope as indicated by slumps, slides and debris flow deposits in the upper D Member of the Mayhah Fm (Figs 23E & F).

The Wasia–Aruma break in the platform sequence is a major unconformity separating the Cenomanian and older Wasia Group from the Coniacian and younger Aruma Group (Fig. 24). Emergence of the platform and development of the unconformity may be related to tectonic activity along the Oman margin and migration of a peripheral bulge over the outer Arabian platform (Robertson, 1987; Patton & O'Connor, 1988). Siliceous sediments of the Qumayrah facies formed at this time in relatively sediment-starved basinal conditions. Both the emergence of the platform and the later development of the Aruma foreland basin may have led to reduced rates of carbonate supply and the resultant siliceous sediments on the continental margin slope.

The Coniacian to Maastrichtian Aruma Group formed within the foreland basin along the outer part of the Arabian carbonate platform and bounded by the Oman Mountains to the north (Figs 24 & 25). The Oman Mountains fold and thrust belt formed during this same interval of time with major tectonism in the Campanian (Fig. 1; Glennie *et al.*, 1973; Glennie, 1974; Robertson, 1987). In the Oman Mountains, a major angular unconformity is associated with the erosion of the uplifted Oman mountain belt and a lacuna (hiatus) extends from deformed Coniacian Qumayrah Fm to the relatively undeformed Maastrichtian Simsima Fm and younger Tertiary deposits (Glennie *et al.*, 1973). Thus, the Aruma foredeep developed in conjunction with emplacement of thrust sheets over the Arabian platform margin in response to tectonic loading.

Evidence for tectonic control on slope sedimentation

In Jurassic time, base-of-slope apron deposits including thick megabreccias indicate that a relatively steep slope existed along the Arabian carbonate platform margin in the Wadi Qumayrah area. Although, such coarse calciclastic sediments can be generated by a variety of slope processes, the thick megabreccias could indicate tectonically steepened slopes and may have been triggered by seismicity (*cf.* Mutti *et al.*, 1984). With time, coarse redeposited limestones became uncommon, suggesting that this area evolved into a basin with gentle ramp-like margins and indicating tectonic quiescence. The margin remained relatively stable until pre-Cenomanian time, when slope-derived slumps and debris flows recorded abrupt steepening of the slope. Steepening or rejuvenation of the slope in the same geographic location as the Jurassic megabreccias suggests structural control. Tectonic steepening of the slope may have been the result of reactivation of N–S trending basement structures. Similar N–S trending basement structures such as the Maradi–Nafun fault zone exist beneath the Arabian carbonate platform farther to the south and may also extend northward toward the Wadi Qumayrah and Sumeini areas (Fig. 1; Tschopp, 1967b; Murris, 1980; Hughes-Clark 1988).

Transition from carbonate to siliceous sedimentation in Cenomanian time occurred in conjunction with down-to-the-north block faulting on the Arabian platform at Fahud oilfield (Figs 23F, 24 & 25; Tschopp, 1967a; Harris & Frost, 1984). Subsidence of the outer part of the passive continental margin platform and slope resulted from its approach toward a north-dipping subduction zone (Robertson, 1987; Patton & O'Connor, 1988). The response of the north Australian platform to attempted subduction beneath the Timor Island arc represents a good modern analogue for the Cretaceous of Oman (Veevers *et al.*, 1978). Similar steepening of the slope and attendant deformation are occurring along the present-day Oman continental margin (R.S. White, Cambridge University, pers. comm., 1988). During the final stages of deformation, major thrust faults between the eastern and western nappes may have formed along pre-existing structures, listric normal faults which developed during Jurassic rifting (Fig. 23H–J; *cf.* Cohen, 1982).

In a more global sense, early Cretaceous drowning of the Arabian platform margin occurred at the same time as the initial opening of the South Atlantic (Dewey *et al.*, 1973). The African plate including the Arabian platform moved eastward relative to Eurasia. With these changes in plate motion, left-lateral and compressive tectonics may have played a major role in the Cenomanian and younger evolution of the continental margin in Oman.

Palaeoceanography and effects of CCD on slope sedimentation

Two major intervals of siliceous sediment occur within the stratigraphic sequence at Wadi Qumayrah: the Tithonian to Valanginian B Member of the Mayhah Fm and the Cenomanian in Coniacian Qumayrah Fm. These siliceous sediments accumulated on the seafloor near or below the CCD. Although tectonically-induced subsidence may have promoted siliceous sedimentation (see above), eustatic sea-level changes and variations in the depth of the CCD also had a significant effect. Figure 26 illustrates the relationship between eustatic sea-level (Haq *et al.*, 1987), sea-level fluctuations on the Arabian platform (Harris *et al.*, 1984), variations in the depth of the CCD (Van Andel, 1975) and the Jurassic to Cretaceous slope sediments of Oman.

Tithonian to Valanginian chert

The Tithonian to Valanginian chert of the B Member formed during a well-documented drowning of the outer portion of the Arabian carbonate platform (Fig. 23C; Glennie, 1974; Harris *et al.*, 1984). The relatively high sea-level for the outer Arabian platform in Oman apparently resulted from tectonic subsidence of the area (Fig. 26; Harris *et al.*, 1984). The timing of the drowning event on the Arabian platform sequence is not precisely dated (Harris *et al.*, 1984; Connally & Scott, 1985; Hughes-Clarke, 1988), but the Tithonian age of chert of the B Member may provide a minimum age for this platform drowning (Watts & Garrison, 1986). In Abu

Dhabi, shallowing-upward sediments of the Habshan Fm suggest that shallow-marine sedimentation became re-established during Valanginian time due to progradation of a carbonate ramp (Connally & Scott, 1985). This correlates with the Valanginian to Hauterivian chert in the youngest B Member immediately below thin-bedded limestone of the D Member.

The CCD at this time is postulated to have been $\approx$ 4 km below sea-level (Fig. 26; Van Andel, 1975). The CCD for recent sediments is typically shallower along continental margins; thus, similarly, the depth of the CCD along the Oman margin could have been relatively shallow compared to estimates shown for open ocean settings. The supply of periplatform ooze from the Arabian platform would also control the position of the CCD. At Jebel Sumeini, red radiolarian chert formed below the local CCD. Near Wadi Qumayrah, the banded cherts represent silicified limestone that accumulated near but above the local CCD. Supply of periplatform carbonate ooze was sufficient for some carbonate to be preserved but early diagenesis lead to silicification by bottom waters enriched in silica and depleted in carbonate.

Cenomanian to Coniacian chert

The Cenomanian to Coniacian Qumayrah Fm abruptly overlies limestone of the Mayhah Fm. Purple siliceous mudstone and chert in this unit formed during the peak of the Cretaceous highstand of sea-level (Fig. 26). The postulated CCD for the Indian Ocean was at its shallowest during Coniacian time at 3.2 km below sea-level (Van Andel, 1975). Thus, an unusually high sea-level and the shallowest known depth of the CCD may have combined to promote siliceous sedimentation. In addition, tectonic factors noted above (subsidence and isolation from the carbonate platform) may have favoured siliceous sediments (Figs 23F & 25). Complex palaeogeography associated with collisional tectonics may have formed small marginal basins which had vigorous circulation (upwelling) with blooms of siliceous plankton resulting in siliceous sedimentation (*cf.* Jenkyns & Winterer, 1982).

In summary, many factors led to favourable palaeoceanographic conditions for siliceous sedimentation along the Oman continental margin during Cenomanian to Coniacian time. Tectonics and subsidence probably played a role, but a high-

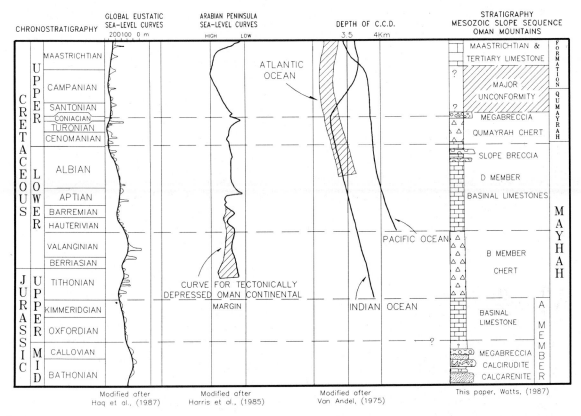

Fig. 26. Palaeoceanography and slope sedimentation along the Oman margin. The figure illustrates the relationship between global and regional sea-level fluctuations, depth of the CCD, and intervals of siliceous sedimentation in the Mayhah and Qumayrah Formations.

stand of sea-level and unusually shallow CCD may have also favoured the accumulation of siliceous sediment.

CONCLUSIONS

The Arabian platform margin slope in the Wadi Qumayrah area formed adjacent to a continental margin that evolved through Jurassic rifting, later passive margin sedimentation and late Cretaceous collisional tectonics. Slope and basinal sedimentation were affected by three major factors:

1 the supply of carbonate sediment from the Arabian platform;

2 tectonism; intervals of subsidence, uplift, steepening of the slope and deformation along the continental margin; and

3 palaeoceanographic effects including variations in eustatic sea-level, depth of the CCD and circulation patterns (upwelling).

Major events which controlled sedimentation on the Arabian platform are also evident in widespread slope and basinal deposits and, therefore, are useful for correlation and interpretation of the evolution of the Oman margin. Thick megabreccias, indicating oversteepening of continental margin slopes and possibly generated by seismicity, are unique and widely occurring cataclysmic deposits that can be correlated between thrust sheets. An interval of Tithonian to Valanginian slope and basinal chert is a widespread marker that can be correlated with a major platform drowning event. Another chert interval is associated with Cenomanian to Coniacian collisional tectonism, changing basin geometries, and possible palaeoceanographic factors such as shallow depth of the CCD and relatively high sea-level.

ACKNOWLEDGEMENTS

The authors are grateful to Sultan Qaboos and the people of Oman for their great hospitality and especially the Ministry of Petroleum and Minerals for authorizing this publication. Greatest thanks is owed to Dr Robert Garrison who supervised Watts PhD thesis and provided much inspiration. The research was funded by Amoco Oman Oil Company through a grant to the Earth Resource Institute (ERI). John Smewing, head of the ERI Oman project, provided excellent support and valuable advice. Many insights were provided by ERI and Amoco geologists working in Oman. The University of California, Santa Cruz provided educational opportunities and financial support through two Amoco Research Fellowships and a Patent Fund grant. Mary Anne McKittrick and Jim Sample were outstanding field assistants. Excellent thin sections were made by Kelly Brown, James Garrison, Miles Grant and Margie Schutt under the supervision of Eugenio Gonzales. James Vigil processed cherts for radiolarians. Drafting was done by Kelly Brown, Jeff Filut, Jeff Rogers, Dale Skinner, Debbie Coccia and Jennette Smith. The paper benefited from reviews by William Irwin, Patrick Lehmann, Chris Kendall, Fred Read, Roland Gangloff and Rick Sarg. The final manuscript was typed by Debbie LaBarre. The Geophysical Institute of the University of Alaska helped with final preparation of the paper.

REFERENCES

BLOME, C.D. & GARRISON, R.E. (1982) *Toward an understanding of the Hawasina Complex: Preliminary report on the biostratigraphy and sedimentology of Hawasina radiolarian cherts and related rocks in the Oman Mountains.* Unpublished company report, Earth Science Resource Institute.

COHEN, C.R. (1982) Models for passive to active continental margin transition; implications for hydrocarbon exploration. *Am. Assoc. Petrol. Geol. Bull.* **66**, 708–718.

COLEMAN, R.G. (1981) Tectonic setting for ophiolite obduction in Oman. *J. Geophy. Res.* **86**, 2497–2508.

CONNALLY, T.C. & SCOTT, R.W. (1985) Carbonate sediment fill of an oceanic shelf, Lower Cretaceous, Arabian Peninsula. In: *Deep-water Carbonates: Buildups, Turbidites, Debris Flows and Chalks—A Core Workshop.* (Eds Crevello, P.D. & Harris, P.M.) Soc. Econ. Paleont. Mineral. Core Workshop 6, p. 266–302.

COOK, H.E. & TAYLOR, M.E. (1977) Comparison of continental slope and shelf environments in the upper Cambrian and lowest Ordovician of Nevada. In: *Deep-Water Carbonate Environments* (Eds Cook, H.E. & Enos, P.) Spec. Publ. Soc. Econ. Paleont. Mineral. 25, pp. 51–81.

COOK, H.E. & MULLINS, H.T. (1983) Basin margin environment. In: *Carbonate Depositional Environments* (Eds Scholle, P.A., Bebout, D.G. & Moore, C.H.) Amer. Assoc. Petrol. Geol. Mem. 33, pp. 540–617.

DAVIES, G.R. (1977) Turbidities, debris sheets and truncation surfaces in Upper Paleozoic deep-water carbonates of the Sverdrup basin. In: *Deep-Water Carbonate Environments* (Eds Cook, H.E. & Enos, P.) Spec. Publ. Econ. Paleont. Mineral. 25, pp. 221–247.

DEWEY, J.F., PITMAN, W.C. RYAN, W.B.F. & BONNIN, J. (1973) Plate tectonics and the evolution of the Alpine System. *Geol. Soc. Am. Bull.* **84**, 3137–3180.

DEMAISON, G.J. & MOORE, G.T. (1980) Anoxic environments and oil source bed genesis. *Am. Assoc. Petrol. Geol. Bull.* **64**(8), 1179–1209.

GARRISON, R.E. & FISCHER, A.G. (1969) Deep-water limestones and radiolarites in the Alpine Jurassic. In: *Depositional Environments in Carbonate Rocks* (Ed. Friedman, G.M.) Soc. Econ. Paleont. Mineral. Spec. Publ. 14, pp. 20–56.

GEALEY, W.K. (1977) Ophiolite obduction and geologic evolution of the Oman Mountains and adjacent areas. *Geol. Soc. Am. Bull.* **88**(8), 1183–1191.

GLENNIE, K.W. BOEUF, M.G.A., HUGHES-CLARK, M.W., MOODY-STUART, M., PILAAR, W.F.H. & REINHART, B.M. (1973) Late Cretaceous nappes in the Oman Mountains and their geologic evolution. *Am. Assoc. Petrol. Geol. Bull.* **57**, 5–27.

GLENNIE, K.W. (1974) *Geology of the Oman Mountains*, Koninklijk Nederlands Geologisch Mijnbouwkundig Genootschap, Verhandelingen. 31, 3 Vols, p. 423.

HAQ., B.U., HARDENBOL, J. & VAIL, P.R. (1987) Chronology of fluctuating sea levels since the Triassic. *Science* **235**, 1156–1167.

HARRIS, P.M., & FROST, S.H. (1984) Middle Cretaceous carbonate reservoirs, Fahud Field and northwest Oman. *Am. Assoc. Petrol. Geol. Bull.* **68** (5), 649–658.

HARRIS, P.M., FROST, S.H., SEIGLIE, G.A. & SCHNEIDERMANN, N. (1984) Regional unconformities and depositional cycles, Cretaceous of the Arabian Peninsula. In: *Interregional Unconformities and Hydrocarbon Accumulation* (Ed. Schlee, J.S.) Am. Assoc. Petrol. Geol. Mem. 36, pp. 67–80.

HASSAN, T.H., MUD, G.C. & TWOMBLEY, B.N. (1975) The stratigraphy and sedimentation of the Thamama Group (Lower Cretaceous) of Abu Dhabi. *9th Arab Petroleum Congress*, Dubai, Paper 107 (B-3), pp. 1–11.

HUGHES-CLARKE, M.W. (1988) Stratigraphy and rock-unit nomenclature in the oil-producing area of interior Oman. *J. Petrol. Geol.* **11**(1), 5–61.

JENKYNS, H.C. & WINTERER, E.L. (1982) Paleoceanography of Mesozoic ribbon radiolarites. *Earth Planet. Sci. Lett.* **60**, 51–375.

JOHNS, D.R., MUTTI, E., ROSELL, J. & SEGURET, M. (1981) Origin of a thick, redeposited carbonate bed in Eocene turbidites of the Hecho Group, South-central Pyrenees, Spain. *Geology* **9**, 161–164.

KENDALL, C.G.ST.C. & SCHLAGER, W. (1981) Carbonates and relative changes in sea level. In: *Carbonate Platforms of the Passive-type Continental Margins, Present and Past* (Eds Cita, M.B. & Ryan, W.B.F.) *Marine Geol.* **44**, 181–212.

KRAUSE, F.F. & OLDERSHAW, A.E. (1979) Submarine carbonate breccia beds: a depositional model for two-layer sediment gravity flows from the Sekwi Formation (Lower Cambrian), MacKenzie Mountains, Northwest Territories, Canada. *Can. J. Earth Sci.* **16**, 189–199.

MULLINS, H.T. & COOK, H.E. (1986) Carbonate apron models. Alternatives to the submarine fan model for paleoenvironmental analysis and hydrocarbon exploration. *Sed. Geol.* **48**(1/2), 37–79.

MURRIS, R.J. (1980) Middle East: Stratigraphic evolution and oil habitat. *Am. Assoc. Petrol. Geol. Bull.* **64**, 597–618.

MUTTI, E., RICCI-LUCCI, F., SEQURET, M. & ZANZUCCHII, G. (1984) Seismoturbidites: a new group of resedimented deposits. *Marine Geol.* **55**, 103–116.

NISBET, E.G. & PRICE, I. (1974) Silicified turbidites: graded cherts as ocean-ridge derived sediments. In: *Pelagic Sediments; on land and under the sea*. (Eds Hsu, K.J. & Jenkyns, H.C.) Spec. Publ. Int. Assoc. Sed. 1, pp. 351–366.

PATTON, T.L. & O'CONNOR, S.J. (1988) Cretaceous flexural history of northern Oman Mountains foredeep, United Arab Emirates. *Am. Assoc. Petrol. Geol. Bull.* **72**(7), 797–808.

READ, J.F. (1982) Carbonate platforms of passive (extensional) continental margins: Types, characteristics and evolution. *Tectonophys.* **81**, 195–212.

READ, J.F. (1984) Carbonate platform facies models. *Am. Assoc. Petrol. Geol. Bull.* **69**(1), 1–21.

ROBERTSON, A.H.F. (1987) The transition from a passive margin to an upper Cretaceous foreland basin related to ophiolite emplacement in the Oman Mountains. *Geol. Soc. Am.* **99**(5), 633–653.

SCHLAGER, W. (1981) The paradox of drowned reefs and carbonate platforms. *Geol. Soc. Am. Bull.* **95**, 675–702.

SCHOLLE, P.A. (1971) Sedimentology of fine-grained deep-water carbonate turbidites, Monte Antola flysch (Upper Cretaceous), northern Apennines, Italy: *Geol. Soc. Am. Bull.* **82**, 629–658.

SCHOLLE, P.A., ARTHUR, M.A. & EKDALE, A.A. (1983) Pelagic Environment. In: *Carbonate Depositional Environments* (Eds Scholle, P.A., Bebout, D.G. & Moore, C.H.) Am. Assoc. Petrol. Geol. Bull. Mem. 33, pp. 540–617.

SEARLE, M.P. (1984) Sequence and style of thrusting and origin of culminations in the northern and central Oman Mountains. *J. Struct. Geol.* **7**(2), 129–143.

SEARLE, M.P., JAMES, N.P., CALON, T.J. & SMEWING, J.D. (1983) Sedimentological and structural evolution of the Arabian continental margin in the Musandam Mountains and Dibba Zone, United Arab Emirates. *Geol. Soc. Am. Bull.* **94**, 1381–1400.

TSCHOPP, R.H. (1967a) Development of the Fahud Field. *Proceedings of the 7th World Petroleum Congress*, Mexico. Vol. 2, pp. 251–255.

TSCHOPP, R.H. (1967b) The general geology of Oman. *Proceedings of the 7th World Petroleum Congress*, Mexico, Vol. 2, pp. 231–241.

VAN ANDEL, T.H. (1975) Mesozoic/Cenozoic calcite compensation depth and the global distribution of calcareous sediments. *Earth Planet. Sci. Lett.* **26**, 187–194.

VEEVERS, J.J., FALVEY, D.A. & ROBINS, S. (1978) Timor trough and Australia. Facies show topographic wave migrated 80 km during last 3 *Myr. Tectonophy.* **45**, 217–227.

WATTS, K.F. (1985) *Evolution of carbonate slope facies along a south Tethyan continental margin: The Mesozoic Sumeini Group and the Qumayrah Facies of the Muti Formation, Oman*. PhD Thesis, University of California, p. 475.

WATTS, K.F. & GARRISON, R.E. (1986) Evolution of a Mesozoic carbonate slope on a south Tethyan continental margin, Sumeini Group, Oman. *Sed. Geol.* **48**(1/2), 107–168.

Index

References to figures appear in *italic type*.
References to tables appear in **bold type**.